T0329843

Wireless Communications Systems

Wireless Communications Systems

An Introduction

Randy L. Haupt
Colorado School of Mines
Department of Electrical Engineering

Registered Office
John Wiley & Sons, Inc., 111 River Street, Hoboken, NJ 07030, USA

Editorial Office
111 River Street, Hoboken, NJ 07030, USA

For details of our global editorial offices, customer services, and more information about Wiley products visit us at www.wiley.com.

Wiley also publishes its books in a variety of electronic formats and by print-on-demand. Some content that appears in standard print versions of this book may not be available in other formats.

Library of Congress Cataloging-in-Publication Data is applied for
9781119419174

Cover Design: Wiley
Cover Image: © Dong Wenjie/Getty Images

To the wonderful girls in my life: Sue Ellen, Bonny, Amy, Adeline, and Rose. You give me love and inspiration

Contents

Preface

This book targets undergraduate and graduate students as well as professionals wanting an introduction to wireless communication systems. Wireless systems pervade all aspects of our lives. I have been an insulin-dependent diabetic most of my life. Recently, I got a continuous blood glucose monitor (Chapter 11) that interfaces with my cell phone via Bluetooth (Appendix D). This monitor dramatically changed my life for the better due to the technologies presented in this book. Even though wireless systems have exponentially expanded over my lifetime, the future looks even brighter. Learning about wireless systems leads to a significant advantage over the uninformed.

My career in wireless systems covers a wide range of different projects in industry, government, and academia. I have taught courses in wireless communications, analog and digital communications, digital signal processing, probability and statistics, antennas, electromagnetics, software-defined radios, electronics, electromagnetic compatibility, optimization, and radar at six different universities and also presented many special topic short courses. I thought that my diverse background would prepare me for writing a book on wireless systems. Writing this book humbled me, however. Universities usually have one or more courses based on each chapter in this book. I learned a lot in the writing process, and I tried to convey very complex information as clear and simple manner as possible with plenty of pictures and examples.

This book has two parts: fundamentals (Chapters 1–5) and system applications (Chapters 6–12). The appendices provide some supplemental information for the reader. I relied on MATLAB for most of the computations and graphics. There are plenty of examples and pictures illustrating many different aspects of wireless technology in our lives. Even though I tried to simplify the concepts, students still need some knowledge in electromagnetics, math through calculus, probability, and linear systems.

The first half of this book delves into theory with many examples of practical applications. Chapter 2 covers data, signals, and digital signal processing. Chapter 3 introduces analog and digital modulation along with multiplexing and multiple access techniques. Chapter 4 has an overview of antennas with emphasis on design for wireless systems. The details on antenna array design are necessary for Chapters 8 and 9 but can be skipped if needed for time constraints. Chapter 5 concerns many aspects of RF propagation from HF to mm waves.

I selected seven examples of wireless applications for the second half of the book. These examples make use of the theory introduced in the first half of the book. Chapter 6 is on satellite communications. My visits to satellite communication facilities and the Smithsonian inspired this chapter. I have worked on satellite projects over my career. Chapter 7 introduces RFID. This technology impacts our daily lives and keeps track of many things that we value. Although I am a novice in this area, I found that a lot of my experience and knowledge in radar useful in writing this chapter. I have written separate books on the material in Chapters 8 and 9. Here, I provide an overview with practical insights into the material. Chapter 10 covers the interesting and confusing topic of multiple input/multiple output (MIMO). I translated this complicated material down to an understandable level for the nonspecialist. I had fun researching the materials for Chapters 11 and 12. Security and health effects of wireless systems concern users and designers.

I teach the first five chapters of the book and some of the systems described in the remaining chapters in my class. I like to have the students do plenty of MATLAB programming and use hardware experiments to enhance the theory. My students do final projects and make presentations in lieu of a final exam. I have students purchase RTL software-defined radios (Appendix F) to have some hardware to test the theory they learn. My goal is to prepare them for a job in the wireless technology sector. This goal makes this book unique.

I have extra material available for instructors who adopt this text in the form of PowerPoint slides, videos, and additional problems. Please contact the publisher to arrange for access.

I owe many thanks to the people who spent time reviewing portions of this book: Sue Haupt, Payam Nayeri, Bonny Turayev, Amy Shockley, Jake Shockley, and Mark Leifer. They pushed me to do better.

Randy L. Haupt
Boulder, CO

Symbols and Acronyms

Symbols

A	low sidelobe taper factor
$\mathbf{A}$	array steering matrix
A_e	effective aperture
A_{LOS}	amplitude of the LOS signal
A_m	amplitude of the message signal
AF	array factor
AF_N	array factor normalized to N
AF_x	x-axis array factor
AF_y	y-axis array factor
AP	antenna pattern
Ap	index of geomagnetic activity
A_p	physical aperture
AR	axial ration
a	one dimension of rectangular slot
a_n	weights
B	bandwidth
b	one dimension of rectangular slot
b_n	bit n
b_n	weights
b_{pn}	parity bit n
BER	bit error rate
C	channel capacity
$\mathbf{C}$	covariance matrix
$C(v)$	Fresnel cosine integral
$\hat{\mathbf{C}}$	sample covariance matrix
$\mathbf{C}_{\mathrm{noise}}$	noise covariance matrix
$\mathbf{C}_{\mathrm{noise}-s}$	noise–signal covariance matrix
$\mathbf{C}_{s-\mathrm{noise}}$	signal–noise covariance matrix
$\mathbf{C}_s$	signal covariance matrix

$\mathbf{C}_{sr}$	receive signal covariance matrix
$\mathbf{C}_{st}$	transmit signal covariance matrix
C_{text}	ciphertext
CL	confidence level
CNR	carrier-to-noise ratio
C/N	carrier-to-noise ratio
c	speed of light
c_p	specific heat
cdf	cumulative distribution function
cdf_{norm}	standard normal CDF
D	antenna directivity
$\mathbf{D}$	singular value diagonal matrix
D	raindrop equivalent spherical diameter
D_{max}	maximum aperture size
D_{out}	outer cylindrical conductor of diameter of coax
D_{TE}	TE diffraction coefficient for a finitely conducting wedge
D_{TM}	TM diffraction coefficient for a finitely conducting wedge
d	distance between antennas
d_h	Hamming distance
d_{ia}	wire diameter
d_{max}	maximum distance
d_p	penetration depth
d_{rh}	distance of receive antenna to horizon
d_{skip}	skip distance
d_{sp}	spacing between turns in a helical antenna
d_{th}	distance of transmit antenna to horizon
d_x	element spacing in x-direction
d_y	element spacing in y-direction
$\vec{E}(t)$	time-dependent electric field
$\mathbf{E}$	error vector
$\mathbf{E}_{antenna}$	electric field of antenna
E_b/N_0	energy per bit
$\mathbf{E}_{diff}$	diffracted electric field
$\mathbf{E}_{GO}$	GO electric field
$\mathbf{E}_i$	incident electric field
E_{LOS}	LOS electric field at receiver
$\mathbf{E}_r$	reflected electric field
E_r	r component of the electric field
E_s	signal energy
E_T	total electric field
$\mathbf{E}_t$	electric field transmitted into medium
E_x	x component of the electric field
E_{xco}	co-polarized electric field in the x-direction

E_{xcross}	cross polarized electric field in the x-direction
E_y	y component of the electric field
E_{yco}	co-polarized electric field in the y-direction
E_{ycross}	cross polarized electric field in the y-direction
E_z	z component of the electric field
E_φ	φ component of the electric field
E_θ	θ component of the electric field
$EB_n(\theta)$	eigenbeam n
EIRP	effective isotropic radiated power
ENOB	effective number of bits
$\hat{e}_i$	polarization vector of incident wave
e_n	bit error n
$\hat{e}_r$	polarization vector of receive antenna
erfc()	complementary error function
F	noise factor
$F(v_F)$	Fresnel integral
F_{LNA}	LNA noise factor
F_{pg}	path gain factor
F_{sp}	frequency spreading factor
F_{tag}	fade margin
f	frequency
f_c	carrier frequency
f_{crit}	critical frequency
f_D	Doppler-shifted frequency
f_{Dmax}	maximum Doppler frequency
f_{hi}	highest frequency in a bandwidth
f_{lo}	lowest frequency in a bandwidth
f_m	frequency of message signal
f_0	resonant frequency
$\mathbf{G}$	generator matrix
G	antenna gain
G_{amp}	amplifier gain
G_{EGC}	EGC diversity gain
G_{MRC}	MRC diversity gain
G_p	processing gain
G_r	gain of receiving antenna
G_{reader}	gain of reader antenna in direction of tag
G_{sd}	selection diversity gain
G_t	gain of transmitting antenna
G_{tag}	gain of tag antenna in the direction of reader
g_n	generating polynomial
$\mathbf{H}$	parity check matrix
$\mathbf{H}$	channel matrix

$\mathbf{H}_i$	incident magnetic field
$\mathbf{H}_n$	$n \times n$ Hadamard matrix
H	entropy
$H(f)$	transfer function
$H_c(f)$	channel transfer function
$H_{eq}(f)$	equalizer transfer function
H_n	screen height
$H_o(f)$	overall transfer function
$\mathbf{H}_r$	reflected magnetic field
H_r	r component of the magnetic field
$H_r(f)$	receive transfer function
$H_{RC}(f)$	raised cosine transfer function
$H_{RRC}(f)$	root-raised cosine transfer function
$\mathbf{H}_t$	magnetic field transmitted into medium
$H_t(f)$	transmit transfer function
H_φ	φ component of the magnetic field
H_θ	θ component of the magnetic field
h	height
$h(t)$	impulse response
h'	virtual height of the ionospheric layer
$h_c(t)$	channel impulse response function
h_{mn}	subchannel impulse response
h_o	obstacle height above ground
h_r	height of receive antenna
$h_{RC}(t)$	raised cosine impulse response
$h_{RRC}(t)$	root-raised cosine impulse response
h_t	height of transmit antenna
I	information
I_{dipole}	dipole current
$\mathbf{I}_N$	$N \times N$ identity matrix
I_0	constant current
$I_0(\xi)$	zeroth order modified Bessel function of the first kind
J	joules
$J_n(\cdot)$	nth order Bessel function
K_L	constraint length
K	kelvin
K	index of geomagnetic activity
Kp	estimated planetary K index
K_r	Rice factor
k	wavenumber
k_B	Boltzmann's constant $= 1.38 \times 10^{-23}$ J/K
k_e	earth enlargement constant
$\hat{\mathbf{k}}_{mp}$	unit propagation vector for the multipath signal

$\hat{\mathbf{k}}_{tr}$	unit propagation vector from transmitter to receiver
k_x	wavenumber in x-direction
k_y	wavenumber in y-direction
k_z	wavenumber in z-direction
k_0	wavenumber in free space
L	loss
L_{block}	blockage loss
L_{dB}	path loss in dB
$L_{diff\,dB}$	diffraction loss in dB
L_{floor}	floor loss
L_{hata}	Hata attenuation
L_{indoor}	indoor propagation loss
L_{rain}	rain loss
L_t	transmission line loss
LUF	lowest usable frequency
L_{wall}	wall loss
l	number of bits in a message
$\mathbf{M}$	vector of messages
M	integer number (quantity)
MUF	maximum usable frequency
mass	mass of object
m_n	message n
$m(t)$	message signal
N	number of elements
N_0	noise power spectral density
$N(D)$	rain drop size distribution
N_a	number of adaptive element
N_{bits}	total number of bits
$N_{cluster}$	number of cells in a cluster
$N_{databits}$	number of data bits
N_{dr}	Marshall and Palmer drop size constant
N_e	electron density
N_{ec}	number of errors corrected
N_{ed}	number of errors detected
N_{frame}	number of frames
N_{hel}	number of turns in the helical antenna
N_k	rank of channel matrix
N_{lev}	number of different quantization levels
N_{mess}	number of messages
N_p	number of parity bits
N_r	number of receive elements
N_s	number of samples of the covariance matrix
N_{samp}	number of samples

N_{sunfade}	maximum number of sun fade days
N_t	number of transmit elements
N_{turn}	number of turns in loop antenna
N_x	number of elements in x-direction
N_y	number of elements in y-direction
NF	noise figure
n	array noise vector
$n(t)$	AWGN with power spectral density $N_0/2$
$\bar{n}$	Taylor array constant
n_a	atmospheric index of refraction
n_i	index of refraction in medium of incident wave
n_t	index of refraction in medium of transmitted wave
P	parity bit generating matrix
P	power
P_a	power absorbed
P_{avg}	average power
P_D	distortion power
P_N	noise power
P_{Namp}	noise power generated by an amplifier
P_{Nin}	input noise power
P_{Nout}	output noise power
P_r	power received
P_{reader}	reader transmit power
P_s	signal power
P_{sin}	input signal power
P_{sout}	output signal power
P_t	transmit power
P_{tag}	power delivered to the tag IC
p	probability
pdf	probability density function
pdf_{norm}	standard normal PDF
P_I	interference power
P_N	noise power
P_{Namp}	noise power generated by an amplifier
P_s	signal power
P_{sin}	input signal power
P_{sout}	output signal power
P_{text}	plaintext
PSD	power spectral density
PSD_B	baseband PSD
PSD_P	bandpass PSD
$Q(\cdot)$	Q function
Q	eigenvector matrix

R	distance
R_{ain}	rainfall rate
R_b	data rate
R_c	code rate
R_e	resistor
R_f	co-channel distance
$\tilde{R}_i$	radius of inscribed circle
R_i	distance to image antenna
R_L	resistive loss
R_{load}	load resistance
R_{LOS}	LOS path
R_M	multipath distance traveled
R_o	radius of circumscribed circle
R_r	radiation resistance
r	distance
r'	distance to source point
r_a	radius of circular aperture
r_c	radius of circular array
r_e	radius of the earth
r_{ec}	apparent earth radius due to refraction
r_ℓ	loop radius
r_{mnp}	length of path p from transmit element m to receive element n
r_n	radius of the nth Fresnel zone ellipse
r_0	distance from center point of diffraction plane
$\mathbf{S}$	syndrome
$S(f)$	Fourier transform of signal
$S(v)$	Fresnel sine integral
SAR	specific absorption rate
S_B	Brussels International Sunspot Number
$S_r(f)$	Fourier transform of received signal
$S_t(f)$	Fourier transform of transmitted signal
S_{wolf}	Wolf number
s	separation between wires
$s(t)$	analog signal
$s[n]$	sampled signal
$\mathbf{s}$	array signal vector
$s_{in}(t)$	input signal
$s_{out}(t)$	output signal
s_p	distance from feed to shorting pin on PIFA
$s_r(t)$	receive signal
$\mathbf{s}_r(t)$	receive signal vector
$\tilde{\mathbf{s}}_r$	recovered data signals at the receiver

$s_{rn}(t)$	signal arriving at receive element n
$s_t(t)$	transmit signal
$\mathbf{s}_t(t)$	transmit signal vector
$\tilde{\mathbf{s}}_t$	transmit data signals before precoding
$\hat{s}_t(t)$	approximation of transmitted signal
$s_{tm}(t)$	signal transmitted from element m
s_{11}	s-parameter, reflection coefficient
SINAD	signal-to-noise and distortion ratio
SINR	signal-to-interference plus noise ratio
SNR	signal-to-noise ratio
$\overline{\text{SNR}}$	average SNR
SNR_{in}	input SNR
SNR_{out}	output SNR
T	period in time
T_0	ambient temperature (290 K)
T_{aext}	temperature of external sources
T_{afeed}	temperature of antenna feed line
T_{aloss}	temperature of antenna losses
T_{ant}	antenna temperature
T_b	bit length
T_c	coherence time
T_e	equivalent noise temperature
T_{line}	transmission line temperature
T_s	symbol length
$T_{sunfade}$	maximum sun fade time
t	time
$\tan\delta_{LF}$	loss factor
$\mathbf{U}$	SVD receive data weights
V	volts
$\mathbf{V}$	SVD transmit data weights
V_{ADCmin}	ADC minimum detectable voltage
V_{ADCmax}	ADC maximum allowed voltage
V_c	carrier voltage
V_D	forward voltage drop
V_{DC}	output DC voltage
V_{load}	voltage across load
V_m	message signal voltage
V_p	peak voltage
V_{rms}	rms voltage
V_{thresh}	threshold voltage
$\mathbf{v}_r$	receiver velocity vector
$\mathbf{v}_t$	transmitter velocity vector
w	width

$\mathbf{w}$	weight vector
w_n	weight at element n
$\mathbf{w}_{\text{opt}}$	optimized weights
$\mathbf{X}$	binary information
X_a	antenna reactance
X_c	decorrelation distance
XPD	cross polarization discrimination
$\hat{x}$	unit vector in x-direction
x_n	x-location of element n
x_0	x-distance in diffraction plane
$\mathbf{Y}$	transmitted codeword
$\hat{y}$	unit vector in y-direction
y_n	y-location of element n
y_0	y-distance in diffraction plane
$\mathbf{Z}$	received codeword
Z_{ant}	antenna impedance
Z_c	characteristic impedance
Z_{in}	input impedance
Z_L^0	load impedance of binary 0
Z_L^1	load impedance of binary 1
Z_0	free space impedance
z	z-transform variable
z_n	x-location of element n
z_n	z-transform of nth zero
$\hat{z}$	unit vector in z-direction
z_0	distance from aperture to diffraction pattern
α	raised cosine filter roll-off factor
β	constant
β_{AM}	amplitude modulation index
β_{FM}	frequency modulation index
β_{PM}	phase modulation index
Γ	reflection coefficient
Γ_d	reflection coefficient from dielectric interface
Γ_g	ground reflection coefficient
$\Gamma_p(m, n)$	reflection coefficient of path p
Γ_{TE}	TE reflection coefficient
Γ_{TM}	TM reflection coefficient
$\Gamma_{0,1}$	reflection coefficient of binary 0,1
γ	power loss due to distance
γ_E	Euler's constant
γ_m	weight for beam m
γ_p	packet throughput
Δ	time difference

Δf	maximum frequency deviation
ΔR	additional length
ΔT	rise in temperature
$\Delta \mathbf{w}[n]$	weight increment
$\Delta \tau$	additional time
$\delta(\cdot)$	delta function
δ_a	aperture efficiency
δ_e	radiation efficiency
δ_n	phase at element n
δ_p	polarization loss factor
ε	permittivity
ε	error
$\varepsilon(t)$	error
ε_r'	real part of permittivity
ε_r''	complex part of permittivity
ε_0	permittivity of free space $= 8.854187817 \times 10^{-12}$ F/m
ε_{eff}	effective permittivity
ε_r	relative permittivity
ζ	GTD distance parameter
η_{ob}	on object gain penalty
η_{PCE}	RF-DC power conversion efficiency
η_{se}	spectral efficiency
η_t	taper efficiency
θ	angle measured from z-axis
θ_{3dB}	3 dB beamwidth
θ_b	Brewster's angle
θ_c	critical angle
θ_D	diffraction angle
θ_i	incident angle
θ_{null1}	location of the first null
θ_r	reflection angle
θ_s	scan angle
θ_t	transmission angle
$\mathbf{\Lambda}_{\text{noise}}$	noise eigenvalues
$\mathbf{\Lambda}_\lambda$	eigenvalue matrix
λ	wavelength
λ_m^{Ψ}	eigenvalues of $\mathbf{\Psi}$
λ_{max}	maximum eigenvalue
λ_{min}	minimum eigenvalue
λ_n	nth eigenvalue
λ_p	number of packets

λ_x	wavelength in x-direction
λ_y	wavelength in y-direction
λ_z	wavelength in z-direction
λ_0	resonant wavelength
$\hat{\lambda}_m$	singular value
μ	permeability
μ_0	permeability of free space $= 4\pi \times 10^{-7} \text{ N/A}^2$
μ	mean
μ_s	signal mean
$\mu_{\chi\text{dB}}$	mean of χ_{dB} in dB
v_F	Fresnel integral input
v_n	noise at element n
ξ	integration variable
ρ	tissue density
ρ_r	Marshall and Palmer drop size constant
ρ_T	threshold voltage normalized to the rms signal level
σ	standard deviation
σ_{MP}	standard deviation of multipath signals
σ_{noise}	noise standard deviation
σ_{noise}^2	noise variance
σ_{rcs}	radar cross section of the tag
σ_s	signal standard deviation
T_{load}	power transmission coefficient
T_{TE}	TE transmission coefficient
T_{TM}	TM transmission coefficient
τ	time delay
τ_{mnp}	time taken for a signal from transmit element m to receive element n via path p
τ_n	time delay at element n
τ_p	packet length
$\mathbf{\Phi}_s$	ESPRIT diagonal matrix
φ	phase
ϕ	angle measured from x-axis
ϕ'	incidence angle, measured from incidence face
ϕ_n	angular location of element n
ϕ_s	scan angle
χ	ratio of received to transmitted power
χ_{dB}	χ in dB
$\mathbf{\Psi}$	estimate of $\mathbf{\Phi}_s$
ψ	phase
ψ_g	angle between ground and signal path

ψ_n	phase of the nth zero
ψ_y	phase of x-component
ψ_z	phase of y-component
Ω	ohms
Υ	volume

Acronyms

3DES	triple data encryption standard
A	ampere
ABS	acrylonitrile, butadiene, and styrene
ACK	acknowledgement
ADC	analog to digital convertor
AES	advanced encryption standard
AF	array factor
AFD	average fade duration
AM	amplitude modulation
AMSAT	Radio Amateur Satellite Corporation
AOA	angle of arrival
AP	access point
AR	axial ratio
ARQ	automatic repeat request
ASCII	American Standard Code for Information Interchange
ASK	amplitude shift keying
AWGN	additive white Gaussian noise
BAT	battery-assisted tag
BER	bit error rate
BFSK	binary frequency shift keying
BPF	bandpass filter
bps	bits per second
BPSK	binary phase shift keying
BSA	basic service area
BSS	basic service set
CBC	cipher block chaining
CDF	cumulative density function
CDM	code division multiplexing
CDMA	code division multiple access
CFB	cipher feedback
CL	confidence level
CMOS	Complementary metal–oxide–semiconductor
CNR	carrier-to-noise ratio
CPS	cyber–physical system

CRC	cyclic redundancy check
CRC	cyclic redundancy check
CSI	channel state information
CSIR	channel state information at the receiver
CSIT	channel state information at the transmitter
CTR	counter
DAC	digital-to-analog converter
DARPA	Defense Advanced Research Projects Agency
dB	decibel
dBm	power relative to 1 milliwatt
DBS	Direct Broadcast Satellite
DC	direct current
DFE	direction finding
DFE	decision-feedback equalizer
DL	downlink
DMI	direct matrix inversion
DOA	direction of arrival
DoS	denial of service
DPSK	differential phase shift keying
DSB	double sideband
DSN	Deep Space Network
DSSS	direct-sequence spread spectrum
ECB	electronic codebook
ECC	Error Correcting Codes
EEPROM	electrically erasable programmable read-only memory
EGC	equal gain combining
EHS	electromagnetic hypersensitivity
EMC	electromagnetic compatibility
EMI	electromagnetic interference
ENOB	effective number of bits
EPC	electronic product code
ESD	electrostatic discharge
ESPRIT	*E*stimation of *S*ignal *P*arameters via *R*otational *I*nvariance *T*echniques
ESS	extended service set
EUI	extended unique identifier
FCC	Federal Communications Commission
FDD	frequency division duplexing
FDM	frequency division multiplexing
FDMA	frequency division multiple access
FDX	full duplex
FFH	fast frequency hopping
FFT	fast Fourier transform

FM	frequency modulation
FM0	type of encoding
FOT	Frequency of Optimum Traffic
FSS	Fixed Satellite Service
Gen 2	generation 2
GEO	geosynchronous earth orbit
GHz	gigahertz
GMSK	Gaussian minimum shift keying
GNSS	Global Navigation Satellite System
GO	geometrical optics
GPS	Global Positioning System
GSM	Global System for Mobile
GTD	geometrical theory of diffraction
HDX	half duplex
HEO	high earth orbit
HF	high frequency
HOW	handover
HPF	high-pass filter
HVAC	heating, ventilation, and air conditioning
Hz	hertz
IARC	International Agency for Research on Cancer
IBSS	independent basic service set
IC	integrated circuit
IDS	intrusion detection system
IEEE	Institute of Electrical and Electronics Engineers
IFF	identify friend or foe
IFFT	inverse fast Fourier transform
IIR	infinite impulse response
IO	input–output
IoT	Internet of Things
IP	Internet protocol
IPv4	Internet protocol version 4
IQ	in phase-quadrature
IR	infrared
ISI	Intersymbol Interference
ISM	industrial, scientific, and medical
ISO	International Organization for Standardization
ITU	International Telecommunication Union
IZ	interrogation zone
J	joules
JPEG	Joint Photographic Experts Group
JPEG-LS	Lossless Joint Photographic Experts Group
K	kelvin

kbps	kilobits per second
kg	kilogram
kHz	kilohertz
LAA	locally administered address
LAN	local area network
LBT	listen before talk
LCR	level crossing rate
LDS	laser direct structuring
LED	light emitting diode
LEO	low earth orbit
LH	left hand
LHCP	left-hand circular polarization
LMS	least mean square
LNA	low noise amplifier
LO	local oscillator
LOS	line of sight
LPDA	log periodic dipole antenna
LPF	low pass filter
LRC	longitudinal redundancy check
LSB	least significant bit
LUF	lowest usable frequency
MAC	media access control
Mbps	megabits per second
Mcps	million chips per second
MEM	maximum entropy method
MEO	medium earth orbit
MHz	megahertz
MID	molded interconnect devices
MIMO	multiple input/multiple output
MISO	multiple input single output
MMSE	minimum mean square error
MOSFET	metal-oxide semiconductor field-effect transistor
MPE	maximum permissible exposure
MPEG	Moving Picture Experts Group
MRC	maximum ratio combining
MRI	magnetic resonance imaging
MSE	mean square error
MSK	minimum shift keying
MUF	maximum usable frequency
MUSIC	*MU*ltiple *SI*gnal *C*lassification
mW	milliwatt
N	Newtons
NASA	National Aeronautics and Space Administration

NCAR	National Center for Atmospheric Research
NCI	National Cancer Institute
NFC	near-field communications
NIC	network interface card
NLM	National Library of Medicine
NRZ	nonreturn to zero
NSA	National Security Agency
NVIS	near vertical incidence
OFB	output feedback
OFDM	orthogonal frequency division multiplexing
OFDMA	orthogonal frequency division multiple access
OOK	on–off keying
OQPSK	offset quadrature phase shift keying
OT	operational technology
OUI	organizationally unique identifier
OWF	Optimum Working Frequency
PC	personal computer
PCB	printed circuit board
PDF	probability density function
PHD	Pisarenko Harmonic Decomposition
PIE	pulse interval encoding
PIFA	planar inverted F antenna
PKC	public key cryptography
PLF	polarization loss factor
PM	phase modulation
PRN	pseudo random noise
PSD	power spectral density
QAM	quadrature amplitude modulation
QPSK	quadrature phase shift keying
RAM	random access memory
RAP	rogue AP
RF	radio frequency
RFI	radio frequency interference
RFID	radio frequency identification
RHCP	right-hand circular polarization
RO	read only
rpm	revolutions per minute
RTF	reader talk first
RW	read write
RZ	return to zero
SAR	specific absorption rate
SBR	shooting and bouncing rays
SEER	Surveillance, Epidemiology, and End Results
SFH	slow frequency hopping

SFU	Solar Flux Units
SIMO	single input multiple output
SIR	signal-to-interference ratio
SISO	single-input single-output
SKC	secret key cryptography
SMI	sample matrix inversion
SSB	single sideband
SSID	service set identifier
STRIDE	spoofing, tampering, repudiation, information disclosure, DoS, and elevation of privilege
SVD	singular value decomposition
TARI	type A reference interval
TDD	time division duplexing
TDM	time division multiplexing
TDMA	time division multiple access
TDRS	Tracking and Data Relay Satellite
TE	transverse electric
TEC	total electron count
TEM	transverse electromagnetic
TID	tag identifier
TLM	telemetry
TM	transverse magnetic
TTF	tag talk first
UAA	universally administered address
UHF	ultra high frequency
UL	uplink
UPC	universal product code
UTD	uniform theory of diffraction
UV	ultraviolet
UWB	ultra wideband
V	volts
VHF	very high frequency
VHP	Visible Human Project
VPN	virtual private network
VRC	vertical redundancy check
W	watts
WBSAR	whole-body average SAR
WEP	Wired Equivalent Privacy
WHO	World Health Organization
WLAN	wireless local area network
WORM	write-once-read-many
WPA	Wi-Fi Protected Access
WPA2	Wi-Fi Protected Access version 2

1

Introduction

At the end of the nineteenth century, "wireless" meant "wireless telegraphy" which eventually became known as radio. Ham radio kept the term "wireless" alive, but obscure, until cell phones resurrected it toward the end of the twentieth century. Most wireless technologies use radio frequencies (RF), but infrared (IR), magnetic, optical, and acoustic systems also enable wireless communication. Wireless systems include a wide range of fixed, mobile, and portable applications. Designing a wireless system involves all the same challenges as a wired system plus the antennas and propagation channel. This chapter begins with a brief history of wireless communications then explains some basic concepts needed for proceeding through the rest of this book. The second half of this book (Chapters 6–12) is devoted to practical applications.

1.1 Historical Development of Wireless Communications

Long distance communications seem easy now, but that was not the case throughout history. In 490 BC, legend says that Philippides ran from Marathon to Athens and announced that the Greeks defeated the Persians in the Battle of Marathon (according to Google Maps about a 44.4 km drive), then dropped dead [1]. That long run became the standard for today's marathon. Current wireless networks deliver that same message in a blink of the eye. People wanted a faster way to communicate over long distances than using a messenger. Several ingenious, low data rate innovations emerged. Figure 1.1 shows four early wireless communication systems that replaced face-to-face delivery of the message: smoke signals, heliographs (mirrors), semaphore (flags), and drums. Weather and limited line of sight hindered most wireless communications. In addition, messages had to be simple and were prone to misinterpretation at the receiver.

The first quantum leap in fast long distance communication occurred in the 1800s with the introduction of electrical circuits that send signals over wires. In

Wireless Communications Systems: An Introduction, First Edition. Randy L. Haupt.
© 2020 John Wiley & Sons, Inc. Published 2020 by John Wiley & Sons, Inc.

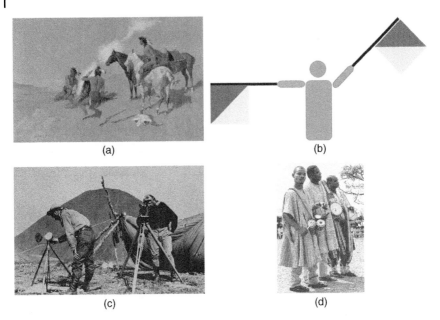

Figure 1.1 Early forms of wireless communications. (a) Smoke signals, (b) semaphore, (c) heliograph, (d) drums.. *Source:* (a) https://commons.wikimedia.org/wiki/File:Frederic_Remington_smoke_signal.jpg. Public domain; (b) Author originated; (c) www.photolib.noaa.gov/historic/c&gs/theb1633.htm. Courtesy of NOAA; (d) https://www.flickr.com/photos/58034970@N00/178631090. Licensed under CC BY 2.0 [3].

the 1830s, Cooke and Wheatstone demonstrated a telegraph system with five magnetic needles that an electric current forced to point at letters and numbers that form a message. Britain adopted this invention for railroad signaling [2]. At the same time, Samuel Morse independently developed the electric telegraph. He collaborated with Gale and Vail to build a telegraph that transmitted an electric signal by pushing an operator key that connects a battery to a wire and sends the electric signal down a wire to a receiver [2]. This simple system required a switch and battery at both ends of a wire. The length of wire and the loss of the signal strength over that wire limited communication distance. Wired communications forced users to established nodes, but significantly increased the data rate as well as made communications independent of weather and line of sight.

Electronic wireless communications began when Maxwell found that all electromagnetic waves travel at the speed of light. He also discovered the relationship between electricity and magnetism [4]. Maxwell's mathematical ideas of electromagnetic wave propagation needed experimental verification, so Heinrich Hertz built and tested the 100 MHz dipole antenna shown in Figure 1.2. In order to increase the radiation intensity in a desired direction, he built the higher gain reflector antenna shown in Figure 1.3. Hertz provided

Figure 1.2 The first radiating dipole designed by Hertz. *Source:* https://en.wikipedia.org/wiki/Heinrich_Hertz#/media/File:Hertz_first_oscillator.png.

Figure 1.3 Hertz designed higher gain reflector antennas. *Source:* https://commons.wikimedia.org/wiki/File:Hertz_spark_gap_transmitter_and_parabolic_antenna.png.

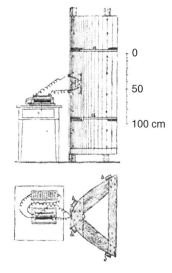

the means for getting the transmitted signal from a wire to the air then back to another wire connected to a receiver. Professor Oliver Lodge demonstrated the reception of wireless Morse code in 1894 using a newly invented "coherer" or receiver [5]. In 1895, Guglielmo Marconi used a more practical setup to demonstrate transmitted signals up to one-half mile [6]. Marconi then tried two new ideas: (i) placing the antenna high off the ground and (ii) grounding the transmitter and receiver. These modifications demonstrated that signals could travel up to 3.2 km and over hills. Marconi received a British patent for radio in 1898 [7] and a US patent a few years later [8]. Around the same time, Tesla tinkered with radiowave propagation and invented radio remote control [9]. He transmitted an RF wave from the apparatus shown in Figure 1.4 that opened and closed switches in order to steer the model boat. Tesla received a US patent for radio in 1898 [10]. A patent battle between Tesla and Marconi continued until after their deaths. In 1943 (six years after Marconi's death and six months after Tesla's death), the US Supreme Court ruled that Tesla was the inventor of radio and not Marconi.

In 1900, Reginald Fessenden demonstrated amplitude-modulation (AM) radio that allowed more than one station to broadcast at the same time and in the same area (as opposed to spark-gap radio, where one transmitter covers the

Figure 1.4 Tesla's apparatus for the remote control of a boat. *Source:* https://en.wikipedia.org/wiki/Nikola_Tesla#/media/File:Tesla_boat1.jpg.

entire bandwidth of the spectrum) [11]. A few years later, Edwin Armstrong patented three important inventions that made today's radio possible: regeneration, superheterodyning, and wide-band frequency modulation (FM) [12]. Regeneration or the use of positive feedback increased the received radio signal amplitude to the point where headphones were no longer needed. The superheterodyne receiver replaced several tuning controls with only one. It made radios more sensitive and selective as well. Wideband FM improved the sound quality and fidelity over AM. Armstrong set the stage for the 1940s when a flurry of inventions made advanced wireless communications possible, including the mobile phone, spread spectrum, and television. In addition, Harry Nyquist's work (Nyquist rate) became the impetus for Claude Shannon to establish the theoretical foundations for modern information theory [13]. Some of the more notable advances in wireless communications appear in Figure 1.5.

1.2 Information

A message contains information that a sender wants the recipient to know. The sender and receiver may be human or not. Some messages are a simple "yes" or "no," while others are quite complicated, such as a movie. Message value depends on the information content. In mathematical terms, the information content of message n is expressed in bits by [14]

$$I_n = \log_2(1/p_n) \text{ bits} \tag{1.1}$$

where p_n is the probability of transmitting message n. Thus, a less likely message has a higher information content than a more likely message. The game of

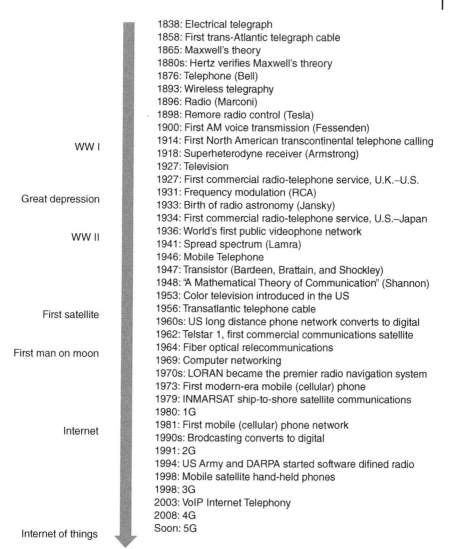

1838: Electrical telegraph
1858: First trans-Atlantic telegraph cable
1865: Maxwell's theory
1880s: Hertz verifies Maxwell's threory
1876: Telephone (Bell)
1893: Wireless telegraphy
1896: Radio (Marconi)
1898: Remore radio control (Tesla)
1900: First AM voice transmission (Fessenden)
1914: First North American transcontinental telephone calling
1918: Superheterodyne receiver (Armstrong)
1927: Television
1927: First commercial radio-telephone service, U.K.–U.S.
1931: Frequency modulation (RCA)
1933: Birth of radio astronomy (Jansky)
1934: First commercial radio-telephone service, U.S.–Japan
1936: World's first public videophone network
1941: Spread spectrum (Lamra)
1946: Mobile Telephone
1947: Transistor (Bardeen, Brattain, and Shockley)
1948: "A Mathematical Theory of Communication" (Shannon)
1953: Color television introduced in the US
1956: Transatlantic telephone cable
1960s: US long distance phone network converts to digital
1962: Telstar 1, first commercial communications satellite
1964: Fiber optical relecommunications
1969: Computer networking
1970s: LORAN became the premier radio navigation system
1973: First modern-era mobile (cellular) phone
1979: INMARSAT ship-to-shore satellite communications
1980: 1G
1981: First mobile (cellular) phone network
1990s: Brodcasting converts to digital
1991: 2G
1994: US Army and DARPA started software difined radio
1998: Mobile satellite hand-held phones
1998: 3G
2003: VoIP Internet Telephony
2008: 4G
Soon: 5G

WW I

Great depression

WW II

First satellite

First man on moon

Internet

Internet of things

Figure 1.5 Timeline for the development of modern wireless systems.

Scrabble uses this concept to assign points to a letter. In Scrabble, players take turns placing tiles with letters and points onto a 15×15 grid of squares in order to form words as in a crossword puzzle [15]. Players receive points on the tiles used to form a word. The letter "Q" has a value of 10, whereas the letter "E" only has a value of 1, because "Q" occurs less frequently in the English language than "E." You know less about a word if it has the letter "E" than if it has the letter "Q." Table 1.1 contains the number of letter tiles and associated points in Scrabble.

Table 1.1 Distribution of letters and points in the game of Scrabble [16].

		Number of tiles							
		1	2	3	4	6	8	9	12
Points	0		[blank]						
	1				L S U	N R T	O	A I	E
	2			G	D				
	3		B C M P						
	4		F H V W Y						
	5	K							
	8	J X							
	10	Q Z							

The average information called entropy (H) equals the information of message n times its probability of occurrence summed over all N_{mess} messages.

$$H = \sum_{n=1}^{N_{mess}} p_n I_n = \sum_{n=1}^{N_{mess}} p_n \log_2(1/p_n) \text{ bits} \tag{1.2}$$

Example

Calculate the information in the first five letters of the English alphabet given the graph in Figure 1.6.

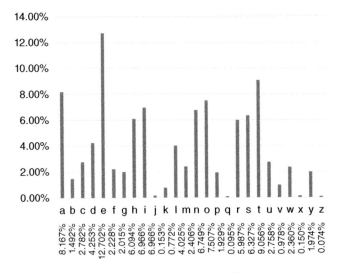

Figure 1.6 Frequency of letters in English text [17].

Solution

Use (1.1) and the values of p_n in Figure 1.6 to generate the following Table 1.2:

Table 1.2 Probability and information associated with the first five letters of the English alphabets.

Letter	A	B	C	D	E
p_n	0.08167	0.01492	0.02782	0.04253	0.12702
I_n	3.6140	6.0666	5.1677	4.5554	2.9769

1.3 Wired Communications

A transmission line or waveguide minimizes the signal loss by forcing the signal into a conduit from the transmitter to the receiver. Even wireless systems have cabling between the transmitter and the antenna or from the antenna to the receiver.

A single wire only carries a DC current (Figure 1.7a). Time varying signals need two paths as shown in Figure 1.7b. At one point along the twin wire transmission line, the current on one wire travels in the opposite direction of the current on the other wire. The fields between the wires add in phase while the fields outside the wires do not. Thus, the signal stays between the wires as it propagates from one end to the other. The copper wires have a protective plastic that keeps the wire separation constant. Twisting the two wires, as shown in Figure 1.7c, reduces coupling from other nearby wires. Cheap two-wire transmission lines work fine at frequencies below 1 GHz. The twin wire transmission line characteristic impedance is given by [18]

$$Z_c = \sqrt{\frac{\mu}{\varepsilon}} \frac{1}{\pi} \cosh^{-1}\left(\frac{s}{d_{ia}}\right) \ \Omega \tag{1.3}$$

where

s = separation between wires
d_{ia} = wire diameter
μ = permeability
$\varepsilon = \varepsilon_r \varepsilon_0$ = permittivity
ε_r = relative permittivity.

In free space, $\mu = \mu_0 = 4\pi \times 10^{-7}$ N/A^2 and $\varepsilon = \varepsilon_0 = 8.854\,187\,817 \times 10^{-12}$ F/m. When all components or lines in an RF circuit have the same impedance, the maximum power reaches the load. For instance, an antenna that has the same impedance as the transmission line receives the maximum possible power.

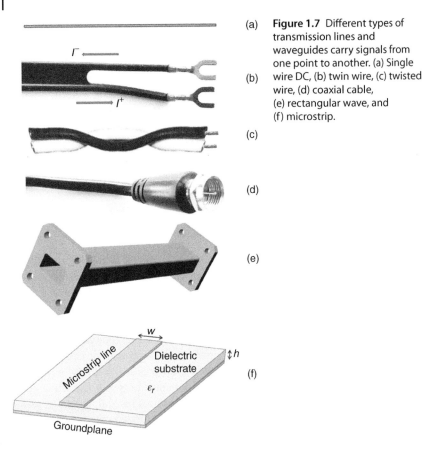

(a)

(b)

(c)

(d)

(e)

(f)

Figure 1.7 Different types of transmission lines and waveguides carry signals from one point to another. (a) Single wire DC, (b) twin wire, (c) twisted wire, (d) coaxial cable, (e) rectangular wave, and (f) microstrip.

Example

Find the wire separation in a twin wire transmission line with an impedance of 75 Ω using wires that are 1 mm in diameter and surrounded by plastic with $\varepsilon_r = 2.2$.

Solution

Substitute the known quantities into (1.3) and solve $75 = \frac{377}{\pi\sqrt{2.2}}\cosh^{-1}\left(\frac{s}{1}\right) \Rightarrow s = 1.462$ mm

Coaxial cable or coax (Figure 1.7d) appears in many communication systems above 50 MHz, including satellite and cellular communication systems. Its advantages include low cost, high bandwidth, and protection from interference. The coax has an inner wire of diameter d_{in} surrounded by an outer cylindrical conductor of diameter D_{out}. A dielectric surrounding the inner wire maintains a constant separation between the two conductors. If the dielectric

Table 1.3 RG-58 coaxial cable attenuation as a function of frequency (dB/m) [19].

Frequency (MHz)	100	500	1000	2500
Loss (dB)	0.125	0.313	0.478	0.87

is air or gas, then dielectric spacers placed at regular intervals maintain a constant separation between the conductors. The current on the inner conductor travels in the opposite direction as the current on the inside of the outer conductor, resulting in the signal propagating in a transverse electromagnetic (TEM) mode where both the electric and magnetic fields are perpendicular to the direction of propagation. The coaxial cable has characteristic impedance given by [19]

$$Z_c = \sqrt{\frac{\mu}{\varepsilon}} \frac{1}{2\pi} \ln \frac{D_{out}}{d_{in}} \tag{1.4}$$

Table 1.3 shows the loss in dB/m of RG-58 coaxial cable. The loss increases with frequency. In contrast, free space loss for wireless systems is independent of frequency.

A waveguide (like the rectangular metal waveguide in Figure 1.7e) contains an electromagnetic wave as it propagates from one end to the other. These reflections form modes that are a function of frequency and the waveguide dimensions. Among other things, the impedance depends on the shape of the waveguide and the mode. Optical fibers rely on variations in the dielectric constant of the glass to contain the signal within the fiber.

A PCB (printed circuit board) consists of a thin dielectric sandwiched between two very thin layers of copper. Microstrip lines have a thin trace etched from the top layer (Figure 1.7f). The line width (w), substrate height (h), and substrate dielectric constant (ε_r) determine the line characteristic impedance [20]. Typically, the impedance is designed to be 50 Ω.

1.4 Spectrum

Signals in wireless communications occupy a designated region of the RF spectrum. The operating frequency depends on regulatory requirements, propagation characteristics, signal attenuation, and available bandwidth. These properties have a distinct impact on key requirements such as the radio link range and the peak throughput as well as on the system capacity. For example more bandwidth allows higher throughput. Throughput depends on the received signal strength. In turn, the received signal strength depends

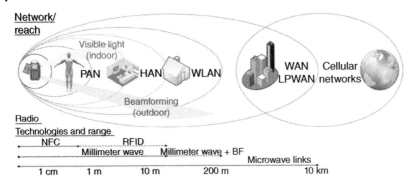

Figure 1.8 Overview of radio technologies. *Source:* Burg *et al.* [21]. Reproduced with permission of IEEE.

on the propagation characteristics (attenuation) and the maximum transmit power. On the one hand, higher frequencies have more available bandwidth, which allows for higher capacity. On the other hand, signal attenuation also increases proportional to the frequency which limits the range at a given transmit power. Higher frequencies are generally more attenuated by obstacles such as walls or windows. Figure 1.8 has an overview of different RF and technologies with their associated radio range.

A wireless system inserts its signal into the frequency spectrum in order to reduce interference with the myriad of other users. A party with many people talking to each other prevents a listener from hearing one particular conversation, because all speakers communicate at baseband. In other words, the frequency components in the signal extend from 0 Hz to some maximum voice frequency. Converting each conversation to an electrical signal does not help unless the different signals differ in time, frequency, coding, or polarization. The message rides on an electrical signal at a higher frequency called a carrier that propagates through the air (wireless) or through a transmission line or waveguide (wired). Frequency and polarization are properties of the carrier while time and coding are properties of the information signal. The transmitter modulates the baseband signal to a higher frequency and the receiver demodulates it.

The radio frequency spectrum extends from about 3 kHz to 300 GHz. Figure 1.9 shows the playground for various wireless applications. The small print precludes reading the designated frequency bands but provides an appreciation for the vast number of applications and the importance of having sufficient bandwidth to perform the desired function. Go to [22] to magnify the small print. Fierce and expensive battles occur between users that want to occupy the same frequency band. Governments auction the rights (licenses) to transmit signals over specific bands in order to efficiently allocate the resource as well as to raise money. Currently, commercial applications have

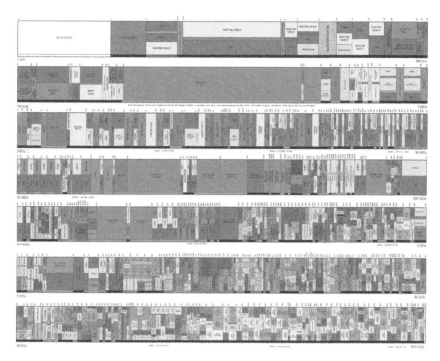

Figure 1.9 United States frequency allocations in the radio spectrum (courtesy of NTIA) [22]. *Source:* www.ntia.doc.gov.

needs that conflict with traditional government allocations, such as military and weather radar bands.

The International Telecommunications Union (ITU) advises national or regional regulatory bodies that assign and regulate licensed and unlicensed frequency bands. Licensed bands cover the majority of the spectrum and require a license for operating wireless systems. While licensing bands are expensive, the exclusive access avoids uncontrolled interference between users of the shared medium to provide reliable quality-of-service. Also, regulations often allow for larger power budgets in licensed bands than in unlicensed bands since interference is better controlled. In addition to the licensed spectrum, some frequency bands exist for use by anybody. These unlicensed parts of the spectrum are known as industrial, scientific, and medical (ISM) bands. Regulations define a set of rules that enables users to coexist. These rules typically restrict the maximum amount of transmit power in order to limit the range of each transmitter and enable spatial reuse of the spectrum.

The severe bandwidth limitations in the microwave spectrum motivate the use of higher (millimeter wave) frequencies at or beyond 28 GHz. Technology initially limited use of millimeter wave frequencies, but CMOS (complementary metal–oxide–semiconductor) processing opened consumer electronics to these frequencies [21]. Another recent push toward millimeter waves was the worldwide availability of almost 7 GHz of bandwidth at the ISM band around 60 GHz. Millimeter waves suffer more loss than microwaves, so they have limited use. Obstacles, including thin walls and windows, highly attenuate millimeter waves. The 60-GHz ISM band lies close to the oxygen absorption frequency that induces even more attenuation.

Radiation levels from wireless systems have government specified limits both in-band and out-of-band. Electromagnetic interference (EMI), also known as radio frequency interference (RFI), occurs when a device transmits signals that interfere with another device. Electromagnetic compatibility (EMC) means that a device does not emit radiation that causes EMI in other devices. EMI results from conducted and radiated emissions, as well as electrostatic discharge (ESD). EMC requires all equipment operating in a common electromagnetic environment to not interfere with each other. Three approaches to EMC include [23]:

1. Suppress emissions at the source.
2. Make the coupling path as inefficient as possible.
3. Make the receptor less susceptible to the EMI.

Simple solutions, such as grounding and shielding, solve many of these issues.

The Federal Communications Commission (FCC) regulates broadcast stations, amateur radio operators, and repeater stations in the United States. In addition, the FCC regulates EMC compliance under Title 47 of the Code of Federal Regulations [24]. Part 15 of these regulations concerns radio frequency devices, including intentional transmitters (e.g. mobile phones) and nonintentional radiators (e.g. PCs and TV receivers). Part 18 concerns equipment operating in the ISM bands. The FCC requirements only relate to radiated and conducted emissions. The FCC has no immunity limits like those associated with European EMC Certification.

1.5 Communication System

This book has two parts. The first part introduces the fundamentals of wireless communications. The block diagram of the wireless system in Figure 1.10 forms an outline for Chapters 2 through 5. Wireless communication starts with information. The information might be data or music. An analog-to-digital converter (ADC) transforms an analog signal into bits. Symbols contain groups of bits. For example the 8-bit ASCII code for the symbol "1" is 00110001. Additional bits added to the code detect and/or correct errors. This "channel coding" allows the receiver to correct errors induced by the channel. The modulator maps the channel encoder output to an analog signal suitable for

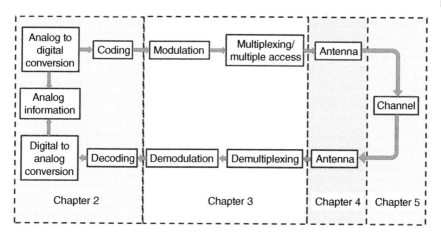

Figure 1.10 Block diagram of a digital wireless communications system.

transmission into the channel. An antenna transmits the signal into the channel at one end and another antenna receives the signal at the other end. A channel is the path taken by the transmitted signal to the receiver. Signals become distorted, noisy, and attenuated in the channel. The demodulator converts the received analog signal into a digital signal that feeds the channel decoder, etc. before arriving at the receiver. Successful signal detection occurs when the signal strength exceeds the receiver threshold and noise and interference did not induce errors that cannot be corrected.

The second half of this book uses the basic information from the first half to cover some practical topics in wireless communications. Chapter 6 introduces satellite communications, while Chapter 7 presents radio frequency identification (RFID). Smart antennas are critical to future advancements in communications, so Chapters 8–10 cover direction finding, adaptive nulling, and multiple input multiple output (MIMO). Security (Chapter 11) and Biological Effects of RF (Chapter 12) are topics of great concern and complete the book. The appendix has several short topics of interest. Many examples and problems in this book use MATLAB, so a few MATLAB hints appear in Appendix A.

Problems

1.1 Calculate the information in the letters A, B, C, D, and E of the English alphabet using the probabilities in Figure 1.6.

1.2 How many bits do the following pieces of information contain? (a) message probability = $\frac{1}{2}$ and (b) message probability = 1.0.

1.3 Calculate the entropy of the English alphabet using Figure 1.6.

1.4 Generate a histogram plot for Scrabble that is similar to Figure 1.6, except the *y*-axis is (a) points and (b) number of letters.

1.5 Calculate the entropy of Scrabble.

1.6 A meter has a read out of $[-5, -3, -1, 0, 1, 3,$ and $5\,\text{V}]$ with corresponding probabilities of [0.05 0.1 0.1 0.15 0.05 0.25 0.3]. Find the (a) entropy and (b) entropy if the output is only represented by three levels $[-4\,\text{V}\ 0\,\text{V}\ 4\,\text{V}]$.

1.7 Find the entropy of a binary code with two symbols, one with probability p and the other with probability $1 - p$.

1.8 Calculate the entropy of the string "asasdgasdgdsg" based on the frequency of occurrence of the letters in the string.

1.9 A codebook has four messages with probabilities of [0.1 0.2 0.3 0.4]. Find the number of bits needs to communicate the message using entropy as the lower bound.

1.10 If four messages have probabilities of [1/8 3/8 3/8 1/8], find the average information per message.

1.11 If five messages have probabilities of [1/2 1/4 1/16 1/16], find the average information per message.

1.12 A code uses a dash that is three times as long as a dot and occurs one in three symbols.
(a) Calculate the information in a dot and a dash.
(b) Calculate the average information of this code.
(c) If a dot lasts 10 ms and the interval between symbols is 10 ms, then calculate the average rate of information transmission.

1.13 Plot the input impedance of a twin wire transmission line vs. s/d_{in} using (1.3). Assume the wires are enclosed in plastic (find permittivity on web).

1.14 Calculate the input impedance for RG-58 cable. Obtain data for the calculation from a company on the web. How does your calculation compare with that given by the company?

1.15 Locate an online calculator for microstrip impedance. Find the impedance of a microstrip line with $\varepsilon_r = 4$, $h = 0.8$, and $w = 1.65$ mm. Assume the microstrip trace is 0.035 mm thick.

References

1 The Editors of Encyclopædia Britannica (2015). Battle of marathon. In: *Encyclopædia Britannica*, Web.
2 http://www.history.com/topics/inventions/telegrapha (accessed 20 May 2016).
3 https://www.flickr.com/photos/58034970@N00/178631090 (accessed 10 February 2019).
4 Maxwell, J.C. (1873). *A Treatise on Electricity and Magnetism*. Oxford: Clarendon Press.
5 https://en.wikipedia.org/wiki/Oliver_Lodge (accessed 25 October 2016).
6 https://en.wikipedia.org/wiki/Guglielmo_Marconi (accessed 25 October 2016).
7 Marconi, G. (1897). Improvements in transmitting electrical impulses and signals, and in apparatus therefor. British Patent No. 12,039. Date of Application 2 June 1896; Complete Specification Left, 2 March 1897; Accepted, 2 July 1897.
8 Marconi, G. (1901). Transmitting electrical impulses and signals and in apparatus, there-for. US Patent RE11,913, filed 1 April 1901; issued 4 June 1901.
9 https://en.wikipedia.org/wiki/Nikola_Tesla (accessed 25 October 2016).
10 Tesla, N. (1900). System of transmission of electrical energy. Issued on, Patent No. 645,576, entitled 20 March 1900.
11 https://www.britannica.com/biography/Reginald-Aubrey-Fessenden (accessed 25 October 2016).
12 https://en.wikipedia.org/wiki/Edwin_Howard_Armstrong (accessed 25 October 2016).
13 Shannon, C.E. (1948). A mathematical theory of communication. *Bell System Technical Journal* 27, pp. 379–423 and 623–656.
14 Johnson, D. (2016). Fundamentals of Electrical Engineering I. http://www.ece.rice.edu/~dhj/courses/elec241/col10040.pdf (accessed 25 May 2016).
15 https://en.wikipedia.org/wiki/Scrabble (accessed 25 May 2016).
16 http://www.wordfind.com/scrabble-letter-values (accessed 25 May 2016).
17 https://en.wikipedia.org/wiki/Letter_frequency (accessed 25 May 2016).
18 Collin, R.E. (1966). *Foundations for Microwave Engineering*. New York: McGraw-Hill.

19 http://www.qsl.net/co8tw/Coax_Calculator.htm (accessed 26 May 2016).

20 Pozar, D.M. (1998). *Microwave Engineering*, 2e. New York: Wiley.

21 Burg, A., Chattopadhyay, A., and Lam, K.Y. (2018). Wireless communication and security issues for Cyber–Physical Systems and the Internet-of-Things. *Proceedings of the IEEE* 106 (1): 38–60.

22 https://www.ntia.doc.gov/files/ntia/publications/january_2016_spectrum_wall_chart.pdf (accessed 10 December 2018).

23 Paul, C.R. (1992). *Introduction to Electromagnetic Compatibility*. New York: Wiley.

24 Ott, H.W. (1988). *Noise Reduction Techniques in Electronic Systems*, 2e. New York: Wiley.

2

Signals and Bits

Information contains facts about something or someone. It takes the form of a message sent from one person or machine to another through a communication system. A message contains symbols known to the transmitter and receiver. For instance, symbols might be letters of the alphabet or words. In wireless communications, symbols are groups of bits. As long as the transmitter and receiver understand what the symbols represent, then intelligible communication occurs. If the message becomes corrupted, then the receiver cannot interpret the correct meaning.

This chapter introduces important concepts about analog signals and their digital representation. Power levels and bandwidth determine how well the signal resists noise and interference as well as determine the speed of communication. A communication system tries to successfully transmit as many messages in the least amount of time. Source coding converts digital baseband signals into symbols with as few bits as possible in order to increase the information transfer rate in terms of bits per second (bps). Channel coding and interleaving protect the data bits from noise and interference, so the received signal has few errors. Sometimes symbols overlap in a message, so steps must be taken to minimize this intersymbol interference (ISI).

2.1 Analog Baseband Signals

Almost all information starts in a low-frequency analog form at baseband, such as a voice signal. For example a sensor measures some physical quantity and outputs an analog signal that corresponds to the information it gathers. Sensors like a thermometer or microphone represents a low output by a low voltage while representing a high output by a high voltage. These voltage signals vary in time like the audio signal in Figure 2.1.

Wireless Communications Systems: An Introduction, First Edition. Randy L. Haupt.
© 2020 John Wiley & Sons, Inc. Published 2020 by John Wiley & Sons, Inc.

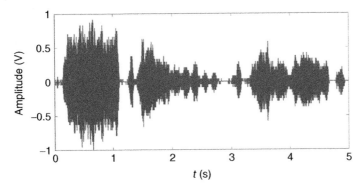

Figure 2.1 Baseband voice signal in the time domain.

A Fourier transform quantifies the frequency content of an analog signal. The signal in the time domain, $s(t)$, has a Fourier transform given by

$$S(f) = \int_{-\infty}^{\infty} s(t)e^{-j2\pi ft}dt \tag{2.1}$$

A spectrum plots $S(f)$ vs. f. The complex power spectral density (PSD) with units of W/Hz has both positive and negative frequencies

$$\text{PSD} = |S(f)|^2 \tag{2.2}$$

Negative frequencies have practical meaning when modulating the signal into a higher frequency band as presented in Chapter 3. Figure 2.2 is the double-sided (positive and negative frequencies) PSD for the signal in Figure 2.1. The unit dBm is dB above one mW.

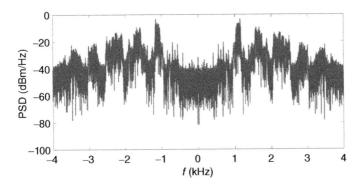

Figure 2.2 Double-sided PSD of the voice signal in Figure 2.1.

The one-sided spectrum of $s(t)$ in Figure 2.3 spectrum $(f \geq 0)$ is found from the two-sided spectrum by

$$\text{PSD} = \begin{cases} |S(0)|^2 & f = 0 \\ |S(f)|^2 + |S(-f)|^2 & f > 0 \end{cases} \quad (2.3)$$

Integrating the PSD over frequency yields the total power. The inverse Fourier transform of the spectrum perfectly recovers the time domain signal.

$$s(t) = \int_{-\infty}^{\infty} S(f) e^{j2\pi ft} df \quad (2.4)$$

In fact, Parseval's theorem states that the power in both the time and frequency representation of a signal are the same:

$$\int_{-\infty}^{\infty} |s(t)|^2 dt = \int_{-\infty}^{\infty} |S(f)|^2 df \quad (2.5)$$

Figures 2.2 and 2.3 have the same PSD at 0 Hz. On either side of 0 Hz, however, the complex PSD is half the amplitude of the plot in Figure 2.3. The ½ factor come from Euler's identity, $\cos(2\pi ft) = (e^{j2\pi ft} + e^{-j2\pi ft})/2$.

The limits of integration of the Fourier and inverse Fourier transforms extend from $-\infty$ to ∞. In practice, a computer calculates the Fourier transform over a finite interval. Figure 2.4 is a time-frequency plot of the signal in Figure 2.1 using the MATLAB command

```
spectrogram(s,kaiser(256,5),220,512,8000,'yaxis')
```

where

s = signal samples
kaiser = window function

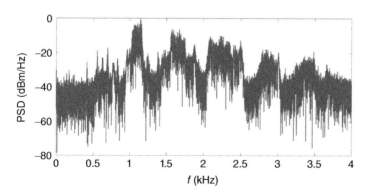

Figure 2.3 Single-sided PSD of the voice signal in Figure 2.1.

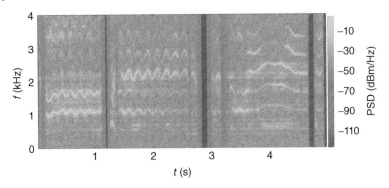

Figure 2.4 Time-frequency plot of the voice signal in Figure 2.1.

220 = number of overlapping samples in adjacent Fourier transforms
512 = number of frequency samples in the Fourier transform
8000 = sampling frequency in Hz.

Figure 2.4 results from plotting the Fourier transform of a finite piece (window) of the time signal in the vertical direction. The window moves in time before computing and graphing another slice of the Fourier transform. This process continues until the window reaches the end of the data. Time-frequency plots reveal how the signal frequency content changes with time.

Example
Show that the power in a 1 ms pulse in the time domain is the same as its power in the frequency domain.

Solution
Assume that the power is measured across a 1 Ω resistor. In the time domain, the power is given by

$$\int_{-\infty}^{\infty} |s(t)|^2 dt = \int_0^1 1^2 dt = 1\,\text{W}$$

The Fourier transform of this pulse is

$$S(f) = \int_0^1 1e^{-j2\pi ft} dt = \frac{e^{-j2\pi f1} - e^{-j2\pi f0}}{-j2\pi f} = \frac{\sin(2\pi f)}{2\pi f} = \text{sinc}(2\pi f)$$

where $\text{sinc}(x) = \frac{\sin \pi x}{\pi x}$. So, the power in the frequency domain is

$$\int_{-\infty}^{\infty} |S(f)|^2 df = \int_{-\infty}^{\infty} |\text{sinc}(2\pi f)|^2 df = 1\,\text{W}$$

The above integral can be solved using MATLAB:

```
integral(@(f)sinc(f).^2,-1000,1000)
```

2.2 Digital Baseband Signals

Analog communication keeps the baseband signal in analog form while digital communication converts the baseband signal to bits via an analog-to-digital converter (ADC). In either case, modulating the baseband signal to a higher radio frequency (RF) (passband) prepares it for transmission. The new passband signal travels through the channel to the receiver that demodulates and decodes it. Successful wireless communication is possible when the channel capacity exceeds the source entropy.

Each digital baseband symbol requires a unique combination of bits in order to distinguish one symbol from another. The sender converts the message into symbols before transmitting them over the wireless channel. The receiver translates the symbols into the original bits that represent the information being transmitted. Successful communication requires the transmitter and receiver know the conversion between symbols and bits.

Figure 2.5 demonstrates the transition from information to message to symbols and finally to a code. A geometrical object has an observable shape. The English word "circle" describes the observed shape. English symbols (letters) form a word that represents the shape. A binary string called ASCII (American Standard Code for Information Interchange) code represents each letter. The resulting binary string has 42 bits and describes the geometric shape called a circle.

At the transmitter, packet and frame bits sandwich the data bits from the ADC. The frame functions like the husk that surrounds the hard shell (the packet) as shown in Figure 2.6. The desirable white coconut meat and milk are like the data. We want the coconut meat and milk (data) which the shell (packet) and husk (frame) encapsulate. A packet consists of data bits along with headers and trailers containing control bits that specify source and destination network addresses, error detection codes, and sequencing information. An IP (internet protocol) packet contains control information such as the source and destination IP addresses. A frame surrounds a packet and typically includes frame synchronization features – a sequence of bits or symbols that indicate to the

O	Information – object shape
Circle	Message – English word
[c i r c l e]	Symbols – English letters
[1100011 1101001 1110010 1100011 1101100 1100101]	Code – ASCII

Figure 2.5 Converting information into code.

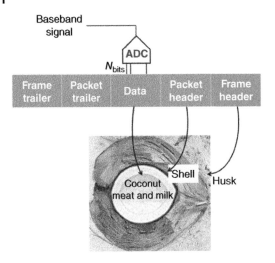

Baseband signal

ADC

N_{bits}

Frame trailer	Packet trailer	Data	Packet header	Frame header

Shell

Coconut meat and milk

Husk

Figure 2.6 A coconut resembles a data frame.

receiver the beginning and end of the data. A receiver strips the headers from the frame to get at the packet. Removing the packet headers reveals the data.

2.3 Source Coding

Source coding converts the data bits into a new binary sequence that efficiently compresses the data before transmission. A source code reduces the redundancy present in the data bits, then represents the data with as few bits as possible. Codebooks at the transmitter map or encode each possible message into a unique sequence of bits (Figure 2.7). The receiver decodes the source code into the original data using the same codebook. Fast communication requires unambiguously representing a message in as few bits as possible.

Lossless coding (entropy coding) perfectly reconstructs the original data from the compressed data. For example the entropy code called Lossless Joint Photographic Experts Group (JPEG-LS) compresses picture and video signals [1]. Lossy coding approximates data with fewer bits, but the receiver cannot perfectly reconstruct the original data. It works well for speech, audio, picture, and video signals that do not need an exact reconstruction. Lossy source coding techniques typically reduce the number of data bits by an order of magnitude compared to lossless source coding. Examples of lossy coding techniques include Joint Photographic Experts Group (JPEG) for still picture coding [2] and Moving Picture Experts Group (MPEG) for video coding [3].

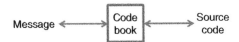

Message ← → Code book ← → Source code

Figure 2.7 A codebook converts between the message and the source code.

Reducing the number of data bits in a message causes that message to transmit faster for a given data rate in bps. If all messages have the same number of bits (fixed length code), then the number of bits in a message is independent of its probability of occurrence. Highly efficient source codes assign short bit sequences to high probability messages and long bit sequences to low probability messages. Morse code, for instance, assigns one "dot" to the most common letter "E." Less common letters have up to four "dashes" and "dots."

Example

The basic ASCII code is an example of a fixed length code with 7 bits for each character. What is the ASCII representation of the message "circle"?

Solution

Use MATLAB to convert the message to ASCII: dec2bin(double ('circle'))

[1100011 1101001 1110010 1100011 1101100 1100101]

Each letter has a binary string of 7 bits. The six symbols or 42 bits comprise the message.

An efficient code minimizes the number of bits representing a message. The function $\ell(m_n)$ counts the number of bits in a binary message m_n. For instance, $\ell(10) = 2$ and $\ell(10001) = 5$. The expected number of bits in a message (overbar on ℓ) is given by [4]

$$\overline{\ell}(\mathbf{M}) = E[\ell(\mathbf{M})] = \sum_{n=1}^{N_{\text{mess}}} \ell(m_n) p_n \tag{2.6}$$

where the $\mathbf{M}$ vector has N_{mess} messages with the probability of the nth message given by p_n. A valid codebook with distinct messages assigned to unique bit sequences has a lower bound on the number of bits required to encode the average message [5]:

$$\overline{\ell}(\mathbf{M}) \geq H(\mathbf{M}) \tag{2.7}$$

where H is the entropy or the average number of bits needed to uniquely communicate the message. Entropy coding assigns fewer bits to more likely messages to get as close to the entropy as possible.

Variable-length codes compress data close to the entropy level with zero error (lossless data compression). A prefix code is a variable-length code that has no other code word for a prefix (initial segment). For example $code_1 = \begin{bmatrix} 1 & 22 & 33 \end{bmatrix}$ is a prefix code – each code word begins with a different number. On the other hand, $code_2 = \begin{bmatrix} 1 & 22 & 122 \end{bmatrix}$ is not, because "1" is a prefix of "1" and "123," so they both begin with the same number "1." A message generated with prefix codes concatenates code words without any special breaks between symbols.

Example
A receiver gets the message [122].
   Is the message unambiguous for $code_1$: $\begin{bmatrix} m_1 = 1 & m_2 = 22 \end{bmatrix}$ or $code_2$: $\begin{bmatrix} m_1 = 1 \\ m_2 = 22 & m_3 = 122 \end{bmatrix}$?

Solution
$code_1$: It is unambiguous because this translates to $m_1 m_2$
$code_2$: It is ambiguous because this may translate to either $m_1 m_2$ or m_3

In 1951, Professor Robert Fano and Claude Shannon were at a dead end in their research to find the most efficient binary source code [6]. In his class at MIT, Fano assigned his student David Huffman a final project to develop a new optimal variable length source code that eluded him and Shannon. The young, ingenious Huffman envisioned an approach using a frequency-sorted binary tree. His technique proved to be the most efficient ever reported at the time. In a single assignment, Huffman bettered his professors (and hopefully got an "A"). Huffman built the tree from the bottom up instead of the accepted top down approach, so he did the inverse of the suboptimal codes of the day. Huffman coding uses a specific method for choosing the representation of each symbol, resulting in a prefix code. The Huffman algorithm has the following steps [7]:

1. Assume there are N_{mess} possible messages.
2. Sort all possible messages in increasing probability.

   $$m_1, m_2, \ldots, m_{N_{mess}}$$

   $$p_1 \leq p_2 \leq \cdots \leq p_{N_{mess}}$$

3. Assign a "1" to the lowest probability message and a "0" to the next lowest probability message (vise versa is fine).

   $$m_1 \rightarrow 1 \quad \text{and} \quad m_2 \rightarrow 0$$

5. The probability of either of these messages occurring equals the sum of the probabilities of the two messages.

   $$m_1 \text{ or } m_2 \rightarrow m_{12} \text{ with } p_{12} = p_1 + p_2$$

6. Replace m_1 or m_2 with m_{12} and $p_1 + p_2$ with p_{12} in the message list.
   If more than one message remains, then go to step 2. Otherwise, output the code words in reverse order.

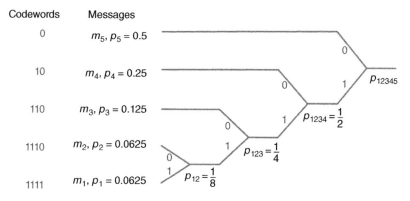

Figure 2.8 Huffman coding example for five messages.

Example

Assume that five messages have the following probabilities:

$$p_1 = 0.0625, \quad p_2 = 0.0625, \quad p_3 = 0.125, \quad p_4 = 0.25, \quad p_5 = 0.5$$

Find a Huffman code that represents these five messages.

Solution

Using the Huffman algorithm, assign $m_1 \rightarrow 1$ and $m_2 \rightarrow 0$ (or $m_1 \rightarrow 0$ and $m_2 \rightarrow 1$) then combine these messages m_1 and $m_2 \rightarrow m_{12}$ and add their probabilities $p_{12} = p_1 + p_2 = 0.0625 + 0.0625 = 0.125$. Next, assign $m_{12} \rightarrow 1$ and $m_3 \rightarrow 0$ then combine these messages m_{12} and $m_3 \rightarrow m_{123}$ and add their probabilities $p_{123} = p_{12} + p_3 = 0.125 + 0.125 = 0.25$. Continue the process as shown in Figure 2.8. Assign a single bit "0" to the highest probability code word, m_5. The next highest probability code word, m_4, is assigned two bits "10." The lowest probability code words have 4 bits each. The receiver knows that the end of a code word occurs after receiving a zero, "0," or four ones, "1111." Thus, the sequence 0101111 can only have one meaning: m_5, m_4, m_1.

A decoder may confuse code words that closely resemble each other. The Hamming distance, d_h, measures the difference between two strings of the same length. It counts the number of errors between the transmitted and received data strings. A binary code also has a Hamming weight which equals the number of nonzero symbols in a code word.

Example

Find the Hamming distance between Symbols 1 and 2 as well as the Hamming weight of Symbol 1.

Solution

Symbol 1	Symbol 2	Hamming distance	Hamming weight of Symbol 1
Adel*ine*	Adel*phi*	3	7
1010*1*010	101*1*010	1	4
123*4*56789	123*5*46789	2	9

2.4 Line Coding

Line coding, also known as baseband modulation, refers to the voltage signals that represent bits. There are two categories of line codes: non-return-to-zero (NRZ) and return-to-zero (RZ). RZ codes have a 0-V level for a portion of every bit interval, while NRZ codes do not.

Figure 2.9 shows some of the many ways to represent bits by voltage levels for NRZ and RZ codes. Unipolar NRZ assigns a high voltage level to a "1" and zero volts to a "0." This approach is also known as on–off keying (OOK). Polar

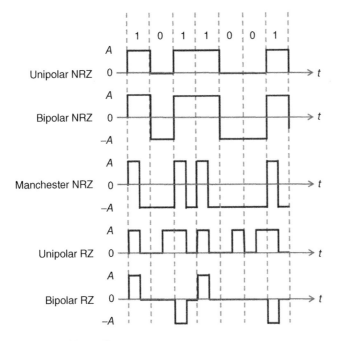

Figure 2.9 Line coding.

NRZ represents a one by A volts and a minus one by $-A$ volts. Manchester NRZ represents a one by A volts in the first half of the bit and $-A$ volts in the second half. In contrast, a zero is represented by $-A$ volts in the first half of the bit and A volts in the second half. Unipolar RZ codes represent a one by A volts in the first half of the bit and 0 V in the second half. A zero is represented by 0 V. Bipolar RZ represents a zero by 0 V and a one by alternating value of A and $-A$.

2.5 Bandwidth

Bandwidth (B) defines a frequency range between a low frequency (f_{lo}) and a high frequency (f_{hi}). Sometimes, the bandwidth is expressed as a ratio f_{hi}/f_{lo}, while other times, it is expressed as a percent and is called fractional bandwidth: $2(f_{hi} - f_{lo})/(f_{hi} + f_{lo}) \times 100\%$. Some set rule defines the high and low frequencies that bound the bandwidth. In most cases, f_{lo} and f_{hi} occur when the gain (or some other power specification) decreases by 3 dB on either side of its peak or center value.

Signal bandwidth contains all significant frequencies in a signal between f_{lo} and f_{hi} [8]. System bandwidth contains all frequencies between f_{lo} and f_{hi} that pass through the system without being attenuated below a specified level. In order to accurately receive the signal, the system bandwidth must be greater than or equal to the signal bandwidth (Figure 2.10), otherwise some signal frequencies attenuate. In general, the impulse response of the system, $h(t)$, or transfer function, $H(f)$, attenuates (or filters) a certain range of frequencies in the input signal, $s_{in}(t)$, as well as adds noise.

A transmitter sends symbols at a known rate. The receiver detects the symbols then reconstructs the transmitted data. If the symbols have more than 1 bit, then the baud rate (symbols/s) equals the bit rate divided by the number of bits per symbol. That is if the symbol has 8 bits, the baud rate is one-eighth of the bit rate. The data rate must be less than twice the bandwidth.

Example
A bit that is T_b long has a power spectrum given by $P_s = T_b \, \text{sinc}^2(fT_b)$. If f_{lo} and f_{hi} are defined to be 3 dB below the peak power, then what is the bandwidth of the signal?

Figure 2.10 System bandwidth should be greater than signal bandwidth.

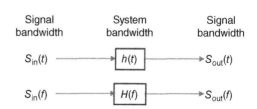

Signal bandwidth   System bandwidth   Signal bandwidth

Solution

The 3 dB points are found from $\mathrm{sinc}^2(fT_b) = 0.5$ and solving the transcendental equation for f.

$$fT_b = \pm 0.443 \Rightarrow B = 2f = 0.886/T_b$$

2.6 Signal Level

At any point in space, the signal level fluctuates over time, making the time average of the signal amplitude an important property. The mean of an analog signal, $s(t)$, over a time period, T, is found by integrating the signal over T.

$$\mu_s = \lim_{T \to \infty} \frac{1}{T} \int_{-T/2}^{T/2} s(t)dt \ \mathrm{V} \tag{2.8}$$

Midpoint integration divides range from $-T/2$ to $T/2$ in (2.8) into equal intervals that are Δ long then multiplies the midpoint value by Δ and sums the result. The mean after sampling $s(t)$ N_{samp} times as shown in Figure 2.11 is

$$\mu_s = \lim_{N_{\mathrm{samp}} \to \infty} \frac{1}{\Delta N_{\mathrm{samp}}} \sum_{n=1}^{N_{\mathrm{samp}}} \Delta s[n] \ \mathrm{V} = \lim_{N_{\mathrm{samp}} \to \infty} \frac{1}{N_{\mathrm{samp}}} \sum_{n=1}^{N_{\mathrm{samp}}} s[n] \ \mathrm{V} \tag{2.9}$$

Note that Δ divides out making the result a function of the samples and the number of samples. Once the signal average is known, then the energy and power are easy to calculate. Energy in analog form is

$$E_s = \lim_{T \to \infty} \int_{-T/2}^{T/2} [s(t) - \mu_s]^2 dt \ \mathrm{J} \tag{2.10}$$

and the energy calculated from the sampled signal is

$$E_s = \lim_{N_{\mathrm{samp}} \to \infty} \sum_{n=1}^{N_{\mathrm{samp}}} (s[n] - \mu_s)^2 \ \mathrm{J} \tag{2.11}$$

The power of the analog signal is

$$P_s = \lim_{T \to \infty} \frac{1}{T} \int_{-T/2}^{T/2} [s(t) - \mu_s]^2 dt \ \mathrm{W} \tag{2.12}$$

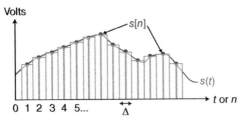

Figure 2.11 A signal $s(t)$ and its samples $s[n]$.

and of the sampled signal is

$$P_s = \lim_{N_{samp} \to \infty} \frac{1}{N_{samp}} \sum_{n=1}^{N_{samp}} (s[n] - \mu_s)^2 \text{ W} \tag{2.13}$$

The mean, power, and energy are important measurable quantities of a signal.

Example

Assume a "0" is 0 V and a "1" is 1 V and both bits are equally likely. Find the signal mean, energy, and power with N_{samp} samples.

Solution

$$\mu_s = \frac{1}{N_{samp}} \sum_{n=1}^{N_{samp}} s[n] = N_{samp} \frac{1}{N_{samp}} \frac{1}{2} = \frac{1}{2} \text{ V}$$

$$E_s = \sum_{n=1}^{N_{samp}} \left(s[n] - \frac{1}{2} \right)^2 = \sum_{n=1}^{N_{samp}} \left(\frac{1}{2} \right)^2 = N_{samp} \frac{1}{4} = 1 \text{ J}$$

$$P_s = \frac{1}{N_{samp}} \sum_{n=1}^{N_{samp}} \left(s[n] - \frac{1}{2} \right)^2 = \frac{1}{N_{samp}} \sum_{n=1}^{N_{samp}} \left(\frac{1}{2} \right)^2 = N_{samp} \frac{1}{N_{samp}} \frac{1}{4} = \frac{1}{4} \text{ W}$$

2.7 Noise and Interference

Noise and interference make the desired signal difficult to detect. Noise comes naturally from heat or electron motion. Interference comes from other signals and may be intentional or unintentional. In either case, the signal transmitter and receiver try to overcome the noise and interference through increased power, superior electronic components, coding, and signal processing. Dependable communication links have a high signal level and a low noise level. The unitless signal to noise ratio serves as a widely used figure of merit.

$$\text{SNR} = \frac{P_s}{P_N} \tag{2.14}$$

where P_s = signal power and P_N = noise power. SNR is usually expressed in dB.

$$\text{SNR} = 10 \log \frac{P_s}{P_N} \tag{2.15}$$

The carrier-to-noise ratio (CNR or C/N) is a version of SNR in which the power in the carrier is divided by the power in the noise.

Thermal noise power arises from vibrations of conduction electrons and holes and is given by

$$P_N = k_B T_0 B \ \text{W} \tag{2.16}$$

with Boltzmann's constant $k_B = 1.38 \times 10^{-23}$ J/K and T_0 is the temperature in kelvin (K). Vibrating charges emit spectral content that falls inside the frequency band of the signals. The thermal noise has a uniform PSD over RF. Unless otherwise specified, T_0 equals the ambient temperature of 290° K. A resistor at room temperature has an average thermal noise power of $P_N = 10 \log(1.38 \times 10^{-23} \times 290 \times 1) = -204 \ \text{dBW/Hz} = -174 \ \text{dBm/Hz}$.

The noise factor, F, of an amplifier indicates the amount of noise that a device adds to a signal passing through it and is defined as the ratio of the input SNR to the output SNR [9].

$$F = \frac{\text{SNR}_{in}}{\text{SNR}_{out}} = \frac{P_{sin}/P_{Nin}}{P_{sout}/P_{Nout}} \tag{2.17}$$

where

$\text{SNR}_{in} = $ input SNR
$\text{SNR}_{out} = $ output SNR
$P_{sin} = $ input signal power
$P_{sout} = $ output signal power
$P_{Nin} = $ input noise power
$P_{Nout} = $ output noise power.

For instance, a perfect amplifier increases the signal and noise by the gain, G_{amp}, so the SNR is the same at the input and output. In the real world, an amplifier adds noise (P_{Namp}) and degrades the output SNR. A low F means that a device does not add much noise.

$$F = \frac{P_{sin}/P_{Nin}}{G_{amp}P_{sin}/(P_{Namp} + G_{amp}P_{Nin})} = \frac{P_{Namp} + G_{amp}P_{Nin}}{G_{amp}P_{Nin}}$$
$$= \frac{P_{Namp} + G_{amp}k_B T_0 B}{G_{amp}k_B T_0 B} \tag{2.18}$$

Note that F is independent of the input signal power. Usually, the noise factor is expressed in dB as the noise figure (NF).

$$\text{NF} = 10 \log F \ \text{dB} \tag{2.19}$$

Both P_{Namp} and P_{Nin} are proportional to bandwidth, so the bandwidths in all the terms in (2.18) divide out and make F independent of bandwidth [10].

Since noise power is proportional to the temperature, temperature is an alternative way to characterize the noise. Consider an amplifier having a noise input power of $P_{in} = kT_{in}B$. An imperfect amplifier adds noise characterized by an equivalent temperature (T_e). As a result, the output noise power of the amplifier is given by

$$P_{\text{out}} = G_{\text{amp}}k(T_{\text{in}} + T_e)B \tag{2.20}$$

which means that the output noise temperature is

$$T_{\text{out}} = G_{\text{amp}}(T_{\text{in}} + T_e) \tag{2.21}$$

Passive components have an equivalent temperature given by

$$T_e = (L - 1)T_0 \tag{2.22}$$

where $L \geq 0$ is the loss with $L = 0$ being lossless. Amplifier effective noise temperature is

$$T_e = (F - 1)T_0 \tag{2.23}$$

The noise factor which for passive components equals the loss. Noise temperatures are not physical temperatures, but theoretical temperatures that produce an equivalent noise power. Often times, satellite communications uses a system temperature to describe the noise performance of a device in place of the *NF* (Chapter 6).

The NF of a cascade of amplifiers (Figure 2.12) results from the NFs and gains/losses of the system components [10]. For one amplifier, the input noise is $k_B T_0 B$ and the output noise is the input noise times the gain plus the noise added by the amplifier (P_{N1}), so the output noise is

$$P_{\text{Nout}} = k_B T_0 B G_1 + P_{N1} \tag{2.24}$$

Adding another amplifier produces output noise given by

$$P_{\text{Nout}} = k_B T_0 B G_1 G_2 + P_{N1}G_2 + P_{N2} \tag{2.25}$$

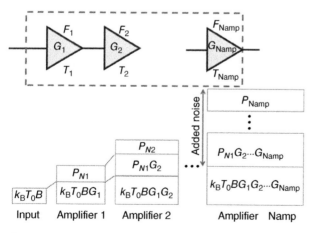

Figure 2.12 Cascaded amplifiers.

If there are N_{amp} devices, then the output noise is

$$P_{Nout} = k_B T_0 B G_1 G_2 \cdots G_{Namp} + P_{N1} G_2 \cdots G_{Namp} + \cdots + P_{Namp} \qquad (2.26)$$

The output signal power (P_{sout}) equals the input signal power (P_{sin}) multiplied by the gains of all the amplifiers.

$$P_{sout} = P_{sin} G_1 G_2 \cdots G_{Namp} \qquad (2.27)$$

Substituting (2.26) and (2.27) into (2.17) yields the equivalent noise factor for cascaded amplifiers.

$$F_{eq} = F_1 + \frac{F_2 - 1}{G_1} + \cdots + \frac{F_{Namp} - 1}{G_1 G_2 \cdots G_{Namp-1}} \qquad (2.28)$$

As long as the first amplifier stage in the receiver has a high gain and low noise factor, then the remaining stages contribute little to the noise that the amplifier adds to the signal. Placing a low noise amplifier (LNA) first minimizes the noise induced by the amplifier chain. The equivalent noise temperature for cascaded amplifiers is given by

$$T_{eq} = T_1 + \frac{T_2}{G_1} + \cdots + \frac{T_{Namp}}{G_1 G_2 \cdots G_{Namp-1}} \qquad (2.29)$$

Example

What is the transmission line noise temperature of a coaxial cable with 1 dB of loss?

Solution

$$T_{line} = T_0(1 - 1/L_t) = 290(1 - 10^{-1/10}) = 77 \text{ K}$$

Additive white Gaussian noise (AWGN) models thermal noise generated within the communication system. Additive means the noise adds to or subtracts from the time domain signal. White implies the noise has constant power across the bandwidth. Gaussian describes the noise power probability density function (PDF) in the time domain. The PDF for AWGN is given by

$$PDF(z) = \frac{1}{\sigma_{noise} \sqrt{2\pi}} e^{-\frac{(z-\mu_N)^2}{2\sigma_N^2}} \qquad (2.30)$$

where μ_N is the mean or average and σ_{noise} is the standard deviation of the noise. The probability that z lies between a and b equals the integral of the PDF between a and b.

Example

A square wave signal has a period of 6.25 ns. Plot the signal and signal plus AWGN having a SNR = 10.

Solution

```
t = (0:0.1:10)';
x = square(t);
y = awgn(x,10,'measured');
fig(1);plot(t,x,'b-',t,y,'r');xlabel('t (ns)');ylabel
    ('Amplitude (V)')
legend('Signal','Signal + AWGN');legend('boxoff')
```

Figure 2.13 is the resulting plot. Electromagnetic interference (EMI) makes the desired signal more difficult to detect. The signal to interference plus noise ratio (SINR) measures EMI.

$$\text{SINR} = \frac{P_s}{P_I + P_N} \tag{2.31}$$

where $P_I =$ interference power. Jamming implies intentional interference. If there is interference but no noise, then the SINR becomes the signal to interference ratio (SIR). When there is no interference, the SINR becomes the SNR.

SNR and CNR characterize analog noise. Digital communications systems frequently use the bit error rate (BER) which equals the number of bit errors divided by the total number of bits over a long interval. Another common digital quality measure in the baseband signal is E_b/N_0 where E_b is the energy per bit (W-s) and N_0 is the noise PSD (W/Hz). In other words, E_b/N_0 is the SNR per bit and serves as a standard of comparison for different modulation methods.

Noise and interference in a digital communication system cause bits to flip from "1" to "0" or "0" to "1." The number of bits impacted depends on the data rate and the time duration of the noise. Errors in a bit stream (Figure 2.14) are either single bit or multi-bit (burst errors). The multi-bit or burst errors do not have to be consecutive bits and have a burst length equal to the number of bits from the first bit error to the last bit error.

Figure 2.13 Signal and signal plus AWGN having an SNR = 10.

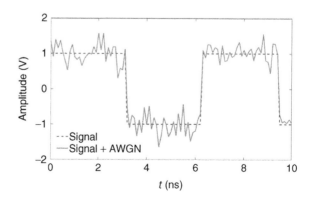

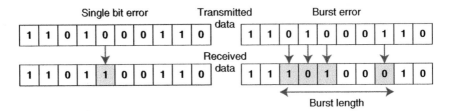

Figure 2.14 Single bit vs. burst errors in a bit stream.

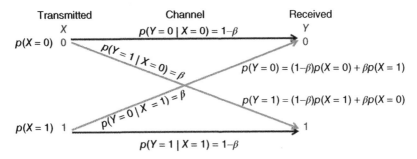

Figure 2.15 Channel-induced bit errors.

The noise and interference in a channel flips a bit with probability p as shown in Figure 2.15. The probability that a transmitted "0" becomes a received "1" or $p(Y=1|X=0)$ equals β. The probability that a bit stays the same, $p(Y=0|X=0)$, equals $1 - \beta$. Thus, the probability of receiving a "1" is given by Bayes theorem.

$$p(Y = 1) = p(Y = 1 \mid X = 1)p(X = 1) + p(Y = 1 \mid X = 0)p(X = 0)$$

$$(2.32)$$

BER is measured by counting the received bit errors in a long received data stream. The number of bits in the received data stream needed to accurately determine the BER depends on the required confidence level (CL) and the desired BER. The CL equals the percentage of tests where the actual BER is less than the desired BER. The CL reaches 100% only when there are an infinite number of bits. The CL is given by [11]

$$CL = 1 - e^{-N_{\text{bits}}\text{BER}} \qquad (2.33)$$

The number of bits that need to be tested for a given BER and CL is

$$N_{\text{bits}} = \frac{-\ln(1 - CL)}{\text{BER}} \qquad (2.34)$$

Example

If the specified BER is 10^{-12} and the required CL is 95%, how many bits need to be tested?

Solution

Use (2.34) to get

$$N_{bits} = \frac{-\ln(1 - 0.95)}{10^{-12}} = 3 \times 10^{12}$$

In an AWGN channel, a normal (Gaussian) PDF models the probability of the noise voltage level. The average value of the noise has the highest probability of occurrence in a normal distribution. The root mean square value of the noise power is the variance or σ_{noise}^2. Typically, the standard normal distribution with a mean of zero and variance of one models noise.

$$PDF_{norm}(z) = \frac{1}{\sqrt{2\pi}} e^{-\frac{z^2}{2}} \tag{2.35}$$

If the distribution has a mean of μ and a variance of σ^2, then its new PDF is given in terms of the standard normal distribution as

$$PDF(x) = PDF_{norm}\left(\frac{x - \mu}{\sigma}\right) / \sigma = \frac{1}{\sigma\sqrt{2\pi}} e^{-\frac{(x-\mu)^2}{2\sigma^2}} \tag{2.36}$$

The cumulative distribution function (CDF) results from integrating the PDF.

$$CDF(x_a) = p(x \leq x_a) = \int_{-\infty}^{x_a} \frac{1}{\sigma\sqrt{2\pi}} e^{-\frac{(x-\mu)^2}{2\sigma^2}} dx \tag{2.37}$$

$$1 - CDF(x_a) = p(x \geq x_a) = \int_{x_a}^{\infty} \frac{1}{\sigma\sqrt{2\pi}} e^{-\frac{(x-\mu)^2}{2\sigma^2}} dx \tag{2.38}$$

and a generic CDF in terms of the unit normal CDF is given by

$$CDF(x) = CDF_{norm}\left(\frac{x - \mu}{\sigma}\right) \tag{2.39}$$

Example

Assume

- a "0" and a "1" are equally likely
- AWGN is zero mean and a variance of σ^2
- $s(t) < 0.5$ the bit is assumed to be "0"
- $s(t) \geq 0.5$ the bit is assumed to be "1."

Find the probability of a bit error.

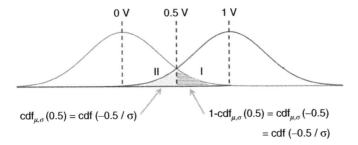

$$\mathrm{cdf}_{\mu,\sigma}(0.5) = \mathrm{cdf}(-0.5 / \sigma)$$

$$1\text{-}\mathrm{cdf}_{\mu,\sigma}(0.5) = \mathrm{cdf}_{\mu,\sigma}(-0.5)$$
$$= \mathrm{cdf}(-0.5 / \sigma)$$

Figure 2.16 Gaussian PDFs representing the voltages associated with the received signals.

Solution
The threshold voltage equals the average voltage of 0.5 V. Noise results in normal distributions centered on 0 and 1 V as shown in Figure 2.16. The standard deviation of the normal distributions define the percent of bits that are misidentified. For instance, the probability that a transmitted "1" is received as a "0" (Type I error or false positive) equals the integral of the PDF with a mean of 1 V that lies below 0.5 V:

$$p(\text{receive } 0 \mid \text{transmit } 1) = \text{CDF}(0.5) = \text{CDF}_{\text{norm}}(-0.5/\sigma)$$

Similarly, the probability that a transmitted "0" is received as a "1" (Type II error or false negative) equals the integral of the PDF with a mean of 1 V that lies below 0.5 V:

$$p(\text{receive } 1 \mid \text{transmit } 0) = 1 - \text{CDF}(0.5) = \text{CDF}(-0.5)$$
$$= \text{CDF}_{\text{norm}}(-0.5/\sigma)$$

Thus, the BER is given by

$$\text{BER} = p(\text{error})$$
$$= p(\text{receive } 0 \mid \text{transmit } 1)p(\text{transmit } 1)$$
$$+ p(\text{receive } 1 \mid \text{transmit } 0)p(\text{transmit } 0)$$
$$= \text{CDF}_{\text{norm}}(-0.5/\sigma)\left(\frac{1}{2}\right) + \text{CDF}_{\text{norm}}(-0.5/\sigma)\left(\frac{1}{2}\right)$$
$$= \text{CDF}_{\text{norm}}(-0.5/\sigma) \tag{2.40}$$

Figure 2.16 shows the Types I and II errors as the shaded areas under the overlapping PDFs.

2.8 Converting Analog to Digital

Converting an analog signal to bits first requires sampling the signal. An ADC samples the analog signal at a rate of f_s which must be at least twice the highest frequency of the analog signal. The ADC output has N_{bits} bits that define the quantized voltage levels. The samples fall into one of

$$N_{lev} = 2^{N_{bits}} \tag{2.41}$$

quantized voltage levels represented by N_{bits} bits. A 10 bit ADC leads to $N_{lev} = 1024$. The ADC's dynamic range extends from the minimum detectable voltage (V_{ADCmin}) to the maximum allowed voltage (V_{ADCmax}). ADC input voltage resolution determines the ADC precision of the analog signal.

$$V_{ADCres} = \frac{V_{ADCmax} - V_{ADCmin}}{N_{lev}} \tag{2.42}$$

Figure 2.17 shows the ADC output in bits for input voltages between $V_{ADCmin} = 0$ to $V_{ADCmax} = V_{max}$.

The minimum change in voltage corresponds to the least significant bit (LSB) in the binary signal. A real ADC has noise and distortion that reduce its theoretical resolution. The effective number of bits (ENOB) specifies the ADC resolution possible in practice with $N_{bits} \geq$ ENOB.

$$ENOB = \frac{SINAD - 1.76}{6.02} \text{ dB} \tag{2.43}$$

Signal-to-noise and distortion (SINAD) is the ratio of the total power: signal (P_s), noise (P_N), and distortion (P_D) to the unwanted power (noise and distortion).

$$SINAD = 10 \log \left(\frac{P_s + P_N + P_D}{P_N + P_D} \right) \tag{2.44}$$

SINAD indicates the ADC dynamic performance, because it includes all components of noise and distortion.

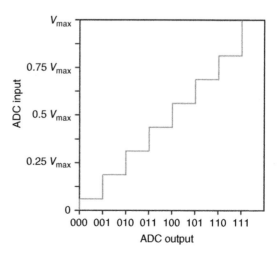

Figure 2.17 The bits assigned to the quantization levels.

Quantization error is the rounding error between the analog input voltage and the quantized output voltage corresponding to the bits. The SNR "6 dB rule" assumes a uniformly distributed quantization error between $-1/2$ LSB and $+1/2$ LSB:

$$\text{SNR} = 20\log(2^{N_{bits}}) \approx 6.02 N_{bits} \text{ dB} \tag{2.45}$$

Applying this equation to a 12 bit ADC leads to a maximum SNR $= 72.24$ dB.

Example
The Texas Instruments TI 32RF45 data sheet [12] shows that it has $f_s = 3$ GHz, SNR $= 60.9$ dB, bandwidth $= 3.2$ GHz, ENOB $= 9.7$ bits, SINAD $= 60.2$ dB, and $V_p = 0.625$ V. Find its noise figure.

Solution
A full-scale input signal has a peak voltage (V_p) that corresponds to the maximum input voltage of the ADC.

$$s_{in}(t) = V_0 \sin(2\pi f t) \tag{2.46}$$

The full-scale power input when the input impedance $50\,\Omega$ is given by

$$P_s = \frac{(0.707 V_0)^2}{Z_{ADC}} = \frac{V_0^2}{100} \tag{2.47}$$

$$= 20\log V_0 - 20 + 30 = 20\log V_0 + 10 \text{ dBm}$$

The signal PSD is given by the signal power divided by the bandwidth.

$$\text{PSD} = 20\log V_0 + 10 - 10\log(f_s/2) \text{ dBm/Hz} \tag{2.48}$$

where f_s is the sampling frequency. Assume that the thermal noise at the input over a 1 Hz bandwidth is -174 dBm/Hz. Usually, the full scale power input is backed off by 1 dBm, so the input SNR is

$$\text{SNR}_{in} = 20\log V_0 + 10 - 1 - 10\log(f_s/2) + 174 \tag{2.49}$$

The output SNR is usually specified and is related to the ENOB

$$\text{SNR}_{out} = 1.76 + 6.02 \text{ ENOB dB} \tag{2.50}$$

The noise figure is calculated from (2.49) and (2.50)

$$\begin{aligned} \text{NF}_{ADC} &= \text{SNR}_{in} - \text{SNR}_{out} \\ &= 20\log V_0 + 9 - 10\log(f_s/2) + 174 - 1.76 - 6.02 N_{bits} \\ &= 20\log V_0 - 10\log(f_s) - 6.02 \text{ ENOB} + 181.24 \text{ dB} \\ &= 24.66 \text{ dB} \end{aligned} \tag{2.51}$$

The data Texas Instruments data sheet show that this ADC has a $NF = 24.7$ dB.

2.9   Channel Coding

An RF channel behaves like a water channel. The RF channel takes the RF signal from the source to its destination. A water channel transfers a well-defined amount of water proportional to the water channel dimensions. Similarly, an RF channel transfers an amount of information proportional to its bandwidth. In reality, the water channel has rocks, turns, and other obstacles that slow the water and induce turbulence. Similarly, the RF channel has obstacles and noise that reduce the channel capacity. Claude Shannon's channel coding theorem defines a maximum channel capacity as

$$C = B\log_2(1 + \text{SNR}) \text{ bps} \tag{2.52}$$

A channel with noise and interference transmits information at a desired error level if the channel capacity is not exceeded. Coding techniques get as close as possible to the Shannon capacity. Spectral efficiency is the data rate (R_b) in bps that can be supported by the bandwidth.

$$\eta_{se} = \frac{R_b}{B} \text{ bps/Hz} \tag{2.53}$$

Example

If the SNR $= 20\,$dB and the bandwidth available $= 4\,$kHz (voice communications) what is the channel capacity and maximum spectral efficiency?

Solution

$$C = 4000 \log_2(1 + 100) = 4000 \log_2(101) = 26.63 \text{ kbit/s}$$
$$\eta_{se} = \frac{26.63}{4} = 6.66 \text{ bps/Hz}$$

A transmitter starts with $N_{databits}$ data bits in a vector $\mathbf{X}$ then transforms them into a code word $\mathbf{Y}$ that has N_{bits} bits with $N_{bits} > N_{databits}$. Channel coding adds bits to the data to combat errors. When the code word passes through the channel, the received code word, $\mathbf{Z}$, may have errors.

$$\mathbf{Z} = \mathbf{Y} + \mathbf{E} = [z_1, \dots, z_{N_{bits}}] \tag{2.54}$$

where the error vector is $\mathbf{E} = [e_1, \dots, e_{N_{bits}}]$. The error vector starts with all zeros. Position where errors occur changes from "0" to "1."

The channel coding enables the receiver to either detect errors or detect and correct errors in the received data bits. A receiver that only detects errors requests the transmitter to resend a message with an error [13]. Automatic repeat request (ARQ where the Q stands for query) requires the receiver to acknowledge receipt of a frame within a timeout period (time period allowed before an acknowledgment [ACK] is received). Otherwise, the transmitter resends the message. Stop-and-wait ARQ transmits one frame at a time. The transmitter does not send another frame until it receives an ACK signal is received by the transmitter. If the ACK does not reach the transmitter before the timeout, the transmitter resends the same frame. Go-Back-N ARQ sends N_{frame} frames even without receiving an ACK from the receiver. If the receiver times out, then the transmitter resends all N_{frame} frames. Selective repeat transmits N_{frame} frames like Go-Back-N ARQ. In this case, the receiver selectively rejects a single frame, which is retransmitted instead of retransmitting all N_{frame} frames.

In order to detect and/or correct errors at the receiver, an algorithm converts the $N_{databits}$ into an N_{bits} code word with $N_{bits} - N_{databits}$ parity bits to detect and/or correct errors. The receiver has a decoding algorithm that recovers the $N_{databits}$ data bits from the code word. The decoding algorithm indicates if an error occurred and possibly where it occurred in the data. The code rate for $N_{databits}$ bits encoded into N_{bits} bits is an indication of the efficiency of the encoding

$$R_c = \frac{N_{databits}}{N_{bits}} \tag{2.55}$$

2.10 Repetition

Repetition breaks the data into blocks of bits and transmits them $1/R_c$ times in a row. For example the data [1001] repeated three times means the transmitter sends [100110011001]. If the receiver recovers [100111011001], then it knows that an error occurred. Repetition efficiency is at most 50% and cannot detect some errors. For instance, if the receiver recovers [110111011101], which has the same error in each location, then no error is detected.

Example
What is the code rate for a message of 8 bits repeated five times? Can you think of an error correction method for the repetition scheme?

Solution

$$R_c = \frac{N_{databits}}{N_{bits}} = \frac{8}{5 \times 8} = 0.2$$

One approach to error correction would be majority rule in which the most common block in the received data is accepted as correct.

2.11 Parity Bits

Parity bits tacked onto the data bits detect whether an error occurred in the transmission. For instance, a code word with 7 data bits $(b_1 \cdots b_7)$ plus 1 parity

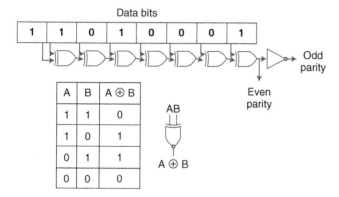

Figure 2.18 Generating even and odd parity with XOR gates.

bit (b_{p1}) take the form

$$[b_1 b_2 b_3 b_4 b_5 b_6 b_7 b_{p1}] \tag{2.56}$$

Even parity sets $b_{p1} = 1$ when the Hamming weight of the data bits is odd and $b_{p1} = 0$ when the Hamming weight is even. Odd parity sets $b_{p1} = 0$ when the Hamming weight of the data bits is odd and $b_{p1} = 1$ when the Hamming weight is even. One parity bit detects, but does not correct, 1 bit error. For example if a transmitter sends the data [1011011] with even parity, then $b_{p1} = 1$. If the receiver gets [11111111], it assumes that the data has no errors – a wrong conclusion due to the 2 bit errors in the received data. If the receiver has 1 bit error [10010111] or 3 bit errors [10010001], for instance, then it correctly assumes the data is corrupted. A parity detection scheme implemented with XOR gates is shown in Figure 2.18.

Example
Show the parity bit value for the following data when even and odd parity is used.

Solution

	Parity bit	
Data	Even parity	Odd parity
00000000	0	1
01011011	1	0
01010101	0	1
00000001	1	0
01001001	1	0
11111111	0	1

2.12 Redundancy Checking

Breaking the data into small blocks then assigning parity bits to the blocks increases the odds of detecting errors over assigning parity bits to all of the data. Figure 2.19 shows two approaches to assigning the parity bits [13]: (i) longitudinal redundancy check (LRC) where a parity bit is found for the bits at location n of each block or (ii) vertical redundancy check (VRC) where a parity bit is derived for each block. For instance, the 16 bits of data in Figure 2.19 are divided into four blocks. An LRC even parity bit is found for each of the 4 bit columns in the blocks. Alternatively, VRC finds an even parity bit for each row in the blocks. Next, those 4 parity bits append to the original data bits before transmitting (Figure 2.19). On reception, the data separates into blocks then either the columns (LRC) or rows (VRC) are summed with the parity bits to see if the result is an even number (for even parity). If not, then errors have corrupted the data. LRC and VRC increase the likelihood of detecting burst errors. One error pattern that LRC cannot detect occurs when 2 bit errors in one data block and 2 bit errors in exactly the same positions in another data block (Figure 2.20). Some systems use a combination of VRC and LRC for improved error detection.

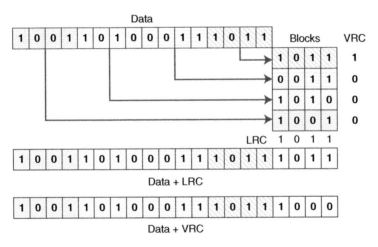

Figure 2.19 LRC and VRC for a 16 data bits.

	Transmitted data									Received data with errors							
	1	1	1	0	0	1	1	0		0	1	1	0	0	1	1	1
	1	1	0	1	1	1	0	1		0	1	0	1	1	1	0	0
LRC	0	0	1	1	1	0	1	1		0	0	1	1	1	0	1	1

Figure 2.20 LRC cannot detect errors when they occur at the same location in two blocks of the data.

Example

If the data 11100111110111010011100110101001 has errors at locations 17, 19, 20, 29, 30, and 31, show that LRC will detect the presence of errors.

Solution

Break the transmit and receive data into four blocks as shown in Figure 2.21. Sum the columns to get the LRC for the transmit data. Do the same for the receive data. The receive bit sum does not equal the LRC, so the data has errors and needs to be retransmitted.

Example

Represent the word "running" in ASCII, then use LRC with even parity and VRC with odd parity to find the parity bits.

Solution

Figure 2.22 shows the results. Note that the ASCII code is a column below the letter, so the LRC and VRC bits are switched accordingly.

Cyclic redundancy check (CRC) error detection [13] assumes the data bits serve as coefficients of a polynomial. For instance, the data bits $X = [10110]$ represents $x^4 + x^2 + x$. Dividing the polynomial by a binary generating polynomial

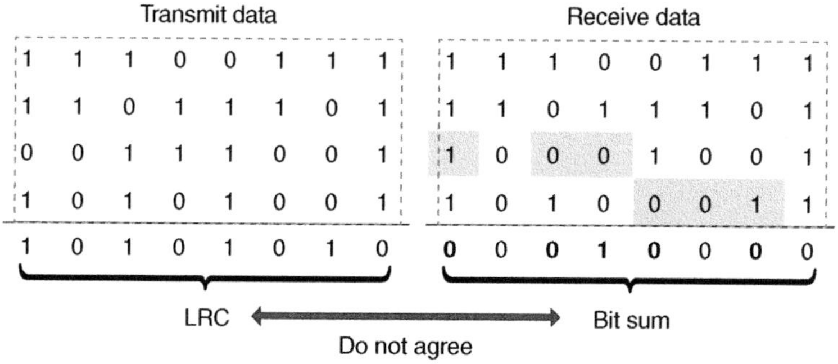

Figure 2.21 Example where LRC finds errors.

Figure 2.22 Using LRC with even parity and VRC with odd parity for the word "running" encoded in ASCII.

bit	r	u	n	n	i	n	g	LRC
0	0	1	0	0	1	0	1	1
1	1	0	1	1	0	1	1	1
2	0	1	1	1	0	1	1	1
3	0	0	1	1	1	1	0	0
4	1	1	0	0	0	0	0	0
5	1	1	1	1	1	1	1	1
6	1	1	1	1	1	1	1	1
VRC	1	0	0	0	1	0	0	

(g_n) results in a quotient (discarded) and a remainder (appended to the data as parity bits to form **Y**). The receiver recovers the data by dividing the received bits, **Z**, by the same generating polynomial. If the remainder is not zero, then an error occurred in the channel. A zero remainder implies that no error occurred. CRC detects the following errors [14]:

- single bit errors
- two bit errors as long as the data polynomial has a factor with at least three terms
- all odd number of errors as long as the polynomial contains the factor $x + 1$
- all burst errors whose length is less than the length of the remainder
- most large burst errors.

Example

Given the data [100100] and the generator [1101], find the 9 transmitted bits using CRC.

Solution

The data polynomial is $x^8 + x^5$ and the generating polynomial is $x^3 + x^2 + 1$. Since the transmitter sends 9 bits with 6 data bits, the remainder has 3 bits. Figure 2.23 shows the quotient and remainder from dividing the data polynomial by the generator polynomial. The quotient is discarded and the remainder is appended to the data prior to transmitting. Upon reception, the code word is divided by the generating polynomial to extract the data as shown in Figure 2.24. Since the remainder is 0 0 0, the received data has no errors.

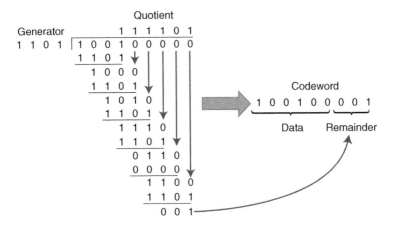

Figure 2.23 Creating the CRC code word for the data polynomial $x^8 + x^5$ and the generating polynomial $x^3 + x^2 + 1$.

Figure 2.24 Recovering the data from the CRC code word. A 0 0 0 remainder means there are no errors.

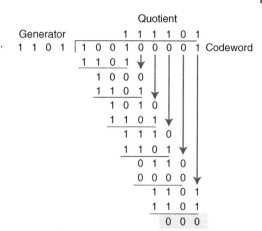

2.13 Error Correcting Codes (ECC)

A receiver recovers the original data from an error correcting code (ECC) even when random errors flip some of the bits. Consequently, the receiver does not have to ask the transmitter to retransmit data when it detects errors. There are two main types of ECCs: block codes and convolutional codes.

2.13.1 Block Codes

An (N_{bits}, N_{dat}) block code maps N_{dat} input bits into N_{bits} output bits as shown in Figure 2.25, where $N_{bits} > N_{dat}$ [22]. Errors are detected and corrected if the Hamming distance exceeds $N_{bits} + 1$ or $d_h \geq N_{ed} + N_{ec} + 1$, where N_{ed} is the number of errors detected, and N_{ec} is the number of errors corrected. Since $N_{ed} \geq N_{ec}$, then $d_h \geq 2N_{ec} + 1$.

A generator matrix, $\mathbf{G}$, encodes binary information with N_{dat} bits, $\mathbf{X} = [x_1, \ldots, x_{N_{dat}}]$, into an N_{bits} code word, $\mathbf{Y} = [y_1, \ldots, y_{N_{bits}}]$, via

$$\mathbf{Y} = \mathbf{XG} \tag{2.57}$$

Figure 2.25 A block code has parity bits for error detection and correction.

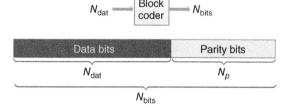

where the generator matrix takes the form

$$
\mathbf{G} = \begin{bmatrix}
1 & 0 & \cdots & 0 & g_{1,N_{\text{dat}}+1} & \cdots & g_{1,N_{\text{bits}}} \\
0 & 1 & \cdots & 0 & g_{2,N_{\text{dat}}+1} & \cdots & g_{2,N_{\text{bits}}} \\
\vdots & \vdots & \ddots & \vdots & \vdots & \ddots & \vdots \\
0 & 0 & \cdots & 1 & g_{N_{\text{dat}},N_{\text{dat}}+1} & \cdots & g_{k,N_{\text{bits}}}
\end{bmatrix} = [\mathbf{I}_{N_{\text{dat}}} \mid \mathbf{P}] \qquad (2.58)
$$

G is partitioned into the $N_{\text{dat}} \times N_{\text{dat}}$ identity matrix, $\mathbf{I}_{N_{\text{dat}}}$, which reproduces the data at the beginning of the code word, and the $N_{\text{dat}} \times N_p$ parity bit generating matrix, **P**, which generates parity bits from the data. The generator matrix creates a systematic code – one that puts **X** in the first N_{dat} symbols of **Y**. Each linear code has a parity check matrix, **H**, that has the orthogonality properties

$$
\mathbf{YH}^T = 0
$$
$$
\mathbf{GH}^T = 0
$$
$$(2.59)$$

The parity check matrix decodes the bits when it takes the form

$$
\mathbf{H} = \begin{bmatrix}
g_{1,N_{\text{dat}}+1} & g_{2,N_{\text{dat}}+1} & \cdots & g_{N_{\text{dat}},N_{\text{dat}}+1} & 1 & \cdots & 0 \\
\vdots & \vdots & \ddots & \vdots & \vdots & \ddots & \vdots \\
g_{1,N_{\text{bits}}} & g_{2,N_{\text{bits}}} & \cdots & g_{N_{\text{dat}},N_{\text{bits}}} & 0 & \cdots & 1
\end{bmatrix} = [\mathbf{P}^T \mid \mathbf{I}_{N_p}]
$$
$$(2.60)$$

H is partitioned by the $N_p \times N_{\text{dat}}$ transpose of the parity bit matrix, $\mathbf{P}^T$, and the $N_p \times N_p$ identity matrix, $\mathbf{I}_{N_p}$ where superscript "T" indicates the transpose of the matrix.

The receiver decodes **Z** into a syndrome, **S**, given by

$$
\mathbf{S} = \mathbf{ZH}^T = (\mathbf{XG} + \mathbf{E})\mathbf{H}^T = \mathbf{X}\overset{0}{\overbrace{\mathbf{GH}^T}} + \mathbf{EH}^T = \mathbf{EH}^T \qquad (2.61)
$$

If the syndrome contains all zeros, then there are no errors. The following example explains how to find the location of the error for a nonzero syndrome.

Example
A block code encodes the data bits $\mathbf{X} = [1011]$ with this generating matrix:

$$
\mathbf{G} = \begin{bmatrix}
1 & 0 & 0 & 0 & 0 & 1 & 1 \\
0 & 1 & 0 & 0 & 1 & 0 & 1 \\
0 & 0 & 1 & 0 & 1 & 1 & 0 \\
0 & 0 & 0 & 1 & 1 & 1 & 1
\end{bmatrix} \qquad (2.62)
$$

and the receiver has a parity check matrix given by

$$
\mathbf{H} = \begin{bmatrix}
0 & 1 & 1 & 1 & 1 & 0 & 0 \\
1 & 0 & 1 & 1 & 0 & 1 & 0 \\
1 & 1 & 0 & 1 & 0 & 0 & 1
\end{bmatrix} \qquad (2.63)
$$

The received data is $\mathbf{Z} = [1011110]$. Find which bit has an error.

Solution

The encoded information equals the data vector multiplied by the generating matrix: $\mathbf{Y} = \mathbf{XG} = [1011010]$. The syndrome is given by

$$\mathbf{S} = \mathbf{ZH}^T = [1011110] \begin{bmatrix} 011 \\ 101 \\ 110 \\ 111 \\ 100 \\ 010 \\ 001 \end{bmatrix} = [100]$$

This syndrome is nonzero indicating that an error occurred in the transmission. The $\mathbf{S}^T$ vector is the same as the fifth column in $\mathbf{H}$, which means that the error occurred in the fifth bit of $\mathbf{Y}$, so $\mathbf{E} = [0000100]$. This approach cannot correct more than one error. The syndrome for two errors is the sum of the syndromes for the individual errors which incorrectly points to a different column in $\mathbf{H}$.

The Reed and Solomon systematic code does a linear mapping of a message to a polynomial [23]. Most two-dimensional bar codes use Reed–Solomon error correction to recover the bar code even when it is damaged. This approach to coding is also used in space missions like Voyager [24].

2.13.2 Convolutional Codes

In 1955, Peter Elias introduced nonsystematic codes – codes that have error detection but do not include the original data [15]. His convolutional code generates a parity bit by performing a mod 2 convolution of the data bits with a generator polynomial having order $K - 1$. The polynomial acts like a window function with constraint length (K). Parity bit i associated with data bit n is calculated from

$$p_i[n] = \left(\sum_{m=0}^{K-1} g_i[m]x[n-m] \right) \bmod 2 \tag{2.64}$$

where mod 2 finds the remainder after division of one binary number by another binary number.

A convolutional code uses generating polynomials to create a code. On average, there are R_c parity bits sent for each data bit. If there are N_{dat} data bits in a block, then $K - 1$ previous data blocks must be retained in order to generate the parity bits. Examples of generator polynomials for rate 1/2 convolutional codes with different constraint lengths are shown in Table 2.1.

Table 2.1 Generator polynomials for $R = 1/2$ convolutional codes [16].

K	g_1	g_2
3	110	111
4	1101	1110
5	11010	11101
6	110101	111011
7	110101	110101
8	110111	1110011
9	110111	111001101
10	110111001	1110011001

Example

A $(N_{bits}, N_{dat}, K) = (2, 1, 2)$ convolutional code has the generating polynomials $g_1 = [111]$ and $g_2 = [101]$. Find the output code word given the input data $\mathbf{X} = [11101]$.

Solution

First, convolve g_1 with $\mathbf{X}$: $g_1 * \mathbf{X} = [111] * [11101] = [1010011]$.
Next, convolve g_1 with $\mathbf{X}$: $g_2 * \mathbf{X} = [101] * [11101] = [1101001]$.
Finally, interleave the bits from the two convolutions to get the code word:

$$\mathbf{Y} = \begin{bmatrix} 11 & 01 & 10 & 01 & 00 & 10 & 11 \end{bmatrix}$$

The code rate is $R_c = 5/14 = 0.357$.

In order to decode the code word, the receiver compares the received bits to a list of possible bit sequences. Decoders, such as the Viterbi algorithm [17], determine the most likely bit sequence that was transmitted. A longer code has better error-correcting performance at the cost of a more complex decoding algorithm and decreased data rate. In the early 1990s, turbo codes revolutionized forward error corrections (FECs) by approaching the Shannon limit. The Cassini Saturn probe, Mars Pathfinder, and Mars Rover used convolutional coding with $R_c = 1/6$ and $K = 15$ and Viterbi decoding [19]. Turbo codes found their first practical application in the Mars Pathfinder mission [21].

2.14 Interleaving

Interleaving mixes up symbols from code words in the same bit stream in order to convert burst errors into single bit errors are detected and corrected using ECC [21]. Consider the case of four code words having four symbols each.

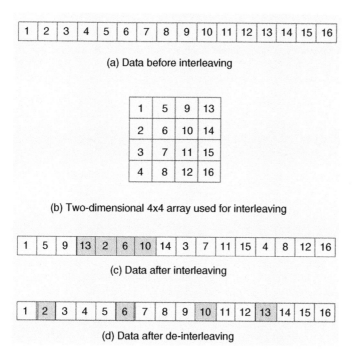

(a) Data before interleaving

(b) Two-dimensional 4x4 array used for interleaving

(c) Data after interleaving

(d) Data after de-interleaving

Figure 2.26 Example of 1-D block interleaving. *Source:* Reprinted by permission of [21]; © 2004 IEEE [17].

The 16 consecutive symbols appear in Figure 2.26a. Block interleaving, first creates a 4×4 2-D array in which each code word occupies a column of the matrix shown in Figure 2.26b. The interleaved code symbols result from writing the code symbols row-by-row from the matrix in Figure 2.26b. The result appears in Figure 2.26c. Assume an error burst affects four consecutive symbols (shaded in Figure 2.26). De-interleaving effectively spreads the burst error over four code words, resulting in one error in each of the four code words as shown in Figure 2.26d. Hence, the burst error appears as random single bit errors.

Example

Create a short ASCII message in MATLAB and use (12,7) cyclic encoding. Corrupt the message with a burst error. Compare the received bits with and without interleaving.

Solution

1. Create message:

```
rndsd=1234; % random seed
word='wireless';
```

```
a=dec2bin(double(word)); % convert to decimal
numsym=length(word); % number of symbols in message
b=reshape(a,numsym*7,1); % binary message in a
                         %column vector
msg=de2bi(str2num(b)); % message is in binary
```

2. Use cyclic coding:

```
n = 12; k = 7; % cyclic code rate
cpoly=cyclpoly(n, k); % create generating polynomial
code = encode(msg,n,k,'cyclic',cpoly); % Encoded data
```

3. Create a 6-bit burst error:

```
errors = zeros(length(code),1); errors(n-2:n+2) =
    [1 1 1 1 1]';
```

4. Interleaving:

```
inter = randintrlv(code,rndsd); % Interleave.
intererr = bitxor(inter,errors); % Include burst error.
deinter = randdeintrlv(intererr,rndsd); % Deinterleave.
decoded = decode(deinter,n,k,'cyclic',cpoly); % Decode.
disp('Interleaving:');
[number,BER] = biterr(msg,decoded) % Error statistics
```

5. No interleaving:

```
code_err = bitxor(code,errors); % Include burst error.
decoded = decode(code_err,n,k,'cyclic',cpoly); % Decode.
disp('No interleaving');
[number,BER] = biterr(msg,decoded) % Error statistics
```

Interleaving has zero errors and a BER = 0.0.
No interleaving has four errors and a BER = 0.0714
Results depend upon the value of rndsd.

2.15 Eye Diagram

An eye diagram consists of overlaying oscilloscope sweeps from consecutive segments of a long data stream. Overlaying positive and negative pulses results in an image that looks like an eye. Ideally, eye diagrams look like rectangular boxes. In the real world, filters round the sharp edges of rectangular pulses. Differences in timing and amplitude from bit to bit cause the eye opening to shrink. Figure 2.27 has two examples of measured eye diagrams. Figure 2.27a has a large eye opening that corresponds to a large SNR. Reducing the SNR, as in eye diagram in Figure 2.27b, causes the eye to close. Optimal sampling occurs

Figure 2.27 Eye diagrams.
(a) Low SNR and jitter and
(b) higher SNR and jitter.

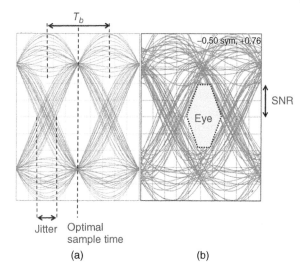

(a)   (b)

in the center of the eye. The start of each bit varies, so the bits do not appear to have a common starting point. This timing variation from bit to bit is called jitter. The eye diagram in Figure 2.27b has more jitter than the eye diagram in Figure 2.27a. The hexagon that roughly outlines the inside of the eye has a vertical opening proportional to the SNR and a horizontal opening relating to the sensitivity of the sampling time. A large eye opening corresponds to a low BER. The steeper the slope of the hexagon sides, the more timing jitter it has.

2.16   Intersymbol Interference

Multipath, noise, and interference distort symbols passing through the channel. The channel does not treat all frequency components the same (i.e. the channel is dispersive). A dispersive channel has an impulse response and noise that changes the transmitted signal by the time it arrives at the receiver.

$$s_r(t) = s_t(t) * h_c(t) + n(t) \tag{2.65}$$

where

$s_r(t)$ = received signal
$s_t(t)$ = transmitted signal
$h_c(t)$ = channel impulse response function
$n(t)$ = AWGN with PSD $N_0/2$.

A bandlimited channel behaves like a low pass filter (LPF) that smears the transmitted signal in time so that adjacent symbols overlap. The spreading of one symbol into an adjacent symbol causes ISI that leads to wrong symbol identification. Delay spread is the difference between the arrival time of the line-of-sight (LOS) signal and the arrival time of the last multipath signal. ISI can be ignored when the symbol duration is much larger than the delay spread.

A signal takes multiple paths of different lengths to the receiver. A longer path usually has more attenuation than a shorter path due to more reflections. The sum of all the multipath signals at the receiver has a long tail compared to the transmitted signal. This tail extends into the adjacent bit causing ISI.

Figure 2.28a displays a sequence of 8 bits originating at the transmitter. Each bit is $T_b = 1$ μs long. These bits pass through a LPF that rounds the sharp edges and spreads the symbols in time. If there is no multipath, the individual pulses arrive at the receiver as shown in Figure 2.28b. The receiver samples the pulses at the center of the pulse or every 1 μs. These orthogonal pulses have no overlap at the sample points, so no ISI occurs when the pulses add together as shown in Figure 2.28c. In this case, 1 V represents a "1" while 0 V represents a "0." Multipath in the channel leads to increasing the pulse duration due to multiple versions of the pulses arriving at different times at the receiver. Figure 2.28d shows the individual received dispersed pulses. Dispersion ruined the pulse orthogonality, so that when one pulse is a "1" the rest are no longer a "0." Adding the dispersed pulses together produces the waveform in Figure 2.28e. The ISI produces 0.4 V at the 2 μs sampling point which is a "0." Now, the sample is about 0.4 V instead of 0 V. Adding noise to this scenario might well push this sample point above 0.5 V, so that it is interpreted as a "1" by the receiver.

Figure 2.28 demonstrates that ISI potentially leads to bit errors. In order to reduce bit errors due to ISI, system designers use the following remedies:

1. Slow down the data rate until the symbol duration is much larger than the delay spread.
2. Use ECC.
3. Use pulse shaping of the symbols in order to minimize the spread into adjacent symbols.
4. Apply an equalizer that reverses the distortion caused by the channel.

Slowing the data rate is usually not an acceptable solution due to the demand for higher data rates. Since ECC has already been presented earlier in the chapter, pulse shaping and the equalizer will be discussed in Section 2.13.

Figure 2.29 diagrams the transmit (H_t), channel (H_c), and receive (H_r) filter transfer functions in a dispersive channel [25]. Let the overall transfer function be represented by

$$H_o(f) = H_t(f)H_c(f)H_r(f) \tag{2.66}$$

Minimizing ISI requires a constant $H_o(f)$ over all frequencies. A constant $H_o(f)$ results from designing H_t and H_r for a given H_c. A matched receive filter has the transfer function:

$$H_r(f) = H_t^*(f)H_c^*(f) \tag{2.67}$$

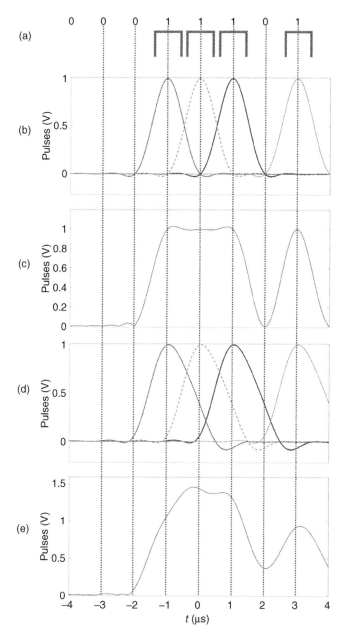

Figure 2.28 ISI example for a digital waveform. (a) Ideal pulses for each bit, (b) individual filtered pulses with no multipath, (c) sum of the received raised-cosine pulses with no multipath, (d) individual received raised-cosine pulses with multipath, and (e) sum of the received raised-cosine pulses with multipath.

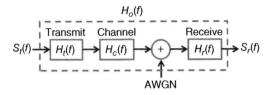

Figure 2.29 Dispersive channel model.

where "*" indicates complex conjugate. Selecting $H_t(f)$ to make $H_r(f)$ constant requires knowledge of the channel transfer function. Three approaches to making $H_r(f)$ constant are

1. A pulse-shaping filter in the transmitter.
2. Pick $H_t(f)$ and design $H_r(f)$ to make $H_o(f)$ constant.
3. Pick a simple filter for $H_r(f)$ then digitize the output. The digital output goes to an equalizer or digital filter that compensates the frequency response. In this way, the digital filter adapts to a changing channel.

Correlation measures the similarity of two things. A high correlation implies that two things are very similar. Correlating the received signal with a matched filter is the same as convolving the received signal with a conjugated time-reversed version of itself. When the matched filter and received signal align, the output of the correlator/matched filter peaks. A matched filter [25, 26] maximizes the receiver SNR in the presence of AWGN. It has an impulse response that is a time reversed complex conjugate of the transmitted signal [27].

2.17 Raised-Cosine Filter

A raised-cosine filter, also known as the Hann window in digital signal processing, shapes a digital pulse at the transmitter in order to reduce ISI. It has a frequency response given by [22]

$$H_{RC}(f) = \begin{cases} 1 & |f| \leq \dfrac{1-\alpha}{2T_s} \\ \dfrac{1}{2}\left\{1 + \cos\left[\dfrac{\pi T_s}{\alpha}\left(|f| - \dfrac{1-\alpha}{2T_s}\right)\right]\right\} & \dfrac{1-\alpha}{2T_s} < |f| \leq \dfrac{1+\alpha}{2T_s} \\ 0 & |f| > \dfrac{1+\alpha}{2T_s} \end{cases}$$

$$\times\, 0 \leq \alpha \leq 1 \tag{2.68}$$

where α is the roll-off factor that determines the bandwidth of the spectrum. This filter has a flat passband with a gain of one and a stop band with a gain of zero. In between, the filter drops off like a cosine function. When $\alpha = 0$, it is a perfect LPF with a spectrum that is flat over the $1/T_s$ bandwidth and zero

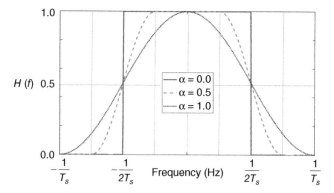

Figure 2.30 Spectrum of a raised-cosine filter.

elsewhere. Figure 2.30 shows the frequency response of a raised-cosine filter. $H_{RC}(f) = 0.5$ for all values of α at $f = 1/2T_s$.

The impulse response is the inverse Fourier transform of (2.68) [22]:

$$h_{RC}(t) = \text{sinc}\left(\frac{t}{T_s}\right)\frac{\cos\left(\frac{\pi \alpha t}{T_s}\right)}{1 - \left(\frac{2\alpha t}{T_s}\right)^2} \tag{2.69}$$

Low values of α produce higher side lobes that extend into the adjacent pulses hence increasing the probability of ISI. The time domain representation of the raised-cosine pulse appears in Figure 2.31. Raised-cosine pulses are zero when $t = nT_s$ for $n \geq 1$.

Figure 2.32 shows a series of raised-cosine pulses spaced $T_s = 1$ μs apart. The peak of one pulse occurs at the nulls of the others. Individual pulses are graphed on the left while their sum is graphed on the right. When $\alpha = 0.0$, pulse sidelobes

Figure 2.31 Raised-cosine filter in time domain.

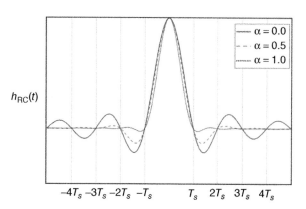

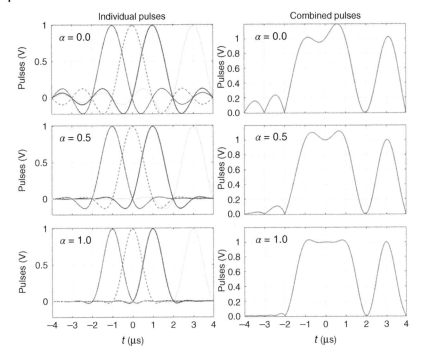

Figure 2.32 Series of raised-cos pulses spaced $T_s = 1$ μs apart.

are high, so the sum of the pulses shows significant distortion. Increasing α decreases the pulse sidelobes and the ISI in their sum.

Example

One way to implement a raised-cosine channel transfer function is to have a root raised-cosine filter (RRC) that has a frequency response of

$$H_{\text{RRC}}(f) = \sqrt{H_{\text{RC}}(f)} \tag{2.70}$$

at the transmitter and receiver, so the product of the two is a raised-cosine filter. Redo Figures 2.30 and 2.31 for the RRC filter.

Solution

The impulse response for the RRC filter is

$$h_{\text{RRC}}(t) = \text{sinc}\left(\frac{t}{T_s}\right) \frac{\sin\left[\frac{\pi(1-\alpha)t}{T_s}\right] + 4\alpha\frac{t}{T_s}\cos\left[\pi\frac{t}{T_s}(1+\alpha)\right]}{\pi\frac{t}{T_s}\left[1 - \left(\frac{4\alpha t}{T_s}\right)^2\right]} \tag{2.71}$$

Figure 2.33 is a plot of the spectrum and Figure 2.34 is a plot of the impulse response.

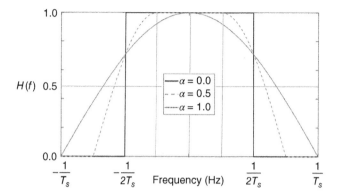

Figure 2.33 Spectrum of a root-raised-cosine filter.

Figure 2.34 Root-raised-cosine filter in time domain.

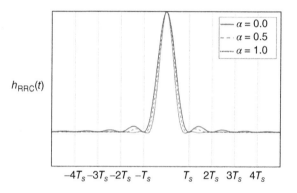

2.18 Equalization

When a high-speed data signal propagates through a channel that has less bandwidth than the signal (lossy channel), it becomes dispersed and picks up noise, so the BER increases. Sharp symbol edges become rounded and spread in time which in turn causes ISI as shown in Figure 2.35a. The zero crossing times depend on the bit amplitudes and lead to significant jitter. In the frequency domain, the lossy channel attenuates high-frequency content to get the reduced spectrum in Figure 2.35b. An equalizer is a high-pass filter (HPF) with a response that is the inverse of the channel response, so it will completely recover the transmitted signal as illustrated in Figure 2.35c [29].

The infinite impulse response (IIR) filter in Figure 2.36 serves as a model for most equalizers. The output of the IIR filter sampled at a period T_s approximates the transmitted signal given by

$$\hat{s}_t(t) = \sum_{m=0}^{M} b_m s_r(t - mT_s) + \sum_{n=1}^{N} a_n \hat{s}_t(t - nT_s) \tag{2.72}$$

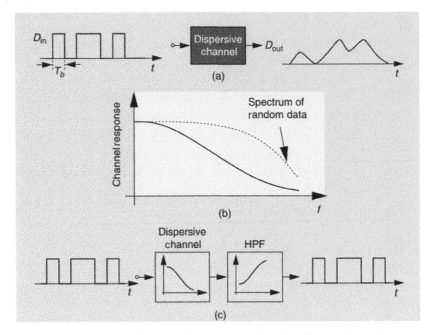

Figure 2.35 Dispersion and equalization in the time domain. (a) Dispersion in a lossy channel, (b) the channel attenuates high-frequency components, and (c) a HPF serves as an equalizer. *Source:* Reprinted by permission of [30]; © 2017 IEEE.

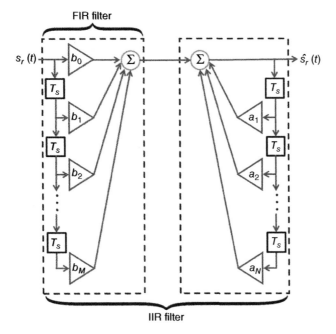

Figure 2.36 Equalizing filter.

where t is sampled. If all the $a_n = 0$ then the filter becomes an all zero filter called a finite impulse response (FIR) filter. It only has feed forward paths. When the feedback is not zero, then the transfer function has poles and the filter becomes IIR. The FIR filter has linear phase and is stable, while the IIR filter has nonlinear phase, and its stability depends on the weights a_n and b_n.

A zero-forcing equalizer [33] has a transfer function that is the inverse of the channel transfer function.

$$H_{eq}(f) = \frac{1}{H_c(f)} \tag{2.73}$$

This FIR filter cancels the ISI, but also amplifies the channel noise at the higher frequencies, because all high-frequency components (signal and noise) get amplified. An FIR filter becomes a good approximation of the inverse of the channel transfer function when the weights, b_m, ($a_n = 0$) are chosen to force the system impulse response to zero at all the delayed samples ($m > 0$).

$$\begin{bmatrix} s_r[0] & s_r[-T_s] & \cdots & s_r[-MT_s] \\ s_r[T_s] & s_r[0] & & \\ & & \ddots & \vdots \\ s_r[MT_s] & \cdots & & s_r[0] \end{bmatrix} \begin{bmatrix} b_0 \\ b_1 \\ \vdots \\ b_M \end{bmatrix} = \begin{bmatrix} 1 \\ 0 \\ \vdots \\ 0 \end{bmatrix} \tag{2.74}$$

which in matrix form is

$$\mathbf{S}_r \mathbf{w} = \mathbf{h}_c \tag{2.75}$$

and the weights are found by

$$\mathbf{w}_{opt} = \mathbf{S}_r^{-1} \mathbf{h}_c \tag{2.76}$$

Example
Figure 2.37 is an example of a received pulse that enters an equalizer. Find the weights for a zero-forcing filter if the samples are $\begin{bmatrix} 0 & 0 & 0.1 & 0.85 & -0.3 & 0.2 & 0 \end{bmatrix}$ at $\begin{bmatrix} -3T_s & -2T & -T & 0 & T & 2T & 3T \end{bmatrix}$.

Figure 2.37 Received pulse example.

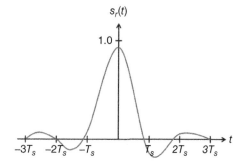

Solution

The ideal filter output for a two tap filter would be $\begin{bmatrix} 1 & 0 & 0 \end{bmatrix}$. The coefficients are found using (2.74)

$$\begin{bmatrix} 0.85 & 0.1 & 0 \\ -0.3 & 0.85 & 0.1 \\ 0.2 & -0.3 & 0.85 \end{bmatrix} \begin{bmatrix} b_0 \\ b_1 \\ b_2 \end{bmatrix} = \begin{bmatrix} 1 \\ 0 \\ 0 \end{bmatrix}$$

Solving for the weights produces $b_0 = 1.128$, $b_1 = 0.412$, $b_2 = -0.120$.

Transmitting a training sequence through the channel to the equalizer at the receiver produces the elements in the $\mathbf{S}_r$ matrix in (2.75). Narrow pulses or pseudo random noise (PRN) make good training sequences with a full spectral content across the bandwidth. PRN looks like noise but is a deterministic signal that repeats with a very long period. It has a larger average power resulting in a higher SNR for the same peak transmitted power of a narrow pulse, so the PRN is usually used. Preset equalization periodically updates the filter weights due to changes in the channel.

An alternative approach, the minimum mean square error (MMSE) equalizer, reduces the mean-square error (MSE) between the equalized and transmitted signal which in turn reduces both ISI and noise. The error is the difference between the estimated data symbol at the equalizer output and the desired data symbol

$$\varepsilon(t) = \hat{s}_t(t) - s_t(t) \tag{2.77}$$

where $\hat{s}_t(t)$ is the equalizer output and s_t is the original transmitted signal. The equalizer minimizes the average square of this error through an appropriate set of weights, b_n.

The error in (2.77) is a random variable, since the received signal is a random variable that changes with time. The expected value of the square of the error is

$$E\{|\varepsilon(t)|^2\} = E\{|s_t(t) - \mathbf{w}^T \mathbf{s}_r(t)|^2\} \tag{2.78}$$

where

$$\mathbf{s}_r(t) = \begin{bmatrix} s_r[0] & s_r[T_s] & \cdots & s_r[MT_s] \end{bmatrix}^T \tag{2.79}$$

Expanding the right side leads to

$$E\{|\varepsilon|^2\} = E\{|s_t(t)|^2\} - 2\mathbf{w}^\dagger E\{s_t(t)\mathbf{s}_r(t)\} + E\{\mathbf{w}^\dagger \mathbf{s}_r(t)\mathbf{s}_r^\dagger(t)\mathbf{w}\} \tag{2.80}$$

where superscript "†" indicates the complex conjugate transpose. For a zero mean signal, the first term equals the signal variance which is a constant. Dropping the expected value, (2.80) becomes

$$E\{|\varepsilon|^2\} \approx \sigma_s^2 - 2\mathbf{w}^\dagger E\{s_t(t)\mathbf{s}_r(t)\} + \mathbf{w}^\dagger \mathbf{C}\mathbf{w} \tag{2.81}$$

where $\mathbf{C}$ is the covariance matrix given by

$$\mathbf{C} = \begin{bmatrix} E\{s_r^\dagger(0)s_r(0)\} & E\{s_r^\dagger(0)s_r(T_s)\} & \cdots & E\{s_r^\dagger(0)s_r(MT_s)\} \\ E\{s_r^\dagger(T_s)s_r(0)\} & E\{s_r(T_s)s_r^\dagger(T_s)\} & & \vdots \\ & & \ddots & \\ E\{s_r^\dagger(MT_s)s_r(0)\} & E\{s_r^\dagger(MT_s)s_r(T_s)\} & \cdots & E\{s_r^\dagger(MT_s)s_r(MT_s)\} \end{bmatrix}$$

(2.82)

Now, take the derivative of (2.81) with respect to the weights and set it to zero to find the optimum solution.

$$E\{\nabla_w|\varepsilon|^2\} = -2s_t(t)E\{\mathbf{s}_r^*(t)\} + 2\mathbf{C}\mathbf{w}_{\text{opt}}^* = 0$$

(2.83)

Solving this equation for the optimum weights results in the Wiener–Hopf solution:

$$\mathbf{w}_{\text{opt}} = \mathbf{C}^{-1}s_t(t)E\{\mathbf{s}_r(t)\}$$

(2.84)

Solving for the optimum weights requires knowing $s_t(t)$ and $\mathbf{s}_r(t)$. Training sequence or predefined sequence of symbols lead to estimates of $\mathbf{s}_r(t)$.

Transmitting a training sequence has several drawbacks:

1. The spectral efficiency goes down, because nonmessage bits increase, thus reducing the data rate.
2. To maintain an acceptable spectral efficiency, the training sequence is short, so the filter does not average enough to significantly reduce noise.
3. If the channel changes after the training sequence is received, then the filter weights are outdated and the BER increases.

Blind equalization omits the training sequence but exploits the statistical properties of the transmitted signal to estimate the channel and data. The weights adjust until the statistical properties of $\hat{s}_t(t)$ match the statistical properties of $s_t(t)$.

In mobile communications, the channel fading changes over time, so equalizers must quickly adapt to these time varying characteristics. An adaptive equalizer continuously changes the weights in order to keep up with a fast changing channel [32]. Many adaptive algorithms exist to update the weights so as to minimize the SNR. A sampling of these algorithms are presented in later chapters.

Linear equalization faces three problems [30]:

1. Lossy channels require significant amplification of high frequency components that in turn significantly amplifies high-frequency noise and corrupts the data.
2. The high amplification requires multiple stages with each stage limiting the bandwidth and requiring considerable power.

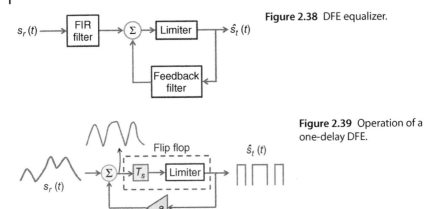

Figure 2.38 DFE equalizer.

Figure 2.39 Operation of a one-delay DFE.

3. Cannot compensate for sharp notches in the channel transfer function, like impedance discontinuities (mismatches) resulting from connectors and other physical interfaces between boards, cables, etc.

Nonlinear equalization with an IIR filter overcomes these problems with feed forward (b_n) and feedback (a_n).

A decision-feedback equalizer (DFE) is a nonlinear equalizer that uses previously detected symbols as feedback to eliminate ISI on current symbols [31] (Figure 2.38). The FIR and feedback filters can take various forms, for example zero-forcing or MMSE. A major drawback of a DFE occurs in low SNR channels where the BER is relatively high. In that case, errors in the feedback quickly cause the overall output error to rise.

Figure 2.39 explains the operation of a simple one delay DFE [30]. Feedback is amplified then subtracted from the received signal. This combination is then delayed by one symbol period. A limiter clips the signal peak in order to eliminate amplitude noise. A flip flop often replaces the combined time delay and limiter. The resulting output looks very similar to the transmitted signal. ISI and noise are dramatically reduced.

Problems

2.1 Record a voice signal for 5 s. Load the signal into MATLAB and plot it like the one in Figure 2.1.

2.2 Plot the PSD of a 5 s voice signal (sig) in MATLAB. Approximate the Fourier transform with the FFT (fast Fourier transform) using sf=abs(fft(sig)). Your plot should be similar to the one in Figure 2.2.

2.3 Plot the spectrogram of a 5 s voice signal in MATLAB using the command that generated Figure 2.4.

2.4 What is the ASCII representation of the message: (a) "three," (b) "radar," (c) "summer"?

2.5 A data source has five symbols that occur with the following frequencies:

Symbol	Frequency
A	24
B	12
C	10
D	8
E	8

Build and label a Huffman code tree.

2.6 A data source has seven symbols that occur with the following frequencies:

Symbol	Frequency
a	37
b	18
c	29
d	13
e	30
f	17
g	6

Build and label a Huffman code tree.

2.7 Write MATLAB code to find the Hamming distance and Hamming weight. Use the strings $a = [1 : 7 \ 0 \ 2]$ and $b = [1 : 9]$ to demonstrate your code.

2.8 Is the code [01 100 101 1110 1111 0011 0001] a prefix code? Is it optimum?

2.9 Find the bandwidth of $P_s = \cos^2[2\pi(f-2)]$ where $1.6 \leq f \leq 2.4$ GHz.

2.10 Find the energy and power of

$$s(t) = \begin{cases} 0 & t < -1 \\ 2 & -1 \leq t \leq 0 \\ 2e^{-t/2} & t > 0 \end{cases}$$

$$s(t) = \begin{cases} 0 & t \leq 0 \\ 1 & t > 0 \end{cases}$$

$$s(t) = \text{sinc}(t)$$

2.11 Find the SNR of a 20 ms rectangular pulse with Gaussian noise ($\sigma_{noise} = 0.000\,01$ V) sampled at 10 kHz for 2 s.

2.12 A signal has an amplitude of 1 V in the presence of Gaussian noise that has an RMS amplitude of 0.5 V. Find the SNR in dB.

2.13 A signal with power 3 mW has noise with power 0.001 mW. What is the SNR in dB?

2.14 Find the SNR when the RMS signal level is 0.707 V, temperature is 300 K, and bandwidth is 50 MHz.

2.15 Show the steps needed to go from (2.28) to (2.29).

2.16 What should the input SNR be for an output SNR $= 10$ dB in a receiver with NF $= 6$ dB and B $= 0.1$ MHz?

2.17 An antenna is connected to an LNA (NF $= 7$ dB, $G = 20$ dB) via a cable (1.5 dB attenuation). The output from the LNA is down converted by a mixer (NF $= 8$ dB, $G = 8$ dB). The output passes through another amplifier (NF $= 6$ dB, $G = 60$ dB).
(a) Find the NF of the entire system.
(b) Find the NF of the entire system if the RF amplifier is placed before the cable with 1.5 dB loss.
(c) Which system is better why?

2.18 A cascade of amplifiers starts with A and ends with (a) C and (b) B.

Amplifier	Gain (dB)	NF (dB)
A	4	2.3
B	12	3
C	20	6

2.19 An RF front end with $G = 20\,dB$ and $NF = 6.5\,dB$ feeds an ADC with $G = 0\,dB$ and $NF = 24\,dB$. Find the system NF.

2.20 Find the SNR of an ADC with (a) 10 bits, (b) 14 bits, and (c) 16 bits.

2.21 Find the capacity of an AWGN channel transmit 10 W at 1 MHz, and power spectral density of noise of 2×10^{-9} W/Hz.

2.22 An analog signal with a bandwidth of 4 kHz is sampled at 1.25 times the Nyquist frequency. Each sample is quantized into 256 levels with equal probability.
(a) What is the information rate?
(b) Is it possible to transmit error-free signals over a channel having AWGN with a 10 kHz bandwidth and $SNR = 20\,dB$.
(c) Find the SNR needed for error-free transmission with AWGN having a 10 kHz bandwidth and $SNR = 20\,dB$.
(d) Compute the required bandwidth to transmit error-free signals over an AWGN channel for a $SNR = 20\,dB$.

2.23 Plot C from (2.52) vs. SNR $(-10\,dB \le SNR \le 20\,dB)$ and B $(1\,kHz \le B \le 1\,GHz)$.

2.24 How much time is needed to test a signal for a BER of 10^{-11}, 10^{-13}, 10^{-15} for data rates (bits/s) 39.813 12 Gbps, 2.488 32 Gbps, 155.52 Mbps?

2.25 Find both the LRC and VRC for the ASCII representation of the word (a) spring, (b) radar, and (c) snowing.

2.26 Find the encoded data using CRC parity bits for the data [100100] for the generating polynomial $x^3 + x + 1$. Check the received data to see if it is error free. Assume that the received word is 100000001. Is this reception error free?

2.27 Find the CRC parity bits for the data [11100110] for the generating polynomial $x^4 + x^3 + 1$.

2.28 Find (a) the 16 possible code words and (b) **H**. Given

$$G = \begin{bmatrix} 1 & 0 & 0 & 0 & 0 & 1 & 1 \\ 0 & 1 & 0 & 0 & 1 & 0 & 1 \\ 0 & 0 & 1 & 0 & 1 & 1 & 0 \\ 0 & 0 & 0 & 1 & 1 & 1 & 1 \end{bmatrix}$$

2.29 Given the data bits [1011], find the parity bits for the convolutional code given $g_1 = [111]$ and $g_2 = [110]$. Assume that all previous bits are 0.

2.30 A message contains the bits [1011]. If

$$H = \begin{bmatrix} 1 & 1 & 0 & 1 & 1 & 0 & 0 \\ 1 & 0 & 1 & 1 & 0 & 1 & 0 \\ 0 & 1 & 1 & 1 & 0 & 0 & 1 \end{bmatrix}$$

then find (a) the code word, (b) the syndrome if the received bits are [1010010], and (c) what bit is in error due to the syndrome in (b)?

2.31 Use MATLAB to plot (2.68) for $\alpha = 0$, 0.5, and 1.0.

2.32 Use MATLAB to plot (2.69) for $\alpha = 0$, 0.5, and 1.0.

2.33 Make a plot of R vs. α for select values of B between 1 MHz and 10 GHz.

2.34 A baseband signal has four levels with $T_s = 100 \, \mu s$.
(a) If the root raised-cosine filter has $\alpha = 0.3$, what is the minimum bandwidth?
(b) How long does it take to transmit 1 000 000 bits?
(c) How many symbol states are needed to transmit the information in half the time given the same bandwidth?

2.35 A satellite receiver operating at 2 GHz has an LNA at 127 K and 20 dB gain followed by an amplifier with $F = 12$ dB and a gain of 80 dB. Find the system noise factor and T_{eq}.

2.36 The first amplifier has a gain of 12 dB and a noise temperature of 28 K. The second amplifier has a noise temperature 200 K. What is T_{eq}?

2.37 An amplifier has a noise temperature of 60 K. What is its noise figure and noise factor?

2.38 An amplifier has noise figure of 1.5 dB. What is its noise temperature?

2.39 Determine the increase in the total noise figure (NF) if an input amplifier with a noise figure of 3 dB is replaced with an amplifier with a noise figure of 6 dB.

2.40 An overall noise figure (NF) of 13 dB indicates that the noise level increases by a factor of what as compared to the increase in signal level?

2.41 Find the weights for a zero-forcing filter if the samples are [0 0.2 0.45 1.0 0.3 −0.25 0] at $[-3T_s\ -2T\ -T\ 0\ T\ 2T\ 3T]$.

2.42 Find the weights for a zero-forcing filter if the samples are [−0.08 0.25] 1 0.3 −0.06] at $[-2T\ -T\ 0\ T\ 2T]$.

References

1 https://en.wikipedia.org/wiki/Lossless_JPEG (accessed 28 October 2016).

2 https://en.wikipedia.org/wiki/JPEG (accessed 28 October 2016).

3 https://en.wikipedia.org/wiki/Moving_Picture_Experts_Group (accessed 28 October 2016).

4 Wiegand, T. and Schwarz, H. (2010). Source coding: part I of fundamentals of source and video coding. *Foundations and Trends in Signal Processing* 4 (1–2): 1–222.

5 Proakis, J.G. and Salehi, M. (1994). *Communication Systems Engineering*. Englewood Cliffs, NJ: Prentice Hall.

6 https://en.wikipedia.org/wiki/Huffman_coding (accessed 28 October 2016).

7 Huffman, D.A. (1932). A method for the construction of minimum redundancy codes. *Proceedings of the IRE* 40 (9): 1098–1101.

8 Haupt, R.L. (2015). *Timed Arrays Wideband and Time Varying Antenna Arrays*. Hoboken, NJ: Wiley.

9 Friis, H.T. (1944). Noise Figures of Radio Receivers. *Proceedings of the IRE* 32 (7): 419–422.

10 (2010). Agilent fundamentals of RF and microwave noise figure measurements, Application Note 57-1.

11 http://www.keysight.com/main/editorial.jspx%3Fckey%3D1481106%26id %3D...&sa=U&ei=p-e8VMmWI8n0UsW6gKAE&ved=0CCAQ9QEwBQ& usg=AFQjCNEO_PZyV0U7VnM9OIg1LF8lmICquw?&cc=US&lc=eng (accessed 7 November 2016).

12 (2016). ADC32RF45 Dual-Channel, 14-Bit, 3.0-GSPS, Analog-to-Digital Converter, Texas Instruments data sheet.

13 Stallings, W. (1994). *Data and Computer Communications*, 4e. New York: Macmillan Publishing Co.

14 Peterson, W.W. and Brown, D.T. (1961). Cyclic codes for error detection. *Proceedings of the IRE* 49 (1): 228–235.

15 Elias, P. (1955). Coding for noisy channels. *IRE Convention Record* 4: 37–46.

16 Peterson, W.W. and Weldon, E.J. (1971). Error Correcting Codes. In: *MIT Press*, Revised 2e.

17 Viterbi, A.J. (1967). Error bounds for convolutional codes and an asymptotically optimum decoding algorithm. *IEEE Transactions on Information Theory* 13 (2): 260–269.

18 https://ocw.mit.edu/courses/electrical-engineering-and-computer-science/6-02-introduction-to-eecs-ii-digital-communication-systems-fall-2012/lecture-slides/MIT6_02F12_lec06.pdf (accessed 14 August 2017).

19 Berrou, C., Glavieux, A., and Thitimajshima, P. (1993). Near Shannon limit error-correcting coding and decoding: turbo-codes. 1. In: *IEEE International Conference on Communications, 1993. ICC '93 Geneva. Technical Program, Conference Record*, Geneva, vol. 2, 1064–1070.

20 Burr, A. (2001). Turbo-codes: the ultimate error control codes? *Electronics & Communication Engineering Journal* 13 (4): 155–165.

21 Shi, Y.Q., Xi, M.Z., Ni, Z.-C., and Ansari, N. (2004). Interleaving for combating bursts of errors. *IEEE Circuits and Systems Magazine* 4 (1): 29–42, First Quarter.

22 Couch, L.W. II, (1993). *Digital and Analog Communication Systems*, 4e. New York: Macmillan Publishing Co.

23 Reed, I.S. and Solomon, G. (1960). Polynomial codes over certain finite fields. *Journal of the Society for Industrial and Applied Mathematics* 8 (2): 300–304.

24 Ludwig, R. and Taylor, J.(2002). Voyager telecommunications, JPL DESCANSO Design and Performance Summary Series, Pasadena, CA.

25 http://wireless.ece.ufl.edu/twong/Notes/Comm/ch4.pdf (accessed 26 June 2018).

26 D. O. North, "An Analysis of the factors which determine signal/noise discrimination in pulsed-carrier systems," in *Proceedings of the IEEE*, vol. 51, no. 7, pp. 1016-1027, July 1963; originally published as RCA Laboratories Report PTR-6C, Jun 1943.

27 Turin, G. (1960). An introduction to matched filters. *IRE Transactions on Information Theory* 6 (3): 311–329.

28 Scholtz, R. (1993). Multiple access with time-hopping impulse modulation. In: *Military Communications Conference, 1993. MILCOM '93. Conference Record. Communications on the Move., IEEE*, Boston, MA, vol. 2, 447–450.

29 Reddy, P.K. and Gadgay, B. (2016). Survey of various adaptive equalizers for wireless communication and its applications. *International Journal of Industrial Electronics and Electrical Engineering*: 83–86.

30 Razavi, B. (2017). The decision-feedback equalizer [A circuit for all seasons]. *IEEE Solid-State Circuits Magazine* 9 (4): 13–132, Fall.

31 Austin, M.E. (1967). Decision-Feedback Equalization for Digital Communication Over Dispersive Channels. *Tech. Rep. 437*, Lincoln Laboratory.

32 Lucky, R.W. (1966). Techniques for adaptive equalization of digital communication systems. *The Bell System Technical Journal* 45 (2): 255–286.

33 Lucky, R.W. (1965). Automatic equalization for digital communication. *The Bell System Technical Journal* 44 (4): 547–588.

3

Passband Signals

Information starts as an analog baseband signal with its lowest frequency close to zero, such as audio, video, sensors, images, etc. Passband signals, on the other hand, are baseband signals elevated to a higher frequency in order to fit into particular slots in the spectrum. Figure 3.1 shows a baseband signal modulated to a passband signal then frequency multiplexed with two other signals, so that they all share the spectrum without interfering with each other. Consider the case of amplitude modulation (AM) radio. A low-pass filtered audio signal has a spectrum from 0 Hz to 10 kHz. Then, a modulator upconverts this signal to a higher frequency that fits into the 20 kHz channel assigned to the AM station. Frequency multiplexing fits all the stations into the spectrum assigned to AM broadcasting. The receiver down-converts or demodulates the signal back to baseband for listening. This chapter introduces different analog and digital modulation schemes as well as several approaches to multiplexing.

3.1 Carrier

The carrier is a single frequency electromagnetic wave (harmonic or tone) represented in rectangular coordinates as:

$$\vec{E}(t) = \hat{x}E_x \cos(2\pi f_c t - k_x x) + \hat{y}E_y \cos(2\pi f_c t - k_y y + \psi_y)$$
$$+ \hat{z}E_z \cos(2\pi f_c t - k_z z + \psi_z) \tag{3.1}$$

where

f_c = carrier frequency
t = time
$c = \lambda/T$ = speed of light
λ = wavelength
T = time period

$$k = \sqrt{k_x^2 + k_y^2 + k_z^2} = \sqrt{\left(\frac{2\pi}{\lambda_x}\right)^2 + \left(\frac{2\pi}{\lambda_y}\right)^2 + \left(\frac{2\pi}{\lambda_z}\right)^2} = \frac{2\pi}{\lambda} = \text{wavenumber}$$

Wireless Communications Systems: An Introduction, First Edition. Randy L. Haupt.
© 2020 John Wiley & Sons, Inc. Published 2020 by John Wiley & Sons, Inc.

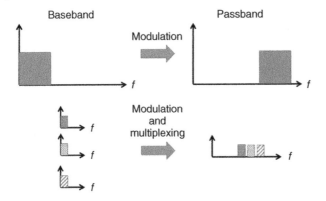

Figure 3.1 Modulation converts the baseband signal to a passband signal. Multiplexing inserts it into the proper place in the spectrum to avoid interference with other signals.

λ_x, λ_y, λ_z = wavelength projected in the x, y, and z directions
$\hat{x}$, $\hat{y}$, and $\hat{z}$ = unit vectors in the x, y, and z directions
E_x, E_y, and E_z = magnitudes of the electric fields in the x, y, and z directions
ψ_y and ψ_z = phases of the y and z components relative to the x component

A carrier contains no information but serves to transport the baseband information signal from the transmitter to the receiver at the desired passband frequencies.

The electric field in (3.1) can be written as:

$$\vec{E}(t) = \mathrm{Re}\{\mathbf{E}e^{j2\pi f_c t}\} \tag{3.2}$$

where $\mathbf{E}$ is the complex steady-state phasor (time independent):

$$\mathbf{E} = \hat{x}E_x e^{-jk_x x} + \hat{y}E_y e^{-jk_y y} e^{j\psi_y} + \hat{z}E_z e^{-jk_z z} e^{j\psi_z} \tag{3.3}$$

Since this chapter only deals with the time-varying part of (3.2), $\mathbf{E}$ is ignored.

3.2 Amplitude-Modulated Signals

AM changes the carrier amplitude in sync with the amplitude of the message signal, $m(t)$. A sinusoidal carrier, for example $V_c \cos(2\pi f_c t)$, multiplies the message to produce double sideband (DSB) modulation:

$$s(t) = m(t)V_c \cos(2\pi f_c t) \tag{3.4}$$

Assume the message signal is a single harmonic at f_m with amplitude V_m/V_c

$$m(t) = \frac{V_m}{V_c} \cos 2\pi f_m t \tag{3.5}$$

Substituting (3.5) into (3.4) yields

$$s(t) = V_m \cos(2\pi f_m t) \cos(2\pi f_c t)$$
$$= \underbrace{\frac{V_m}{2} \cos(2\pi (f_c - f_m)t)}_{\text{lower sideband}} + \underbrace{\frac{V_m}{2} \cos(2\pi (f_c + f_m)t)}_{\text{upper sideband}} \qquad (3.6)$$

As seen in (3.6), the resulting signal has two frequency bands on either side of f_c but the carrier disappeared, hence the name double sideband modulation. The lower sideband lies f_m below the carrier, and the upper sideband lies f_m above the carrier. Figure 3.2 shows how $\pm m(t)$ bounds the modulated carrier signal when $V_m = V_c = 1.0$ V and $f_c = 300$ and $f_m = 20$ kHz.

DSB modulation requires synchronized detection (transmitter and receiver aligned in frequency and phase). A local oscillator (LO) in the receiver reproduces an exact copy of the transmit carrier that multiplies the received signal. This multiplication results in a replica at baseband and another replica at twice the carrier frequency. A low pass filter (LPF) discards the replica at $2f_c$ but keeps the desired baseband signal. Figure 3.3 diagrams the process as well as shows the mathematical derivation.

Figure 3.2 DSB AM signal.

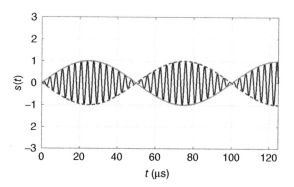

$$m(t) \cos(2\pi f_c t + \varphi) = \frac{1}{2} m(t) \cos \varphi + \frac{1}{2} m(t) \cos(4\pi f_c t + \varphi)$$

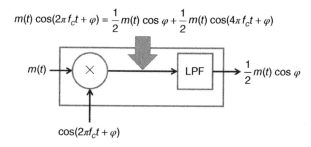

Figure 3.3 Synchronous detection of a DSB signal.

Adding a carrier to (3.6) produces an AM signal that does not require synchronous detection.

$$s(t) = V_c [1 + m(t)] \cos(2\pi f_c t) \tag{3.7}$$

This modulated signal has an envelope with the same shape as the message. As a result, the receiver consists of a simple envelope detector that recovers $m(t)$, unlike DSB modulation. Small variations in the carrier frequency do not impact signal reception. An AM modulation of the message signal in (3.5) has the following representation:

$$\begin{aligned}
s(t) &= V_c[1 + m(t)] \cos(2\pi f_c t) \\
&= V_c \cos(2\pi f_c t) + V_m \cos(2\pi f_m t) \cos(2\pi f_c t) \\
&= \underbrace{V_c \cos(2\pi f_c t)}_{\text{carrier}} + \underbrace{\frac{V_m}{2} \cos(2\pi (f_c - f_m)t)}_{\text{lower sideband}} + \underbrace{\frac{V_m}{2} \cos(2\pi (f_c + f_m)t)}_{\text{upper sideband}} \quad (3.8)
\end{aligned}$$

Figure 3.4 shows the AM spectrum with a carrier and two copies of the message at the harmonics at $f_c \pm f_m$. Adding the carrier takes power away from the message signal which in turn decreases the signal to noise ratio (SNR). The components of (3.8) appear in the time-frequency plot in Figure 3.5.

An AM envelope detector has a diode that passes only the positive portion of the waveform then a shunt capacitor that strips off the high-frequency components and leaves the baseband signal. Figure 3.6 shows an AM signal passing through an envelope detector. This asynchronous detection does not require an LO to recover the baseband signal.

Applying Euler's identity to (3.8) produces a complex AM spectrum.

$$\begin{aligned}
s(t) &= \frac{V_c}{2}[e^{j2\pi f_c t} + e^{-j2\pi f_c t}] + \frac{V_m}{4} \left[e^{j2\pi(f_c - f_m)t} + e^{-j2\pi(f_c - f_m)t} \right] \\
&\quad + \frac{V_m}{4} \left[e^{j2\pi(f_c + f_m)t} + e^{-j2\pi(f_c + f_m)t} \right] \tag{3.9}
\end{aligned}$$

Figure 3.4 AM spectrum.

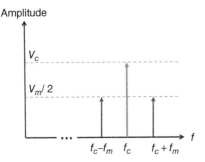

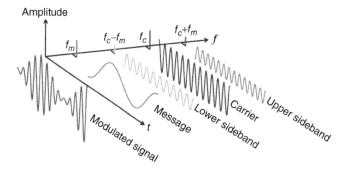

Figure 3.5 Time–frequency plot of an AM signal components.

Figure 3.6 Envelope demodulation of an AM signal.

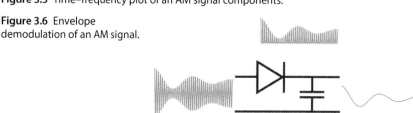

This equation may be rewritten as

$$s(t) = \frac{V_c}{2} e^{j2\pi f_c t} \left[1 + \frac{\beta_{AM}}{2} \left(e^{j2\pi f_m t} + e^{-j2\pi f_m t} \right) \right]$$
$$+ \frac{V_c}{2} e^{-j2\pi f_c t} \left[1 + \frac{\beta_{AM}}{2} \left(e^{j2\pi f_m t} + e^{-j2\pi f_m t} \right) \right] \tag{3.10}$$

where the AM modulation index is

$$\beta_{AM} = \frac{V_m}{V_c} \tag{3.11}$$

Figure 3.7 shows the complex spectrum for the AM signal in (3.9). The spectral components of the one-sided spectrum in Figure 3.4 split into positive- and negative-frequency components with half the amplitude.

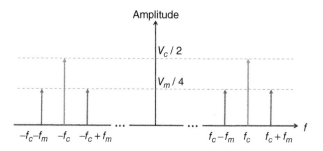

Figure 3.7 AM complex spectrum.

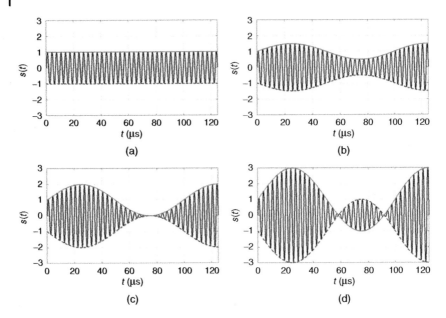

Figure 3.8 Effect of the modulation index on the time domain AM signal. (a) $\beta_{AM} = 0.0$. (b) $\beta_{AM} = 0.5$. (c) $\beta_{AM} = 1.0$. (d) $\beta_{AM} = 2.0$.

The modulation index determines the bounds of the AM signal envelope. Figure 3.8 shows a plot of (3.10) for four different values of β_{AM}. When $\beta_{AM} = 0$, only the carrier wave exists. When $0 < \beta_{AM} < 1.0$, the carrier signal has higher amplitude than the message signal, and the message signal clearly forms an envelope that defines the carrier amplitude. Figure 3.8b shows the modulated signal when $\beta_{AM} = 0.5$. When $\beta_{AM} = 1.0$, the positive envelope touches the negative envelope (Figure 3.8c). Beyond this point, envelope detection does not perfectly retrieve $m(t)$. When $\beta_{AM} > 1.0$ (e.g. $\beta_{AM} = 2.0$ in Figure 3.8d), over modulation occurs. Putting more power into the message than the carrier makes sense, but the envelope of the modulated signal becomes distorted and the carrier experiences a 180° phase shift. Overmodulation introduces spurious harmonics into the signal that cause radiation outside the signal bandwidth.

The Fourier transform of (3.8) yields the frequency domain representation of the complex AM spectrum.

$$S(f) = \frac{V_c}{2}\left[\delta(f - f_c) + \delta(f + f_c)\right]$$
$$+ \frac{V_m}{4}\left[\delta(f - f_c + f_m) + \delta(f + f_c - f_m)\right]$$
$$+ \frac{V_m}{4}\left[\delta(f - f_c - f_m) + \delta(f + f_c + f_m)\right] \quad (3.12)$$

The AM signal power at a resistor R_e is

$$P = \frac{1}{R_e} \left[\frac{V_c^2}{2} + \frac{V_m^2}{4} \right] \tag{3.13}$$

Substituting $V_m = \beta_{AM} V_c$ into (3.13) yields

$$P = \frac{1}{2R_e} [V_c^2 + 0.5\beta_{AM}^2 V_c^2] \tag{3.14}$$

Substituting the carrier power, $P_c = \frac{V_c^2}{2R_e}$, into (3.14) produces

$$P = P_c \left[1 + 0.5\beta_{AM}^2 \right] \tag{3.15}$$

Example

Use MATLAB to modulate a 300 kHz carrier with a voice signal using $\beta_{AM} = 0.5$.

Solution

Figure 3.9 shows a small part of a voice signal modulated with the 300 kHz carrier. The following code captured the voice signal into the vector "song" and normalizes the amplitude. The key commands to find the modulated signal are

```
load music; % this loads the voice file "music" into
    myRecording
song=myRecording'; % the vector "song" contains the music
song=song/max(song); % normalize the amplitude
t=[0:1250]*1e-6; % time in microseconds
car= cos(2*pi*300e3*t); % carrier
bAM=0.5; % modulation index
vAM=[1+bAM*song]*car; % AM signal
```

The AM radio frequency band extends from 535 to 1605 kHz as shown in Figure 3.10. Carrier frequencies occur at 10 kHz intervals from 540 to

Figure 3.9 AM voice signal.

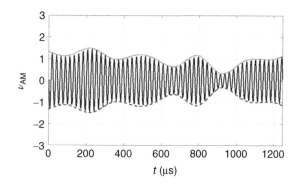

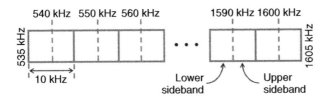

Figure 3.10 The AM radio spectrum.

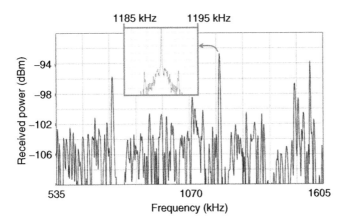

Figure 3.11 Measured broadcast AM spectrum near Boulder, CO with expanded view from 1185 to 1195 kHz.

1600 kHz. Two sidebands and the carrier for transmitting voice signals lie within the 10 kHz bandwidth. The measured broadcast AM spectrum from 1185 to 1195 kHz near Boulder, CO appears in Figure 3.11. The inset box shows an expanded frequency view about the 1190 kHz carrier frequency (1185–1195 kHz) where the radio station resides. Note the sidebands on either side of the carrier.

AM is not efficient, because the carrier consumes power, and the information resides in both the upper and the lower sidebands. Single sideband (SSB) modulation increases efficiency by only transmitting one sideband of AM (usually the upper sideband). This approach halves the AM bandwidth while still transmitting the entire message signal. SSB suppressed carrier modulation removes the carrier to reduce the transmitted power. In order to make the signal easier to detect, many SSB schemes have some remnant of the carrier for receiver synchronization. Simple envelope detection does not work for SSB signals.

Digital AM or amplitude shift keying (ASK) represents binary symbols by discrete amplitude steps. The simplest form of ASK have two states: the presence of a carrier wave that indicates a "1," and its absence that indicates a "0."

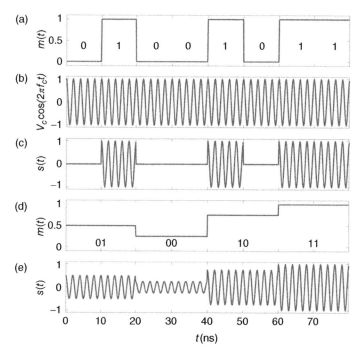

Figure 3.12 ASK modulation. (a) 8 bits, (b) carrier, (c) OOK, (d) symbols with 2 bits, and (e) 4-ASK.

Figure 3.12a shows an on–off keying (OOK) bit stream multiplying a carrier in Figure 3.12b to get the ASK signal in Figure 3.12c. Using more than two amplitude states to represent symbols requires a high *SNR* in order to distinguish the different symbols. *M*-ASK modulation uses $M = 2^{N_{bits}}$ amplitude levels to represent symbols that have N_{bits} bits. Figure 3.12d shows four different symbols that multiply the carrier to get the 4-ASK signal in Figure 3.12e. The measured power received as a function of frequency for a 2 GHz ASK signal with a symbol rate of 200 kbps appears in Figure 3.13. The carrier causes the spike at 2 GHz.

AM has the advantages of being simple and cheap to implement. The disadvantages include low power efficiency, low bandwidth efficiency, and high sensitivity to noise. As a result, AM works well for narrowband applications that tolerate noise as in the transmission of audio signals. AM consumes a lot of spectrum and over half of its power lies with the carrier which has no information. DSB does not waste power in a carrier but has the same bandwidth as AM and requires synchronous detection. SSB modulation has half the bandwidth of AM, puts all power in one information band, has less noise due to smaller bandwidth, and prevents selective fading (carrier and sidebands arriving at different times) but requires synchronous detection.

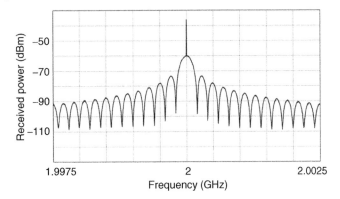

Figure 3.13 Measured spectrum of a 2-ASK signal with a carrier frequency of 2 GHz and a symbol rate of 200 kbps.

3.3 Frequency-Modulated Signals

Frequency modulation (FM) varies the frequency of the carrier in proportion to $m(t)$. The voltage waveform of an FM signal looks like

$$s(t) = V_c \cos\left[2\pi f_c t + 2\pi \Delta f \int_0^t m(\tau)d\tau\right] \tag{3.16}$$

where Δf is the maximum frequency deviation from f_c. As in AM, assume $m(t)$ is a single harmonic at f_m

$$m(t) = V_m \cos 2\pi f_m t \tag{3.17}$$

Substituting $m(t)$ into (3.16) and integrating produces

$$s(t) = V_c \cos[2\pi f_c t + \beta_{FM} \sin(2\pi f_m t)] \tag{3.18}$$

where the FM modulation index is

$$\beta_{FM} = \frac{\Delta f}{f_m} \tag{3.19}$$

and $\Delta f \propto A_m$.

Example
Given a single harmonic message signal at 1 GHz and a carrier at 10 GHz, plot the FM modulated signal for half a period of the message when $\beta_{FM} = 0.2$, 1.0, 2.0, and 5.0 and $V_c = V_m = 1V$.

Solution
Figure 3.14 shows the plots of $s(t) = \cos[2\pi 10^7 t + \beta_{FM} \sin(2\pi 10^6 t)]$.

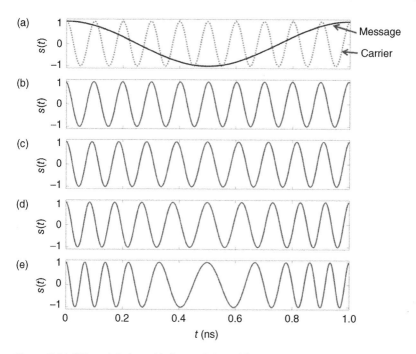

Figure 3.14 FM modulation with $f_c = 10$ GHz and $f_m = 1$ GHz. (a) Message signal superimposed on carrier, (b) $\beta_{FM} = 0.2$, (c) $\beta_{FM} = 1.0$, (d) $\beta_{FM} = 2.0$, and (e) $\beta_{FM} = 5.0$.

Rewriting (3.18) in terms of Bessel functions results in an expression having an infinite number of harmonics in the FM spectrum.

$$s(t) = V_c \sum_{n=-\infty}^{\infty} J_n(\beta_{FM}) \cos\left[2\pi(f + nf_m)_c t\right], \tag{3.20}$$

where $J_n(\beta_{FM})$ is an nth order Bessel function evaluated at β_{FM} (Figure 3.15). Next, taking the Fourier transform of (3.20) results in

$$S(f) = \frac{V_c}{2} \sum_{n=-\infty}^{\infty} J_n(\beta_{FM}) \left[\delta(f - f_c - nf_m) + \delta(f + f_c + nf_m)\right] \tag{3.21}$$

Figure 3.15 Bessel functions of the first kind.

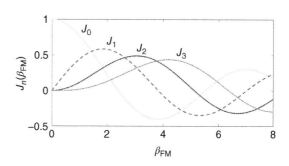

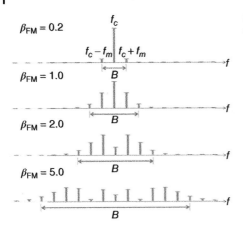

Figure 3.16 FM bandwidth as a function of β_{FM}.

The peak amplitude of $J_n(\beta_{FM})$ decreases with increasing n. As β_{FM} increases, the number of effective sidebands in the FM signal increases, which in turn, increases the bandwidth. When $\beta_{FM} = 0.25$ only one significant sideband exists. Narrowband FM has a carrier, an upper sideband, and a lower sideband similar to AM. Wideband FM occurs when $\beta_{FM} > 1.0$. If $\beta_{FM} = 5$, then the spectrum has eight significant sidebands. Figure 3.16 shows the FM spectrum that correspond to the modulated signals in Figure 3.14.

An FM signal that varies in frequency from $f_c - \Delta f$ to $f_c + \Delta f$ has a bandwidth given by Carson's rule [1]:

$$B = 2\left(\Delta f + f_m\right) = 2\left(\beta_{FM} + 1\right)f_m \tag{3.22}$$

where f_m is the highest frequency in the modulating signal. In North America, $\Delta f = 75\,\text{kHz}$ for commercial FM broadcasting in order to prevent adjacent stations from interfering with one another. FM radio station carrier frequencies always end in 0.1, 0.3, 0.5, 0.7, or 0.9 MHz (88.1 MHz but not 88.0 MHz). If the highest frequency in an audio signal is $f_m = 15\,\text{kHz}$, then $\beta_{FM} = 5$. The bandwidth is $B = 2(5 + 1)5\,\text{kHz} = 180\,\text{kHz}$ which is close to the 200 kHz allocated channel bandwidth. Figure 3.17 shows the measured broadcast FM spectrum from 87.5 to 108.0 MHz near Boulder, CO. The inset box has an expanded frequency view at the 87.5 MHz carrier frequency of a radio spectrum.

Example
Use MATLAB to frequency modulate a 20 kHz carrier with a voice signal and $\Delta f = 10\,\text{kHz}$.

Solution
Substitute into (3.16) to get $v(t) = \cos\left[40\pi t + 20\pi \int_0^t m(\tau)d\tau\right]$ with t in ms. Figure 3.18 is a MATLAB plot of the voice signal superimposed on the modulated signal.

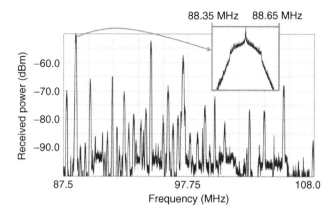

Figure 3.17 Measured broadcast FM spectrum near Boulder, CO.

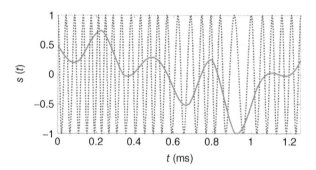

Figure 3.18 Solid line is voice signal and dotted line is the FM-modulated carrier.

Digital FM, called frequency shift keying (FSK), represents symbols by distinct frequencies. Binary frequency shift keying (BFSK) has two possible waveforms:

$$s(t) = \begin{cases} A_c \cos\left(2\pi f_{hi} t\right) & \text{for a "1"} \\ A_c \cos\left(2\pi f_{lo} t\right) & \text{for a "0"} \end{cases} \tag{3.23}$$

where $f_{hi} = f_c + \Delta f$ and $f_{lo} = f_c - \Delta f$. Figure 3.19 shows a BFSK signal corresponding to a bit stream of 01001011 with each bit 1 μs long. In this example, $f_{lo} = 3.9$ MHz and $f_{hi} = 5.2$ MHz. M-FSK (multiple frequency shift keying) uses M different frequencies to represent M symbols, where each symbol has $\log_2 M$ bits. Figure 3.20 shows the power received as a function of frequency for a 2 GHz FSK signal with a symbol rate of 200 kbps.

FM has the advantage that amplitude variations due to noise do not degrade signal detection. Most noise impacts the signal amplitude and not the frequency or phase. FM performance benefits from using nonlinear amplifiers

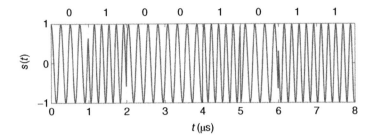

Figure 3.19 BFSK signal.

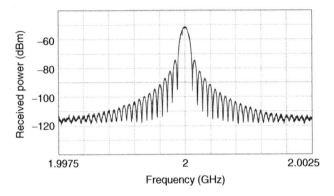

Figure 3.20 Measured spectrum of an FSK signal with a carrier frequency of 2 GHz and a symbol rate of 200 kbps.

that have a higher efficiency than the AM linear amplifiers. FM has the disadvantage of complicated modulation and demodulation circuits compared to AM.

3.4 Phase-Modulated Signals

Phase modulation (PM) changes the carrier phase in proportion to $m(t)$. A PM signal takes the form

$$s(t) = V_c \cos \left[2\pi f_c t + \beta_{PM} m(t) \right] \tag{3.24}$$

where β_{PM} (rad/V), the PM index, equals the peak phase deviation and $\beta_{PM} \propto V_m$. Note that (3.16) and (3.24) look very similar. In fact, FM falls into the PM category. Carson's rule also gives the bandwidth for PM signals as

$$B = 2(\beta_{PM} + 1)f_m \tag{3.25}$$

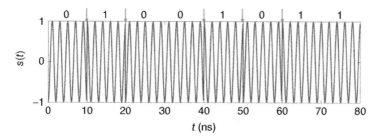

Figure 3.21 BPSK where the 2 bit values are 180° out of phase.

In digital modulation, PM represents binary symbols with unique phase values. Binary phase shift keying (BPSK) has two phase states with a "1" 180° out of phase with a "0." For instance, a BPSK signal has the following representation:

$$
s(t) = \begin{cases} A_c \cos\left(2\pi f_c t\right) & \text{for a "1"} \\[2mm] -A_c \cos\left(2\pi f_c t\right) & \text{for a "0"} \end{cases} \tag{3.26}
$$

An example of an 8-bit BPSK signal appears in Figure 3.21. No phase change occurs when consecutive bits remain the same. A constellation diagram displays symbol samples in the real-imaginary or in-phase-quadrature (IQ) plane. Figure 3.22 shows the measured constellation and eye diagram plots for BPSK as the transmitted signal decreases but the noise stays the same. The eye shrinks as the SNR decreases relative to the noise. Ellipses encircle symbols in the constellation diagram. As the SNR decreases, the ellipses increase in size to the point where they overlap. Distinguishing symbols becomes more and more difficult with decreasing SNR.

Quadrature phase shift keying (QPSK) maps four symbols containing 2 bits each into four unique carrier phases. The phase state to symbol mapping is given by

$$
s(t) = \begin{cases} A_c \cos\left(2\pi f_c t - 45^0\right) & \text{for "00"} \\[2mm] A_c \cos\left(2\pi f_c t - 135^0\right) & \text{for "01"} \\[2mm] A_c \cos\left(2\pi f_c t - 315^0\right) & \text{for "10"} \\[2mm] A_c \cos\left(2\pi f_c t - 225^0\right) & \text{for "11"} \end{cases} \tag{3.27}
$$

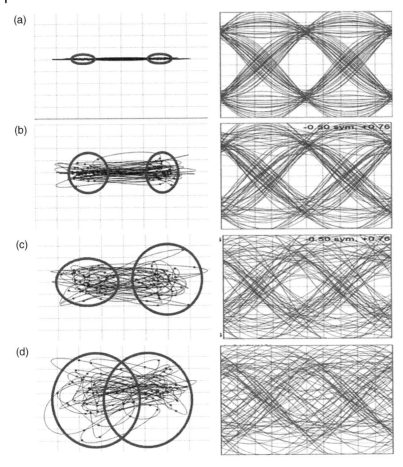

Figure 3.22 Measured constellation plot (left) and eye diagram (right) for BPSK as the transmitted signal level decreases: (a) 10 dBm, (b) −35 dBm, (c) −40 dBm, and (d) −43 dBm.

Using the trigonometric identity $\cos(x+y) = \cos x \cos y - \sin x \sin y$, (3.27) becomes

$$s(t) = \begin{cases} 0.707A_c \left[\cos\left(2\pi f_c t\right) + \sin\left(2\pi f_c t\right)\right] & \text{for "00"} \\ 0.707A_c \left[-\cos\left(2\pi f_c t\right) + \sin\left(2\pi f_c t\right)\right] & \text{for "01"} \\ 0.707A_c \left[\cos\left(2\pi f_c t\right) - \sin\left(2\pi f_c t\right)\right] & \text{for "10"} \\ 0.707A_c \left[-\cos\left(2\pi f_c t\right) - \sin\left(2\pi f_c t\right)\right] & \text{for "11"} \end{cases} \tag{3.28}$$

If "cos" is "in-phase," then "sin" is "quadrature" or 90° out-of-phase. Figure 3.23 shows a plot of the constellation diagram for (3.28). Arrows point from the

Figure 3.23 IQ plot of a QPSK signal.

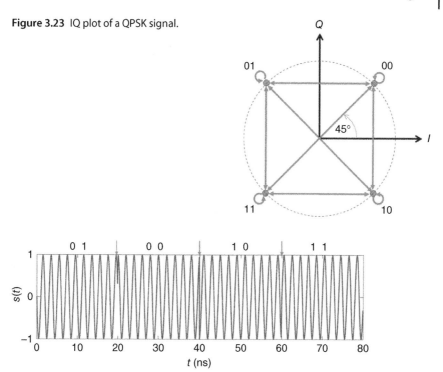

Figure 3.24 QPSK with four phase changes for the four possible symbols.

present phase state to the next possible phase state. QPSK allows any of the four phase states to follow the current phase state. The QPSK signal in Figure 3.24 has four symbols "01," "00," "10," and "11." Arrows start at the current state and point at possible future states. The phase changes do not produce a smooth transition between symbols in the time domain.

A QPSK signal has up to a 180° phase change between symbols (e.g. 00 → 11). In a practical system, large changes in phase produce significant unwanted spectral components outside the desired bandwidth. Alternative forms of QPSK limit phase changes to less than 180° phase changes between states. As an example, offset quadrature phase shift keying (OQPSK) delays the quadrature portion of the signal or the odd bits by 1 bit-period (half a symbol-period) relative to the even bits, so that the in-phase and quadrature components never change at the same time. This approach limits the maximum phase shift between symbols to only 90°. Figure 3.25 shows an IQ plot of an OQPSK signal. In this case, three instead of four phase states possibly follow the current phase state. No symbol sequence has a path through the origin, which corresponds to a 180° phase change. Generating the OQPSK symbols resemble QPSK, except for the 1-bit time delay (T_b) in the quadrature channel shown in Figure 3.26.

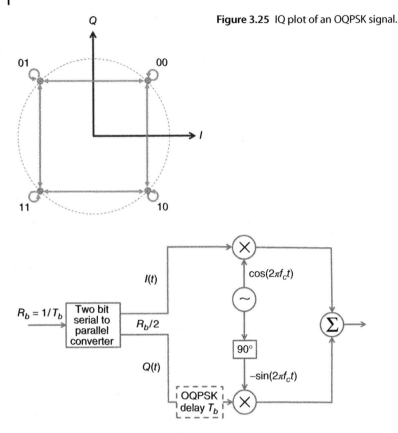

Figure 3.25 IQ plot of an OQPSK signal.

Figure 3.26 QPSK and OQPSK modulation. OQPSK has a 1 bit time delay in the quadrature path.

The I and Q channels have half the bit rate $(R_b/2)$ of the input data (R_b). Figure 3.27 shows the output I and Q bits for QPSK vs. OQPSK when the input symbols are "10," "11," "00," "01," and "11." QPSK and OQPSK have identical in-phase bits. In QPSK, the output phase changes as much as 180° with every symbol. The 1-bit delay in the quadrature channel of OQPSK, on the other hand, causes the output phase to change with every bit (twice as fast as QPSK).

$\pi/4$-QPSK also reduces the phase change between symbols by using two identical constellations with one rotated 45° relative to the other (Figure 3.28). The four open circles in the constellation plot correspond to odd-numbered symbols 1, 3, 5, ..., while the closed circles correspond to even-numbered symbols 2, 4, 6, ... In Figure 3.28, the message starts with symbol, 01, on the open circle constellation. The second symbol, 00, is on the closed circle constellation. The sequence moves to symbol, 10, then finally to 11. $\pi/4$-QPSK has a maximum phase shift of 135°. A plot of the resulting $\pi/4$-QPSK signal

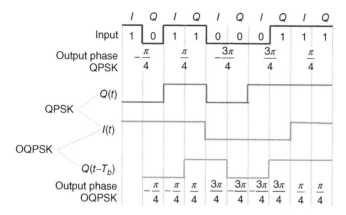

Figure 3.27 Plots of the in-phase (V_I and, quadrature (V_Q) parts of the QPSK and OQPSK signals.

Figure 3.28 IQ diagram for $\pi/4$ QPSK with the bit sequence of 01 to 00 to 10 to 11.

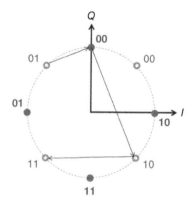

for the symbol sequence "01," "00," "10," and "11" appears in Figure 3.29. Demodulation of $\pi/4$-QPSK consists of differential detection in which the phase of the previous symbol serves as a reference for the phase of the current symbol. In the presence of multipath and fading, $\pi/4$-QPSK out performs OQPSK [2].

Minimum shift keying (MSK) is the same as continuous-phase FSK with $\beta_{FM} = 0.5$. MSK has no phase discontinuity during the frequency change between a logical 1 and 0 as seen by the waveform plot in Figure 3.30. The MSK frequency spectrum decreases much faster than the QPSK spectrum, so it is more efficient. Gaussian minimum shift keying (GMSK) first passes the bits through a Gaussian filter before entering the MSK modulator. A Gaussian filter has a smooth transfer function with a spectrum that falls off even faster than the MSK spectrum and has no zero crossings. Its transfer function and

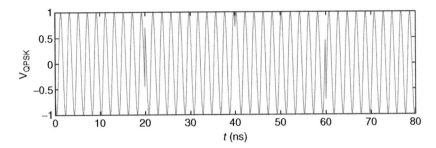

Figure 3.29 Plot of a $\pi/4$ QPSK signal.

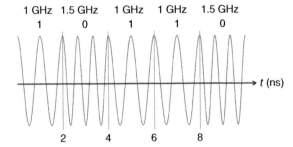

Figure 3.30 MSK modulation.

impulse response are given by

$$H(f) = e^{-\left(\frac{0.5887f}{B}\right)^2} \tag{3.29}$$

$$h(t) = \frac{3}{B}e^{-(5.3365Bt)^2} \tag{3.30}$$

In the time domain, it minimizes the rise and fall times and has no overshoot to a step function input.

Differential phase shift keying (DPSK) uses relative phase rather than absolute phase to represent symbols. Detecting a phase change is easier than accurately determining the phase of the received signal, because the demodulator does not need a coherent carrier. When DPSK transmits a "1," the phase changes by 180° while transmitting a "0" has no phase change. Figure 3.31 shows an example of a DPSK signal. Note the phase changes at 4 and 6 ns.

3.5 Quadrature Amplitude Modulation

Quadrature amplitude modulation (QAM) combines ASK and PSK in order to generate more distinguishable and longer symbols than either ASK or PSK alone. A symbol change results in an amplitude and/or PM of the carrier. Points

Figure 3.31 DPSK signal example.

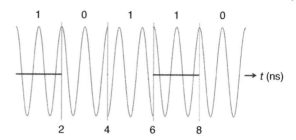

Figure 3.32 IQ diagram for a 16-QAM signal.

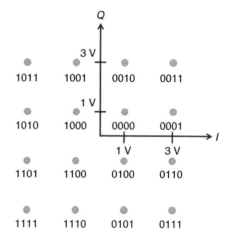

on the constellation diagram have radii that are proportional to the amplitude of the modulated signal, while the angular location of the points correspond to the phase of the modulated signal. As the signal changes symbols (moves from one point to another on the constellation diagram), amplitude and PM occurs. Figure 3.32 displays a constellation diagram for a 16-QAM signal. The symbol 0011 has the same phase as 0000 but a higher amplitude. In contrast, the symbol 1011 has the same amplitude as 0011 but a different phase.

Adding noise to the signal creates scatter plots about the constellation points as shown in Figure 3.33. For instance, if the SNR = 16 dB ($Eb/N0$ = 10 dB), then 16-QAM results in the constellation in Figure 3.33a with a BER = 0.002. Adding more noise so that SNR reduces to 10 dB ($Eb/N0$ = 4 dB), results in a greater spread in the scatter plots about the 16 constellation points as shown in Figure 3.33b. Its BER increases to 0.0766. In addition, intersymbol interference (ISI) occurs because a point on the constellation diagram representing one symbol moves into the detection region of another symbol. Grid lines separate symbol detection regions. Points that lie within a 2 by 2 V box are interpreted as the symbol in the center of the box (indicated by a "+"). The ISI in Figure 3.33b is noticeably greater than in Figure 3.33a, because many points from one symbol lie within the box of another symbol.

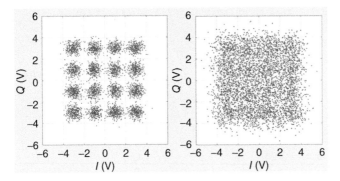

Figure 3.33 Constellation diagrams for 16-QAM. (a) SNR = 16 dB and (b) SNR = 10 dB.

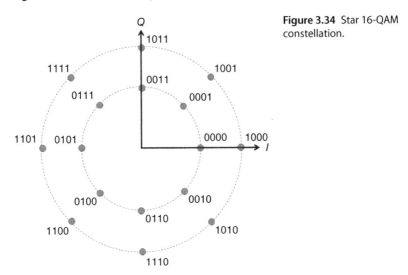

Figure 3.34 Star 16-QAM constellation.

QAM constellation points do not have to be in a rectangular grid as in Figure 3.32. One variation, called star QAM [3] is PSK with two or more amplitude levels (Figure 3.34). Star QAM minimizes error rates for a given transmission power. The star 16-QAM symbols with the same amplitude in Figure 3.32 are 45° apart in phase.

3.6 Power Spectral Density of Digital Signals

A baseband signal defined over an interval T has a normalized average power of

$$P_{avg} = \lim_{T \to \infty} \frac{1}{T} \int_{-T/2}^{T/2} s^2(t) dt \; \text{W} \tag{3.31}$$

Parseval's theorem says that the integral of the square of a signal equals the integral of the square of its transform ($S(f) = \mathfrak{I}\{s(t)\}$). Applying this theorem to (3.31) produces

$$P_{avg} = \lim_{T \to \infty} \frac{1}{T} \int_{-\infty}^{\infty} |S(f)|^2 df \quad W \tag{3.32}$$

The PSD is the integrand in (3.32)

$$PSD_B(f) = \lim_{T \to \infty} \frac{|S(f)|^2}{T} \quad W/Hz \tag{3.33}$$

Normalized average power in the frequency domain is the PSD integrated over all frequencies.

$$P_{avg} = \int_{-\infty}^{\infty} PSD(f) df \quad W \tag{3.34}$$

Table 3.1 lists the baseband PSD expressions for several digital modulation schemes. All the equations are in terms of T_b even though most modulation techniques deal with symbols where the symbol period is

$$T_s = T_b \log_2 M \tag{3.35}$$

The energy per bit equals the square of the carrier amplitude times the bit period

$$E_b = A_c^2 T_b \tag{3.36}$$

The PSD for the bandpass signal is written in terms of the baseband PSD as

$$PSD_P = \frac{1}{4} \left[PSD_B(f - f_c) + PSD_B(f + f_c) \right] \tag{3.37}$$

Table 3.1 Baseband PSD for selected modulations. Reminder: $\text{sinc}(x) = \sin(\pi x)/(\pi x)$.

Modulation	PSD_B
ASK	$\dfrac{E_b}{2}[\delta(f)/T_b + \text{sinc}^2(fT_b)]$
M-ASK	$\dfrac{E_b}{2}[\delta(f)/T_b + \text{sinc}^2(fT_b)]$
M-FSK	$\displaystyle\sum_{m=1}^{M} \left\{ \dfrac{E_b}{4M}\text{sinc}^2[(f - f_m)T_b] + \dfrac{E_b}{4M}\text{sinc}^2[(f + f_m)T_b] \right\}$
BPSK	$E_b \text{sinc}^2(fT_b)$
QPSK	$2E_b \text{sinc}^2(2fT_b)$
M-PSK	$E_b \log_2(M)\text{sinc}^2(nfT_b \log_2(M))$
OQPSK	$2E_b \text{sinc}^2(2fT_b)$
M-QAM	$\dfrac{E_b}{2}\text{sinc}^2(fT_b \log_2 M)$

This formula allows the simple conversion of PSD_B to a bandpass signal modulated by a carrier.

The spectral or bandwidth efficiency of a modulation scheme is the bit rate divided by the bandwidth.

$$\eta_{se} = \frac{R_b}{B} = \frac{T_b \log_2 M}{2/T_b} = 0.5\log_2 M \text{ bits/s/Hz} \tag{3.38}$$

Spectral efficiency increases as M increases, but symbols get closer together and require a higher SNR. Increasing the symbol energy increases the SNR.

3.7 BER of Digital Signals

Closed form expressions exist for the *BER* associated with some common digital modulation techniques [4]. ASK has the following BER:

$$BER = Q\left(\sqrt{\frac{E_b}{N_0}}\right) \tag{3.39}$$

BPSK, QPSK, and OQPSK have the same BER:

$$BER = Q\left(\sqrt{\frac{2E_b}{N_0}}\right) \tag{3.40}$$

where the Q and the complementary error functions are defined as

$$Q(z) = \frac{1}{2}\text{erfc}\left(\frac{z}{\sqrt{2}}\right) \tag{3.41}$$

$$\text{erfc}(z) = \frac{2}{\sqrt{\pi}}\int_z^\infty e^{-\xi^2 2}d\xi \tag{3.42}$$

Other modulation schemes have more complicated expressions that apply under special circumstances, so they are not listed here.

3.8 Multiplexing in Time and Frequency

A simplex wireless system conveys information one way – from a transmitter to a receiver. Radio stations transmit music to user radios that receive it and cannot transmit back. No communication occurs in the reverse direction.

In contrast, a duplex wireless system allows transmission and reception by all devices on the network. A full duplex system allows users to simultaneously transmit and receive, like a telephone. Frequency division duplexing (FDD) transmits at one frequency and receives at another frequency. Satellite communications use a C band uplink at 5.925–6.425 GHz and a downlink at 3.7–4.2 GHz. A half-duplex system means that only one device transmits at a time. In a half-duplex system like walkie-talkie radios, one person says "over" when finishing a transmission in order to let the other person know that the channel is open for transmission. Time division duplexing (TDD) has designated times for transmit and receive signals.

Multiplexing combines multiple signals over a shared medium. The process of multiplexing divides a communication channel into a number of subchannels with each message assigned to a different subchannel. Demultiplexing extracts the multiplexed signals from the different subchannels. Frequency division multiplexing (FDM) divides the spectrum into bands and assigns signals to the bands. Time division multiplexing (TDM) divides time into slots then assigns signals to time slots. This section introduces FDM and TDM before explaining the more versatile multiple access schemes.

3.8.1 Frequency Division Multiplexing

FDM divides a specified frequency band into nonoverlapping subbands with each subband having a carrier frequency. It places multiple independent signals in a designated spectrum (e.g. AM or FM radio) or splits a high-data-rate signal into multiple lower-data-rate signals that exist in parallel. All the subbands combine to form a composite signal that occupies a designated portion of the spectrum.

Figure 3.35 shows a diagram of an FDM system that has M signals modulated into subbands that occupy the composite signal bandwidth. Either each subband has its own transmitter at a unique carrier frequency, or a single transmitter sends the entire composite signal. At the receiving end, the composite

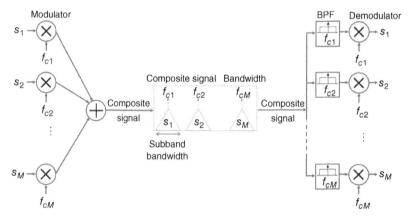

Figure 3.35 Frequency division multiplexing with M subbands.

signal passes through a bandpass filter (BPF) centered at the carrier frequency of one subband. The filtered signal is then demodulated to recover the baseband signal. A good example of FDM is the television UHF Band from 470 to 806 MHz. Each station, numbered 14–69, has a 6 MHz slot. Radio astronomy fills a gap at station 37, 608–614 MHz.

Orthogonal frequency division multiplexing (OFDM) divides a high-data-rate signal into a number of low-data-rate signals then transmits them in parallel by modulating with subcarriers spaced $1/T_s$ apart in order to make all the low-data-rate signal spectra orthogonal. An orthogonal subcarrier has a peak in its spectrum when all the other subcarrier spectra equal zero in order to reduce adjacent channel interference. The subcarrier bands do not have to be contiguous. A conventional digital modulation scheme (such as QPSK, 16-QAM, etc.) modulates the carrier with a low-symbol-rate signal. The combined data rates of the subcarriers result in an overall data rate similar to conventional single-carrier modulation schemes with an equivalent bandwidth.

Figure 3.36 shows three symbols transmitted in time with guard bands between symbols. The OFDM representation in the frequency domain is the inverse fast Fourier transform (IFFT) of the symbol. Each IFFT bin in the frequency domain corresponds to an orthogonal subcarrier spectrum. A subcarrier signal occupies a null-to-null bandwidth equal to $2/T_s$. OFDM usually has several channels with guard bands between channels, so it can handle many data streams.

3.8.2 Time Division Multiplexing

TDM combines several low-bit-rate signals into one high-speed bit stream by assigning signals to specific time slots. Synchronous TDM assigns each signal to a predefined slot recognized by the receiver. All signals have the same sampling

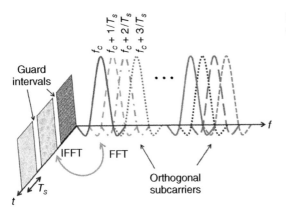

Figure 3.36 Time–frequency diagram for OFDM.

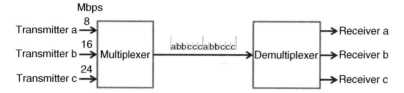

Figure 3.37 Example of synchronous TDM.

rate, so all data streams have the same bit rate which allows the transmitter and the receiver to perfectly synchronize with the slot period. If the signal data rate goes down, then slots go unused and efficiency drops.

Example
Given three signals with bit rates of 8, 16, and 24 Mbps, use TDM to combine them into one channel.

Solution
The TDM channel has a transmission rate of 48 Mbps (sum of the bit rates of the three signals). The ratio of the three data rates is 8 : 16 : 24. Thus, the time slots are allocated in proportion to 1 : 2 : 3. The ratio sums to 6, which corresponds to the minimum length of the slot assignment period. The slot assignment is illustrated in Figure 3.37.

Statistical TDM does not have synchronized time slots with fixed assignments for each different data stream. Instead, the system dynamically assigns time slots based on the past, current, and predicted data rates of the different signals [5]. Allocating time slots based on need allows more users to participate. This approach promises to be more efficient but requires some intelligent processing.

3.8.3 Multiple Access

In TDM/FDM, known users have assigned time slots/subbands. Multiplexing works well for a fixed number of users. In a wireless system where users come and go, like a cellular network, dynamic assignment using multiple access techniques proves to be much more efficient.

The multiple access approach in packet radio, allows several transmitters to send bursts of data packets to a receiver at any time. Each packet competes for the receiver's attention. In an early research project on radio access to computer systems, Professor Norman Abramson of the University of Hawaii invented a simple multiple access approach called ALOHA in which many remote terminals share one radio channel without central control [6]. The receiver responds with an acknowledgement once it successfully receives a packet. If two packets

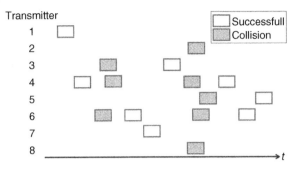

Figure 3.38 ALOHA packets as a function of time. If packets overlap, then they are retransmitted.

from different transmitters overlap, then a collision occurs and the packets are not received. If the transmitter does not get an acknowledgement after a short time, then it resends the packet. Figure 3.38 shows an example where eight transmitters randomly send packets. Packets that do not overlap are successfully received. Transmitters must resend packets that collide.

Assume that the users of an ALOHA system generate random packets at a rate that follows a Poisson process [7] in which λ_p packets that are τ_p long are transmitted per unit of time. The probability of no collisions means that no other transmission occurs over the period $t - \tau_p$ to $t + \tau_p$. The probability of no collisions equals the probability that no other packets transmit over the interval $t - \tau_p$ to $t + \tau_p$ and is given by

$$p(\text{no collisions}) = e^{-2\lambda_p \tau_p} \tag{3.43}$$

The ratio of the average rate of successfully transmitted packets (λ_p) divided by the channel packet rate and is a unitless quantity called throughput:

$$\gamma_p = \lambda_p \tau_p e^{-2\lambda_p \tau_p} \tag{3.44}$$

ALOHA's throughput increases when a user checks the channel availability before transmitting. Another improvement to ALOHA forces all packets into established time slots rather than random time slots. Slotted ALOHA reduces collisions due to small overlaps of packets. Collisions still occur when two or more packets occupy a time slot as shown in Figure 3.39. Collisions in slotted ALOHA only occur over the interval t to $t + \tau_p$, so the probability of no packet collisions in an interval is given by [8]

$$p(\text{no collisions}) = e^{-\lambda_p \tau_p} \tag{3.45}$$

with a corresponding throughput of

$$\gamma_p = \lambda_p \tau_p e^{-\lambda_p \tau_p} \tag{3.46}$$

Figure 3.40 shows a plot of the throughput as a function of load ($\lambda_p \tau_p$) for ALOHA and slotted ALOHA. A small load means that few users transmit packets over the channel. As the load increases, more transmitters send

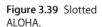 **Figure 3.39** Slotted
ALOHA.

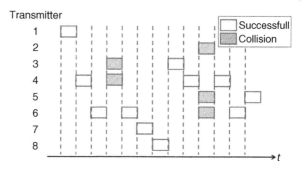

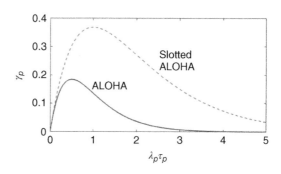

Figure 3.40 Throughput as a
function of load for ALOHA and
slotted ALOHA.

packets that result in more collisions. ALOHA has a peak throughput of 0.184 when $\lambda_p \tau_p = 0.5$ which means that successful transmission only occurs 18.4% of the time. Slotted ALOHA improves throughput with a peak of 0.368 when $\lambda_p \tau_p = 1.0$ which means that successful transmission occurs 36.8% of the time. In either case, these approaches are extremely inefficient for transmitting data over a wireless network. An Aloha network needs some central management, or it becomes overloaded and the throughput falls to zero. Although novel at the time, more efficient approaches to multiple access are currently available.

Frequency division multiple access (FDMA) is FDM that allows users access to any open subband in the channel. Orthogonal frequency-division multiple access (OFDMA), a multi-user version of OFDM, dynamically assigns subcarriers to users based on the availability of subbands. Figure 3.41 illustrates the difference between OFDM and OFDMA. OFDM assigns all the subcarriers to a user in a given time slot whereas OFDMA assigns subcarriers in a given time slot to multiple users. Some OFDMA time slots have no users assigned to subcarriers due to lack of demand.

Time division multiple access (TDMA) shares a single-carrier frequency with several users where users have nonoverlapping time slots. TDMA data transmission occurs in bursts and has low battery consumption, because the transmitter only turns on when transmitting. A user gets a time slot only when actively transmitting.

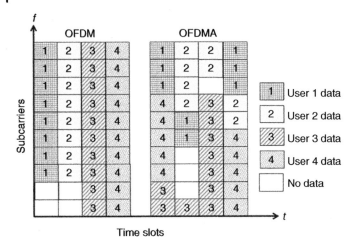

Figure 3.41 OFDMA spectrum vs. OFDM slot assignment.

3.9 Spread Spectrum

Spread spectrum forces a signal with bandwidth B to occupy a larger bandwidth of BG_p, where the processing gain $G_p > 1.0$. The extra bandwidth spent on spread spectrum overcomes several problems encountered in the channel.

Spread spectrum originated during World War II, when the actress Hedy Lamarr learned how easily the enemy jammed and steered away radio-controlled torpedoes [9]. She originated the idea of hopping the torpedo control signals from one frequency to another to avoid jamming. In order to do so, the frequency hopping requires synchronization between the transmitter and the receiver on the torpedo. Her pianist friend, George Antheil, helped her develop a synchronization process based on the idea of a miniaturized player-piano. Their patent for frequency-hopping spread spectrum [10] was too difficult to implement at the time, so the first practical application did not occur until 1962 on Navy ships during the Cuban missile crisis. Since then, the military developed many different spread-spectrum techniques which eventually became common in the commercial sector (e.g. Bluetooth – see Appendix IV).

The channel-capacity theorem illustrates the tradeoff made when converting a narrowband signal into a wideband signal. For a constant channel capacity, decreasing the SNR requires increasing the bandwidth. If $C = 10$ Mbs and SNR $= 20$ dB, then $B = 66$ MHz. To maintain the same capacity, if the SNR decreased to -10 dB, then B must increase to 3.2 GHz. Thus, a high-power narrowband signal spreads into a low-power wideband signal while maintaining the same channel capacity. Burying the signal into the noise is very attractive for military or spy applications. It allows spectrum sharing if the signal just appears

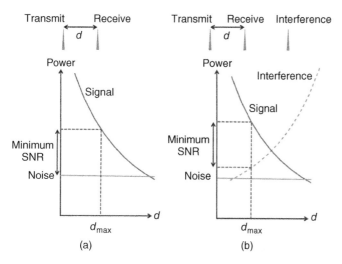

Figure 3.42 Signal, interference, and noise power levels in a channel. (a) A minimum SNR limits d_{max} when only the signal and noise are present and (b) a minimum SINR limits d_{max} when the signal, noise, and interference are present.

to be noise. Spread-spectrum transmit power levels are similar to narrowband signal power levels but have a much lower spectral power density. This means that spread spectrum and narrowband communication systems will not interfere with each other when operating in the same frequency band.

3.9.1 Interference

Power decreases as the separation distance between the transmitter and receiver increases. Noise remains relatively constant, though, so at a maximum distance, d_{max}, the SNR drops below the minimum level for reliable communication. Figure 3.42a shows a plot of the signal power decay vs. distance from the transmitter as well as the constant noise floor or receiver sensitivity. The signal power level that yields the minimum SNR dictates d_{max} as indicated. An interfering signal also experiences a drop in power with distance from the receiver as shown in Figure 3.42b. The interference reduces d_{max}, because its power exceeds that of the noise. A minimum SINR then determines a reduced d_{max} rather than a minimum SNR.

3.9.2 Frequency-Hopping Spread Spectrum

Frequency hopping requires synchronization between the receiver and the transmitter. The hopping sequence comes from a pseudo random noise (PRN) code that has a spectrum similar to a random sequence of bits. The PRN code determines the order of the frequencies transmitted. The receiver decodes

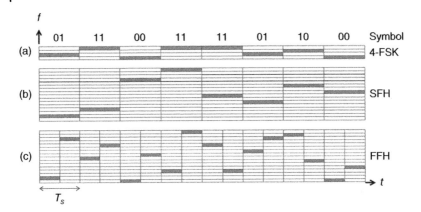

Figure 3.43 Frequency-hopping spread spectrum. (a) 4-FSK signal, (b) 4-FSK signal with SFH, and (c) 4-FSK signal with FFH.

the signal by following the same PRN sequence as the transmitter. In order to initialize the synchronization process, the frequency hopper starts at a fixed frequency before hopping to other predetermined frequencies. As long as the jammer does not interfere with the start frequency, the frequency hopping occurs over a much broader bandwidth than the jamming signal, so the desired signal usually avoids the jamming. The bandwidth of a frequency-hopped signal equals the number of available frequency slots times the bandwidth of a slot. Certain bands can be avoided if they contain potential interference.

Slow frequency hopping (SFH) means that at least one symbol transmits in every frequency subband. Fast frequency hopping (FFH) means that more than one frequency hop occurs during a symbol. SFH can use coherent detection, while FFH cannot, so SFH is more commonly used. Figure 3.43a has the symbols for a 4-FSK. If this FSK signal slow hops between four frequency channels at the rate of one change in frequency per symbol then Figure 3.43b shows one possible hopping sequence. If the fast hopping occurs twice in one symbol, then Figure 3.43c shows a possible hopping sequence. In both cases, the frequency hopping bandwidth is four times that of the original signal.

An FM signal with a 100 kHz bandwidth that hops over a 100 MHz bandwidth has 1000 possible subbands. For equally likely subbands, the spread-spectrum signal interferes with any narrowband receiver that has a 100-kHz bandwidth at one of the hops only 1/1000 of the time. As a result, the average interference power is 1000 times less than the transmitted power. This low level of interference does not interfere with an FM radio as long as the hopping is fast enough. Frequency hopping within the high bandwidth of a television signal might cause the loss of one of the picture lines that viewers would notice, so avoid frequency hopping in the television spectrum.

Figure 3.44 Time hopping a data pulse within established time slots.

Time-hopping spread-spectrum positions a pulsed RF carrier within a time interval in accordance with a PRN code sequence (Figure 3.44) [11]. Time hopping combined with frequency hopping creates a TDMA spread-spectrum system.

3.9.3 Direct-Sequence Spread Spectrum

In order to spread the spectrum of a narrowband signal, a PRN sequence first modulates the code word as shown in Figure 3.45. One bit in the PRN sequence, called a chip, has a period of T_{cp}. The modulated data stream now has a bandwidth increase equal to the spreading factor: $F_{sp} = T_b/T_{cp}$. After spreading, the data modulates a carrier frequency then transmits to the receiver.

Figure 3.45 illustrates the interference rejection capability of direct-sequence spread spectrum (DSSS). A narrowband high-power density interference mixes with the low-power density spread-spectrum signal in the channel. The receiver down converts the signal and interference to baseband using the same PRN sequence as the transmitter. The down conversion de-spreads the baseband signal but spreads the interference. In order for the de-spreading to work, the PRN synchronizes with the baseband spread-spectrum signal. The synchronizer locks onto the strongest baseband multipath signal in order to align its chips with the PRN. A matched filter or correlator identifies the bits at the receiver in Figure 3.45. Interference power spreads over a bandwidth that is F_{sp} times bigger than the original interference bandwidth. Since the power density

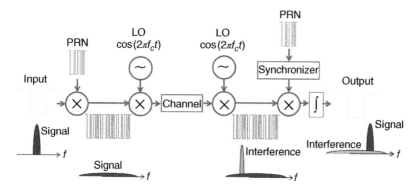

Figure 3.45 Narrowband interference in the channel has little impact on the received signal in a DSSS system.

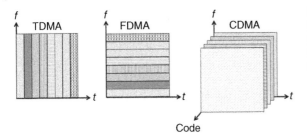

Figure 3.46 TDMA and FDMA lie in the time–frequency plane. CDMA adds another dimension called code.

of the interference decreases by F_{sp}, it has little impact on the desired signal. The processing gain is the increase in signal power density after de-spreading.

3.9.4 Code Division Multiple Access (CDMA)

Code division multiple access (CDMA) is a multiple user version of DSSS. Each user has a different PRN code, so that a receiver has access to one signal in a channel when many signals occupy the same bandwidth. Figure 3.46 illustrates the difference between TDMA, FDMA, and CDMA. TDMA signals exist in time slots while FDMA signals exist in frequency bands. CDMA signals, on the other hand, fill up the time–frequency plane but are separated in a third dimension called code.

CDMA comes in two forms: synchronous and asynchronous. Synchronous CDMA or code division multiplexing (CDM) requires all signals in the channel to precisely align in time. This works with point to multi-point communication, such as transmission from a base station to mobile users. Synchronous CDMA uses orthogonal codes (e.g. Walsh codes) to spread the signals. Asynchronous CDMA, on the other hand, allows signals to arrive randomly as with communication between a mobile unit and a base station. No orthogonal codes exist that spread signals with arbitrary starting points. The orthogonal codes used in synchronous CDMA do not correlate well with asynchronous CDMA signals, so PRN codes or Gold codes [10] are used instead.

The Walsh codes used in CDM come from rows of Hadamard matrices. The elements of a Hadamard matrix are either 1 or -1. Its rows are mutually orthogonal and $\mathbf{H}_n \mathbf{H}_n^T = 2^n \mathbf{I}_{2^n}$. A Hadamard matrix results from a recursive relationship that starts with the initial matrices:

$$\mathbf{H}_0 = [1] \quad \text{and} \quad \mathbf{H}_1 = \begin{bmatrix} 1 & 1 \\ 1 & -1 \end{bmatrix} \tag{3.47}$$

then creates larger square matrices using the formula

$$\mathbf{H}_{n+1} = \begin{bmatrix} \mathbf{H}_n & \mathbf{H}_n \\ \mathbf{H}_n & -\mathbf{H}_n \end{bmatrix} \quad \text{for a } 2^{n+1} \times 2^{n+1} \text{ Hadamard matrix} \tag{3.48}$$

Example

Generate H_2 and H_3 using MATLAB.

Solution

The MATLAB commands are `hadamard(4)` and `hadamard(8)` where the argument is the number of rows/columns.

$$H_2 = \begin{bmatrix} 1 & 1 & 1 & 1 \\ 1 & -1 & 1 & -1 \\ 1 & 1 & -1 & -1 \\ 1 & -1 & -1 & 1 \end{bmatrix} \quad H_3 = \begin{bmatrix} 1 & 1 & 1 & 1 & 1 & 1 & 1 & 1 \\ 1 & -1 & 1 & -1 & 1 & -1 & 1 & -1 \\ 1 & 1 & -1 & -1 & 1 & 1 & -1 & -1 \\ 1 & -1 & -1 & 1 & 1 & -1 & -1 & 1 \\ 1 & 1 & 1 & 1 & -1 & -1 & -1 & -1 \\ 1 & -1 & 1 & -1 & -1 & 1 & -1 & 1 \\ 1 & 1 & -1 & -1 & -1 & -1 & 1 & 1 \\ 1 & -1 & -1 & 1 & -1 & 1 & 1 & -1 \end{bmatrix}$$

Asynchronous CDMA, uses a PRN code or Gold code to spread the signal. For instance, GPS signals use Gold codes. These codes provide a better correlation than orthogonal codes when the bits are not perfectly aligned. On the other hand, the asynchronous codes raise the noise floor, because the cross-correlation of the codes is not zero like it is with the orthogonal codes.

Example

Three receivers each have different 4-bit PRN codes:

PRN code A $= [1 \; -1 \; -1 \; -1]$
PRN code B $= \begin{bmatrix} -1 & -1 & 1 & 1 \end{bmatrix}$
PRN code C $= \begin{bmatrix} 1 & 1 & -1 & 1 \end{bmatrix}$.

A transmitter sends the message 1 0 0 1 using PRN A to encode it. Show that a receiver using PRN A will receive the message while a receiver using PRN B or C will not.

Solution

Assuming that a 1 bit is encoded by "1" and a 0 bit by "−1," then the encoded message sent by the transmitter is given by

$$\begin{bmatrix} 1 & -1 & -1 & 1 \end{bmatrix} = \begin{bmatrix} 1 & -1 & -1 & -1 & -1 & 1 & 1 & 1 & -1 & 1 & 1 & 1 & 1 & -1 & -1 & -1 \end{bmatrix}$$

The message and codes are graphed in Figure 3.47. The receiver using PRN A takes the dot product of the encoded sequence with A, 4 bits at a time to get $\begin{bmatrix} 4 & -4 & -4 & 4 \end{bmatrix}$. The other two receivers take the dot product with B to yield $\begin{bmatrix} 2 & 2 & 2 & 2 \end{bmatrix}$ and C to yield $\begin{bmatrix} 0 & 0 & 0 & 0 \end{bmatrix}$. Note that decoding with B produces a

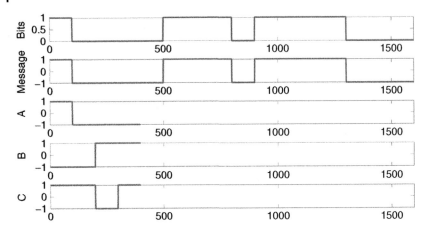

Figure 3.47 Plots of the bits in the message, the signal, and the codes for the three transmitters.

nonzero output. If these codes were orthogonal, then this dot product would be $\begin{bmatrix} 0 & 0 & 0 & 0 \end{bmatrix}$. In this case, the dot product with B appears to be noise.

Problems

3.1 If $V_c = 10$ V and $V_m = 5$ V, find:
(a) The modulation coefficient
(b) The Percent modulation.

3.2 If an AM signal has a peak of 20 V and a minimum positive amplitude of 6 V, determine:
(a) The modulation coefficient
(b) The carrier amplitude.

3.3 For an envelope with a maximum of 30 V and a positive minimum of 10 V, determine:
(a) The unmodulated carrier amplitude
(b) The peak change in the amplitude of the envelope
(c) The modulation coefficient
(d) The percent modulation.

3.4 Describe the following expression for an amplitude-modulated wave in terms of frequency content and voltage amplitude:

$$s(t) = 6 \sin 2\pi 300 \times 10^3 t - 2 \cos 2\pi 312 \times 10^3 t + 2 \cos 2\pi 288 \times 10^3 t$$

3.5 For $V_c = 12\,\text{V}$ and $\beta_{AM} = 0.5$, determine the following:
(a) The percent modulation
(b) The peak voltages of the carrier and the side frequencies
(c) The maximum amplitude of the envelope, V_{max}
(d) The minimum amplitude of the envelope, V_{min}
(e) Plot the AM envelope (label all pertinent voltages).

3.6 For a modulation coefficient $\beta_{AM} = 0.4$ and a peak carrier power $P_c = 400\,\text{W}$, determine:
(a) The peak power in each sideband
(b) The total sideband power
(c) The total transmitted power.

3.7 Use MATLAB to amplitude modulate a 300 kHz carrier with a voice signal. Use your own voice file or the one provided. Plot the time domain signal when $\beta_{AM} = 0.33, 0.67, 1.0, 1.33$.

3.8 An ASK signal transmits data at 28.8 kbps over a channel with a bandwidth from 300 to 3400 Hz.
(a) How many symbol states are needed?
(b) How many symbol states are needed if the channel passband goes from 0 to 3400 Hz and baseband signaling was used.
(c) What is the theoretical capacity if the SNR = 33 dB?

3.9 Determine the percent modulation for a television broadcast station with $\Delta f = 50\,\text{kHz}$ when $f_m = 30\,\text{kHz}$.

3.10 Determine the number of side frequencies produced for the following FM modulation indices: 0.25, 0.5, 1.0, 2.0, 5.0, and 10.

3.11 For an FM modulator with $\beta_{FM} = 5$, modulating signal $= 2\sin(27\pi 5 \times 10^3 t)$, and a carrier at 400 kHz, determine the following:
(a) The number of sets of significant sidebands
(b) The sideband amplitudes.

3.12 For an FM transmitter with 80 kHz carrier swing, determine the frequency deviation. If the amplitude of the modulating signal decreases by a factor of 4, determine the new frequency deviation.

3.13 A modulated signal takes the form $s(t) = 100\cos[2\pi(10 \times 10^6)t + 4\sin 2\pi(10^3)t]$
(a) Find the modulation index and bandwidth if the signal is FM.
(b) Repeat (a) if f_m is doubled.

(c) Find the modulation index and bandwidth if the signal is PM.

(d) Repeat (c) if f_m is doubled.

3.14 Use MATLAB to frequency modulate a 300 kHz carrier with a voice signal. Use your own voice file. Plot the time domain signal when $\beta_{FM} = 0.67$, 1.33, 2.67.

3.15 What happens to the demodulated DSB signal when the receiver LO has a frequency of $f_1 = f_c + \delta f$ instead of f_c?

3.16 Create 10 000 I and Q bits. Plot points in I–Q plane for (a) SNR = 6 dB, (b) SNR = 10 dB, and (c) SNR = 20 dB.

3.17 A video signal with a bandwidth from 0 to 2 MHz is sampled at four times the highest frequency using a 16 bit ADC. This data is modulated to 16-QAM with a raised cosine filter having $\alpha = 0.5$. What is the transmitted video signal bandwidth?

3.18 Random data is BPSK modulated and sent at 4800 bps over a bandpass channel. Find the transmission bandwidth such that the frequency spectrum is below 35 dB outside of the band.

3.19 An hour of temperature data sampled at 50 kHz and quantized to 16 bits is stored in memory. How many bits are stored in memory?

3.20 Plot the normalized analytical PSD in dB/Hz for (a) ASK, (b) BPSK, (c) QPSK, (d) 4-QAM, and (e) 16-QAM when $T_b = 1\,\mu s$.

3.21 Plot the BER for QPSK vs. $Eb/N0$.

3.22 A source uses the same average power to transmit ASK and PSK. Does ASK or PSK have a faster data rate for a given bit error probability at the receiver.

3.23 Plot the throughput as a function of load for ALOHA and slotted ALOHOA.

3.24 A DSSS system has four codes based on the 4×4 Hadamard matrix. Generate 4, 16 random bit sequences where a one is represented by "+1" and a zero is represented by a "−1." Spread the bit sequence n using row n of the Hadamard matrix. The output is a $4 \times 16 = 64$ bit sequence. Add the four spread messages together into 1 total bit sequence. Show row *n* of the Hadamard matrix recovers bit sequence *n* from the total bit sequence.

3.25 Repeat Problem 21 but add noise to the bit sequences. Can the bit sequences be recovered when there is a low SNR?

3.26 GPS uses Gold codes for spreading the signal. Use MATLAB to demonstrate transmission and reception of a binary signal using Gold codes.

References

1 Carson, J.R. (1963). Notes on the theory of modulation. *Proceedings of the IEEE* 51 (6): 893–896.

2 http://www.rfwireless-world.com/Terminology/QPSK-vs-OQPSK-vs-pi-4QPSK.html (accessed 18 December 2018).

3 Hanzo, L.L., Ng, S.X., Keller, T., and Webb, W. (2004). *Star QAM Schemes for Rayleigh Fading Channels*, 307–335. Wiley-IEEE Press.

4 https://www.mathworks.com/help/comm/ug/bit-error-rate-ber.html (accessed 30 July 2019).

5 Win, M.Z. and Scholtz, R.A. (2000). Ultra -wide bandwidth time-hopping spread-spectrum impulse radio for wireless multiple-access communications. *IEEE Transactions on Communications* 48: 679–691.

6 Abramson, N. (1970). The ALOHA system – another alternative for computer communications. *Proc. Fall Joint Computer Conf.*, AFIPS Press 37: 281–285.

7 Gold, R. (1967). Optimal binary sequences for spread spectrum multiplexing (Corresp.). *IEEE Transactions on Information Theory* 13 (4): 619–621.

8 Goldsmith, A. (2013). *Wireless Communications*. New York: Cambridge University Press.

9 https://www.youtube.com/watch?v=NI8nOa9BvjY (accessed 27 June 2018).

10 Markey, H.K. and Antheil, G. (1942). Secret communication system. US Patent 2, 292,387, issued 11 August 1942.

11 Molisch, A.F., Zhang, J., and Miyake, M. (2003). Time hopping and frequency hopping in ultrawideband systems. In: *2003 IEEE Pacific Rim Conference on Communications Computers and Signal Processing (PACRIM 2003) (Cat. No.03CH37490)*, vol. 2, 541–544. Victoria, BC, Canada: IEEE.

4

Antennas

Signals move charges. When charges accelerate, they radiate. A charge moving on a curved path or bouncing off the end of a wire accelerates/decelerates and creates a time-changing electromagnetic field that propagates away at the speed of light. Antennas confine charges to accelerate in ways that cause signals to radiate in desirable directions.

Transmit antennas radiate a modulated signal into one end of a channel that travels to the receive antenna at the other end of the channel. The antenna's ability to receive or transmit a signal depends on the antenna size, pointing direction, polarization, and frequency response. This chapter introduces basic antenna design principles for a wireless system.

4.1 Signal Properties that Influence Antenna Design

The transmitter and receiver along with the channel characteristics influence the antenna design specifications. As explained in Chapter 2, the signal has a direction, a spectrum, and a polarization. The antenna in a wireless system must have enough gain to increase the signal power, operate in the same spectrum as the signal, and match the signal's polarization.

4.1.1 Impedance

Like any circuit element, an antenna has an impedance represented by a complex number written as

$$Z_{\text{ant}} = R_r + R_L + jX_a \qquad (4.1)$$

The resistive loss (R_L) due to lossy material in the antenna converts some of the wireless energy into heat. Radiation resistance (R_r) and antenna reactance (X_a), on the other hand, result from the antenna geometry and material composition. Thus, any signal arriving at the transmit antenna terminals reflects, converts to heat, or radiates. The impedance usually determines the antenna's operating

Wireless Communications Systems: An Introduction, First Edition. Randy L. Haupt.
© 2020 John Wiley & Sons, Inc. Published 2020 by John Wiley & Sons, Inc.

bandwidth. An antenna's impedance should match that of the feedline over the operating bandwidth in order to insure maximum power transfer between the transmitter and the antenna or the antenna and the receiver.

4.1.2 Gain

An antenna is a spatial filter that concentrates a signal in a specific direction as shown in Figure 4.1. This spatial filter response, the antenna pattern, has a main beam which is the spatial passband over a defined angular region, as well as sidelobes and nulls due to diffraction in the stopband.

An isotropic point source radiates equally in all directions from a single point in space (dashed line in Figure 4.1). It serves as a standard for comparing antennas that magnify the signal in some directions over others. Directivity describes an antenna's ability to concentrate power in one direction more than other directions. The directivity of any antenna equals the maximum radiated power density divided by the average radiated power density which equals the power density radiated by an isotropic point source:

$$D = \frac{4\pi}{\int_0^{2\pi} \int_0^{\pi} |AP(\theta, \phi)|^2 \sin\theta d\theta d\phi} \tag{4.2}$$

where $AP(\theta, \phi)$ is the normalized antenna pattern (peak value of 1).

The antenna gain combines directivity with antenna dissipative losses. Gain is defined as the radiation intensity in a given direction divided by the radiation intensity of an isotropic point source accepting the same power. Gain does not include losses arising from impedance and polarization mismatches and is independent of its wireless system. Gain relates to the directivity by

$$G(\theta, \phi) = \delta_e D(\theta, \phi) \tag{4.3}$$

where δ_e is the radiation efficiency (ratio of the power radiated by the antenna to the power input to the antenna). Realized gain, yet another descriptor of the antenna radiation, includes mismatch loss between the feed and the antenna. The term "gain" without any angular dependence, G, means the maximum gain of the antenna and is usually expressed in decibels ($10 \log G$ in dB).

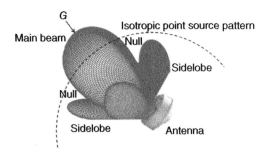

Figure 4.1 The antenna pattern is the response as a function of angle.

The effective (or equivalent) isotropic radiated power (EIRP) of a transmitter multiplies the gain of a transmitting antenna (G_t) by the net power delivered to the transmitting antenna (P_t).

$$\text{EIRP} = P_t G_t \tag{4.4}$$

EIRP describes how effectively a transmitter concentrates power in a given direction.

The effective area (A_e) of a receiving antenna describes its ability to collect radio frequency (RF) signals. Effective area is the ratio of the available power at the receiving antenna output to the power flux density of a plane wave incident on the antenna. It relates to the gain of the receiving antenna by

$$G = \frac{4\pi A_e}{\lambda^2} \tag{4.5}$$

For many antennas, the effective aperture is proportional to the physical aperture (A_p):

$$A_e = \delta_a A_p \tag{4.6}$$

where δ_a is an aperture efficiency and $0 \le \delta_a \le 1.0$.

Example
Find the directivity of an antenna with $AP(\theta) = \cos\theta$ in the upper half plane $(\theta \ge 0)$ and zero in the lower half plane.

Solution
Substitute into (4.2) to get

$$D = \frac{4\pi}{\int_0^{2\pi} \int_0^{\pi} \cos\theta \sin\theta d\theta d\phi} = \frac{4\pi}{\pi} = 4$$

or 6 dB.

4.1.3 Polarization

Polarization defines the direction of the electric field vector as a function of time. If a plane wave travels in the z-direction, the electric field lies in the x–y plane and takes the form:

$$\vec{E}(t) = E_{x0} \cos(\omega t - kz)\hat{x} + E_{y0} \cos(\omega t - kz + \Psi_y)\hat{y} \tag{4.7}$$

This electric field vector traces an ellipse in the x–y plane over one period. Figure 4.2 shows the electric field rotation for left hand and right hand elliptical polarization. Pointing the right thumb in the direction of wave propagation, fingers curl in the direction of the E field trajectory when the wave is right hand

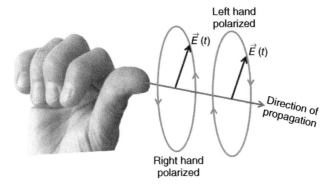

Figure 4.2 Rotation of the electric field for right-hand and left-hand polarization.

polarized (RHP) or $180° < \Psi_y < 360°$. Otherwise, it is left hand polarized (LHP) or $0° < \Psi_y < 180°$.

Axial ratio (AR) describes the shape of the polarization ellipse of an antenna where

$$AR = \frac{\text{Length of major axis}}{\text{Length of minor axis}} \tag{4.8}$$

An AR of 1 (0 dB) indicates circular polarization. Linear polarization has an AR = ∞ because the minor axis is zero, and the amplitude of the peak of the electric field vector lies along a straight line. *Circular polarization* occurs when $E_{x0} = E_{y0}$, and the $\hat{x}$ and $\hat{y}$ components are 90° out of phase. An antenna has the same polarization, hence AR, whether it transmits or receives.

A receive antenna that does not have the same polarization as the incoming electromagnetic wave will not receive the maximum possible power. Polarization loss factor (PLF) accounts for the polarization mismatch between an incident wave and an antenna's polarization and is given by

$$\delta_p = |\hat{e}_i \cdot \hat{e}_r^*|^2 \tag{4.9}$$

where

$\hat{e}_i$ = polarization vector of incident wave = $\dfrac{\mathbf{E}_i}{|\mathbf{E}_i|}$

$\hat{e}_r$ = polarization vector of receive antenna = $\dfrac{\mathbf{E}_{\text{antenna}}}{|\mathbf{E}_{\text{antenna}}|}$

$\mathbf{E}_i$ = incident electric field

$\mathbf{E}_{\text{antenna}}$ = electric field of antenna.

PLF indicates the reduction in the received power due to the polarization mismatch. A receive antenna with the same polarization as the transmit antenna receives all the power when the two antennas directly face each other. Antenna polarization changes with angle off boresight.

Example

What is the polarization of the following fields?

E_x	E_y	Ψ_y	Solution :
1	0	45°	x linear
0.707	0.707	0	linear 45° from x-axis
0.707	0.707	90°	LHP circular
0.867	0.5	90°	elliptical

4.1.4 Bandwidth

Antennas serve as bandpass frequency filters that operate best from a low frequency (f_{lo}) to a high frequency (f_{hi}). Outside this range of frequencies, the antenna performs at a degraded level that does not meet acceptable standards. The values of f_{lo} and f_{hi} depend on one or more of the following specifications:

1. *Antenna gain*: The upper and lower frequency limits define the region where the antenna gain stays above -3 dB or half of the peak gain at the center frequency.
2. *SWR*: The upper and lower frequency limits define the region where the SWR is less than 2 or the reflection coefficient <-10 dB. In other words, at least 90% of the power goes to the antenna.
3. *Polarization*: A circularly polarized antenna has a 0 dB AR at the center frequency. The high and low frequencies where the AR increases to 3 dB marks the polarization bandwidth.

A wideband antenna has a bandwidth greater than 10% otherwise, the antenna is narrow band. In 1990, a Defense Advanced Research Projects Agency (DARPA) Ultra-Wideband Radar Review Panel defined ultra-wide band (UWB) as any system where $B \geq 25\%$ when f_{hi} and f_{lo} are 20 dB below the peak power density [1]. Later in 2002, the Federal Communications Commission (FCC) defined UWB as $B \geq 25\%$ or $B \geq 15$ GHz [2]. These two government agencies used different definitions for f_{hi} and f_{lo}.

Example

An amplitude modulation (AM) antenna operates between $f_{lo} = 540$ and $f_{hi} = 1600$ kHz. What is its bandwidth?

Solution

$f_c = 1070$ kHz $B = f_{hi} - f_{lo} = 1060$ kHz, $B = \frac{f_{hi} - f_{lo}}{f_c} \times 100 = 1060/1070 \times 100 = 99.065\%$ and $B = \frac{f_{hi}}{f_{lo}} = 1600/540 = 2.963$.

4.2 Common Antennas

Wireless systems require many types of antennas with a wide range of capabilities that have tradeoffs depending upon the application. This section categorizes antennas as either point sources, wire antennas, aperture antennas, or microstrip antennas. These arbitrary groupings cover most antennas of importance in wireless communications.

4.2.1 Point Sources

An antenna appears to radiate from a single point when observed at a great distance. For example a star looks like a point source of light on earth even though it is physically huge. The phase center of an antenna looks like a point where the radiation originates. It lies at the center of an imaginary sphere of equal phase radiating from the antenna. Point sources serve as approximations for modeling the phase interactions of multiple antennas.

An antenna in Figure 4.3 transmits a spherical wave front from a single point. The wavefront propagates to receive antenna 1 at a distance R_1 then on to receive antenna 2 at a distance R_2. The phase difference between the center and edge of antenna 1 is greater than at antenna 2 ($\Delta R_1 > \Delta R_2$). According to the Institute of Electrical and Electronics Engineers (IEEE), the antenna far field starts when the separation distance is

$$R = 2D_{max}^2/\lambda \tag{4.10}$$

At this distance, the incident wave approximates a plane wave ($\Delta R \leq \lambda/16 \Rightarrow k\Delta R \approx 0$) [3].

Example
Derive (4.10) given the geometry in Figure 4.3.

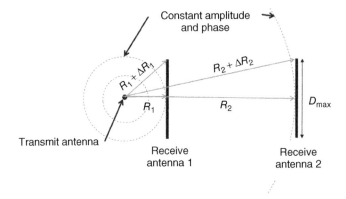

Figure 4.3 Receive antenna in the near field and the far field.

Solution

$$(D_{max}/2)^2 + R^2 = (R + \Delta R)^2$$

$$D^2_{max}/4 + \cancel{R^2} = \cancel{R^2} + 2R\lambda/16 + \cancel{(\lambda/16)^2}^{\;0}$$

$$R = \frac{2D^2_{max}}{\lambda}$$

4.2.2   Wire Antennas

A twin-wire transmission line (left side of Figure 4.4) has current flowing in opposite directions on the two wires. The currents on these wires have electric fields that point in the same direction between the wires but in opposite directions everywhere else. Consequently, the fields outside the wires cancel but the fields between the wires add and propagate. Bending the ends of a two-wired transmission line by 90°, as shown in right side of Figure 4.4, forces the currents on the bent ends to flow in the same direction along the z-axis. Now, the fields radiated by the bent wires radiate. This simple antenna is called a dipole.

The most common type of dipole is half of a wavelength long ($\ell = \lambda/2$). It has a standing wave at the resonant frequency, reminiscent of a vibrating string. If the dipole lies along the z-axis and has a current I_{dipole} flowing in the wires, then the electric field at a distance r in the far field is [4]

$$E_\theta = jZ_0I_{dipole}\frac{e^{-jkr}}{2\pi r}\frac{\cos\left(\frac{k\ell}{2}\cos\theta\right) - \cos(k\ell/2)}{\sin\theta} \qquad (4.11)$$

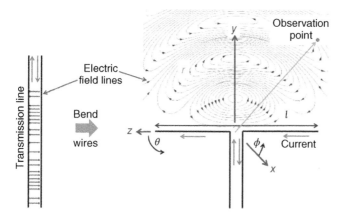

Figure 4.4 Bending the ends of a transmission line creates a dipole that radiates an electric field.

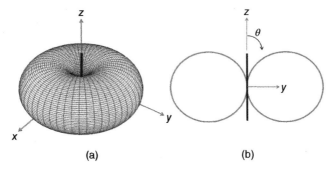

Figure 4.5 Magnitude of the electric field of a half-wavelength dipole. (a) 3D pattern and (b) cut of 3D pattern in the y–z plane showing variation in θ.

where $Z_0 = 377\Omega$ = characteristic impedance of free space. Its antenna pattern is the dipole response at one frequency as a function of angle. The normalized dipole antenna pattern (AP) has a peak of one.

$$AP(\theta) = \frac{\cos\left(\frac{k\ell}{2}\cos\theta\right) - \cos(k\ell/2)}{\sin\theta} \tag{4.12}$$

Since the dipole current only flows in the z-direction, the electric field is z- or θ-polarized depending upon whether the coordinate system is rectangular or spherical. AP has a maximum at $\theta = 90°$, and nulls arise when AP = 0. Nulls in the AP occur at $\theta = 0°$ and $\theta = 180°$. Since the dipole is symmetric in the ϕ direction, (4.12) is a function of θ and independent of ϕ as shown by the 3D antenna pattern and the pattern cut in Figure 4.5.

Half wavelength dipoles have a 73Ω input impedance at resonance. A feed line having an impedance of 73Ω insures maximum power transfer. An impedance match insures maximum power transfer when the feed line and antenna impedances differ. Dipoles made from thin wires have a very narrow impedance bandwidth, while fat wire versions have a much wider bandwidth.

Example
Find the directivity of a half-wavelength ($\ell = \lambda/2$) dipole. Assume the power density equals the magnitude of the electric field squared.

Solution
Substitute (4.12) into (4.2) to calculate the directivity. Since S_{rmax} is in the numerator and denominator, all the constants divide out leaving

$$D = \frac{4\pi}{\int_0^{2\pi}\int_0^{\pi}\left[\frac{\cos\left(\frac{\pi}{2}\cos\theta\right)}{\sin\theta}\right]^2 \sin\theta \, d\theta \, d\phi} = 1.643 = 2.16\,\text{dB} \tag{4.13}$$

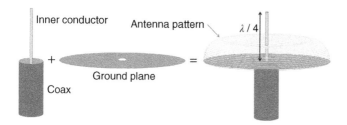

Figure 4.6 Making a monopole from a coaxial cable and a circular ground plane with a hole in the middle.

Extending a coaxial cable center conductor $\lambda/4$ through a hole in a ground plane (Figure 4.6) and soldering the outer conductor to the ground plane forms a simple monopole. A monopole has half the input impedance of a half wavelength dipole ($Z_{in} = 36.5 + j21.25\,\Omega$) but twice the directivity, since it only radiates in the upper hemisphere ($0 \le \phi < 360°$ and $0 \le \theta < 90°$). A monopole antenna pattern looks like half of a donut and is omnidirectional in azimuth (ϕ). Practical monopoles rarely look like the ideal one in Figure 4.6. The monopole on top of the car in Figure 4.7a is not perpendicular to the ground plane in order to make it more aerodynamic. Common radio monopoles rarely have any noticeable ground plane like the one in Figure 4.7b. Broadcast monopoles (Figure 4.7c) have a wire ground screen buried a few centimeters below the surface that serves as the ground plane. The impedance of a nonideal monopole differs from the theoretical perfect monopole. A receive antenna tolerates an impedance mismatch better than a transmit antenna, because high-power reflections from the transmit antenna may harm sensitive electronics.

The folded dipole (Figure 4.8) joins the two ends of a dipole to form a squashed loop with dimensions $\ell = \lambda/2$ long and $d \ll \ell$. Its 292 Ω impedance

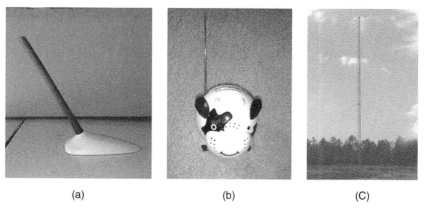

Figure 4.7 Examples of monopole antennas. (a) Car, (b) AM/FM radio, and (c) HF communications.

Figure 4.8 Folded dipole in Tekapo, New Zealand.

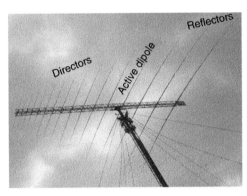

Figure 4.9 An HF Yagi–Uda antenna with 1 active dipole, 3 reflectors, and 12 directors.

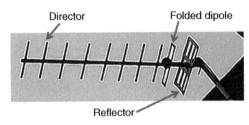

Figure 4.10 Folded dipole has high impedance that is significantly lowered by the presence of parasitic elements.

nearly matches a 300-Ω transmission lines [5]. Folding the dipole increases its bandwidth.

A Yagi–Uda antenna has one active dipole that lies in the same plane as many parasitic (passive) dipoles of different lengths (Figure 4.9). This design increases the gain and bandwidth of a single dipole antenna [6]. The parasitic elements reduce the impedance of the active element, so using a high impedance dipole, like the folded dipole, for the active element counters this impedance reduction (Figure 4.10). Passive elements longer than the resonant length have an inductive impedance (current lags voltage) that reflects a signal. Directors are shorter, capacitive (current leads voltage), passive dipoles. By placing one or two reflectors on one side of the active element and array directors on the other side, the phase distribution across the elements points the beam away from the reflectors and toward the directors.

A log periodic dipole array (LPDA) looks similar to a Yagi [7] but has all its elements connected to the feed (no passive elements). The half wavelength dipoles have resonant frequencies, lengths, diameters, and spacings that are logarithmically proportional. Elements much shorter than a half wavelength are too capacitive to radiate, while elements much longer than a half-wavelength are too inductive to radiate, so nearly all of the radiation comes from the resonant element and the two adjacent elements. Figure 4.11 shows a vertically polarized HF (high frequency) log periodic monopole array made from monopoles instead of dipoles. An LPDA operates over a very wide bandwidth (Yagi is narrow band) but has a lower gain compared to the Yagi.

Ham radio operators invent ingenious alternatives to the very tall monopole at HF frequencies where a wavelength is between 10 and 100 m. For instance, the inverted L antenna bends a monopole by 90° so that most of the wire runs parallel to the ground (Figure 4.12a). Ham radio operators like the inverted

Figure 4.11 HF log periodic monopole array.

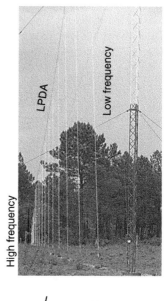

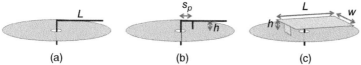

Figure 4.12 Diagrams of (a) inverted L antenna, (b) inverted F antenna, and (c) PIFA

L antenna, because trees and buildings support the long antenna instead of a tall tower. Shorting the arm to the ground plane at a distance s_p from the feed point decreases the length of the L antenna at the resonant frequency (keeps it out of your neighbor's yard). Now, this inverted F antenna looks like the letter F rotated 90° (Figure 4.12b). Decreasing s_p reduces the input impedance [8, 9]. The planar inverted-F antenna (PIFA) (Figure 4.12c) is a popular handset antenna made from a small $L \times w$ metal patch. The feed is at the center of the w dimension and along the L-dimension, while the shorting post is placed along the w-dimension. A rule of thumb for the PIFA resonant frequency is [10]

$$f_0 = \frac{c}{4(w + L)} \tag{4.14}$$

Some PIFA designs perform well in multiple frequency bands [11]. Figure 4.13 shows a patented dual band PIFA. The high- and low-band antennas have their arms wound to reduce their length, so that they fit inside a handset.

A small wire loop antenna has a circumference less than one-tenth of a wavelength. Current around a small loop is approximately constant. The magnetic field close to a small loop dominates the electric field, so it is often called a magnetic field probe in the near field. A small loop has very low input impedance that requires a match to a feed line such as a coaxial cable. This large mismatch makes the small loop an adequate receive antenna but a poor transmit antenna. Thus, the small loop is typically used as a receive antenna and not a transmit antenna (too much power reflected back to a transmitter).

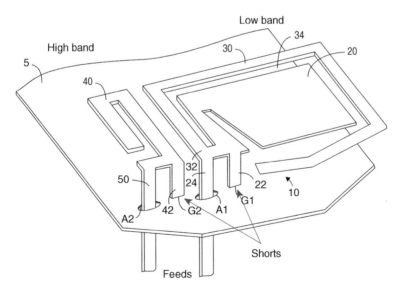

Figure 4.13 Diagram of a two band PIFA [12].

Figure 4.14 Diagram of a loop antenna in the x–y plane.

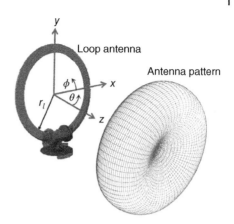

The small loop antenna in Figure 4.14 has electromagnetic fields given by [4]

$$E_r = E_\theta = H_\phi = 0$$

$$E_\phi = \frac{Z_0(k_0 r_\ell)^2 I_0 \sin\theta}{4r}\left[1 + \frac{1}{jk_0 r}\right]e^{-jk_0 r}$$

$$H_r = j\frac{k_0 r_\ell^2 I_0 \cos\theta}{2r^2}\left[1 + \frac{1}{jk_0 r}\right]e^{-jk_0 r}$$

$$H_\theta = -\frac{(k_0 r_\ell)^2 I_0 \sin\theta}{4r}\left[1 + \frac{1}{jk_0 r} - \frac{1}{(k_0 r)^2}\right]e^{-jk_0 r} \tag{4.15}$$

In the far field, the $1/r^2$ and $1/r^3$ terms quickly become very small a short distance from the loop, so only the E_ϕ and H_θ terms remain. A loop is linearly polarized in the x–y plane and has a directivity of 1.5. Its electric field goes to zero at $\theta = 0°$, so the antenna pattern has a null perpendicular to the plane containing the loop as seen by the antenna pattern in Figure 4.14.

Adding multiple turns to the loop increases its input impedance in order to better match to the feed line. If the small loop of radius r_ℓ has N_{turn} turns (Figure 4.15a), then its radiation resistance depends on the number of turns squared.

$$R_r = 31.171\pi^2 N_{\text{turn}}^2 \left(\frac{r_\ell}{\lambda}\right)^4 \tag{4.16}$$

Adding turns increases the loop impedance until it matches its feed line. AM radio receivers often use multi-turn loop antennas with a ferrite core to increase the radiation efficiency (Figure 4.15b).

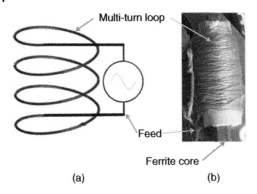

Multi-turn loop

Feed

Ferrite core

(a)   (b)

Figure 4.15 Multi-turn loops (a) antenna with $N_{turn} = 4$ and (b) AM radio multi-turn loop with ferrite core.

Example

Design a multi-turn loop antenna that has a radiation resistance of 50 Ω at 800 MHz. Assume the loop has a radius of 3 cm and the wavelength is 37.5 cm.

Solution

Substitute the given data into (4.16) to get

$$50 = 31.171\pi^2 N_{turn}^2 \left(\frac{3}{37.5}\right)^4 \Rightarrow N_{turn} = 63$$

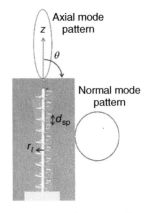

Axial mode pattern

Normal mode pattern

Figure 4.16 The two modes of a helical antenna.

A helical antenna extends the center conductor of a coaxial cable then twists it into the shape of a cork screw of radius r_ℓ with spacing d_{sp} as shown in Figure 4.16. Unlike the multi-turn loop, only one end of the helical antenna connects to the feed. If the total wire length is much less than a wavelength, then it behaves like a monopole and operates in the normal mode. The normal mode helix has a maximum gain perpendicular to the axis of the helix (like a monopole) and is linearly polarized with the electric field parallel to its axis. A normal mode helix connected to a monopole produces an antenna length shorter than $\lambda/4$ as shown in Figure 4.17.

An axial mode helix ($d_{sp} \approx \lambda/4$ and $r_\ell \approx 1/k$) has circular polarization with a peak gain along its axis. The AR at $\theta = 0°$ for an axial mode helix with N_{hel} turns is approximately [13]

$$AR = \frac{2N_{hel} + 1}{2N_{hel}} \tag{4.17}$$

Monopole   Helix

Figure 4.17 Monopole shortened by a helix [14].

An approximate formula for the directivity axial mode helix is [13]

$$D = 12 \left(\frac{2\pi r_\ell}{\lambda} \right)^2 N_{hel} \left(\frac{d_{sp}}{\lambda} \right) \tag{4.18}$$

As θ increases, the directivity decreases, and the antenna becomes elliptically polarized. The impedance of the helical antenna depends upon its feed. One approximate formula for an axial mode helical antenna radiation resistance with an axial feed is [13]

$$R_r = 280\pi \frac{r_\ell}{\lambda} \tag{4.19}$$

Example

Design an axial mode helical antenna that has circular polarization at 3 GHz with directivity of 6 dB and a radiation resistance of 50 Ω.

Solution

The wavelength at 3 GHz is 10 cm.

$$R_r = 280\pi \frac{r_\ell}{10} = 50 \Rightarrow r_\ell = 0.5684 \text{ cm}$$

Substituting into (4.18) and solving for the number of turns results in

$$D = 4 = 12 \left(\frac{2\pi \times 0.5684}{10} \right)^2 N_{hel} \left(\frac{d_{sp}}{10} \right) \Rightarrow N_{hel} d_{sp} = 26.1343$$

Assume that $d_{sp} = 3$ cm then

$$N_{hel} = 9$$

4.2.3 Aperture Antennas

An aperture refers to an opening that radiates/collects electromagnetic radiation. An $a \times b$ rectangular slot with a uniform electric field has a relative diffraction pattern at a point (x_0, y_0) in a plane that is z_0 away from the aperture is given by [15]

$$AP \simeq sinc \left(\frac{a x_0}{\lambda z_0} \right) sin\, c \left(\frac{b y_0}{\lambda z_0} \right) \tag{4.20}$$

Figure 4.18 shows the relative antenna pattern of a rectangular aperture when $a > b$. As a/λ and b/λ get large, their respective sinc pattern main beams get narrower. Its diffraction pattern is the Fourier transform of a uniform electric field in the rectangular aperture. The first sidelobe is 13.26 dB below the main beam, and the first null is at $\theta_{null1} = 57.3° /(a/\lambda)$. The point on the main beam that is 3 dB below the peak is at $\theta_{3dB} = 50.8° /(a/\lambda)$, so the 3 dB beamwidth in the x–z plane is $\theta_{3dB} = 101.6° /(a/\lambda)$ and in the y–z plane is $\theta_{3dB} = 101.6° /(b/\lambda)$.

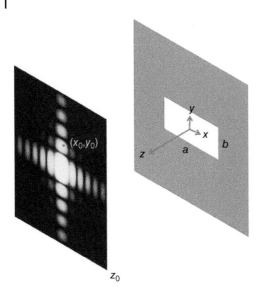

Figure 4.18 Rectangular aperture and its diffraction pattern.

A circular aperture with radius r_a and a uniform electric field has the following relative diffraction pattern at a distance $r_0 = \sqrt{x_0^2 + y_0^2}$ from a point $z = z_0$ from the aperture [15]:

$$\text{AP} \simeq \frac{J_1\left(\frac{kr_a r_0}{z_0}\right)}{\frac{kr_a r_0}{z_0}} \tag{4.21}$$

Its first sidelobe is 17.6 dB below the main beam while the first null is at $\theta_{\text{null}} = 35°\,/(a/\lambda)$. The 3 dB beamwidth is $\theta_{3\text{dB}} = 29.2°\,/(r_a/\lambda)$. In reality, the fields in the apertures have a polarization and an amplitude and phase variation (mode), so their antenna patterns are more complicated [4]. The expressions in (4.20) and (4.21) take the form of a typical antenna pattern by converting the variables to spherical coordinates. The directivity depends on the aperture area: $A_p = ab$ for a rectangular aperture and $A_p = \pi r_a^2$ for a circular aperture.

$$D \approx \frac{4\pi A_p}{\lambda^2} \tag{4.22}$$

Example
Assume that a 12 cm × 6 cm rectangular aperture with a 10 GHz uniform field projects onto a flat screen in the far field. Plot the relative diffraction pattern over a 100 cm by 100 cm area that is centered on the aperture.

Solution

The MATLAB code is given below and the resulting pattern is shown in Figure 4.19.

```
f=10e9; lam=3e10/f;
a=12; b=6;
z0= 2*(a^2+b^2)/lam;
x0=-100:.5:100; y0=x0;
[xx,yy]=meshgrid(x0,y0);
AP=sinc(a*xx/(lam*z0)).*sinc(b*yy/(lam*z0));
figure(1);mesh(x0,y0,abs(AP))
xlabel(' x (cm)');ylabel('y (cm)');zlabel('|AP|')
```

A leaky feeder communication system uses a coaxial cable with holes or apertures in the outer conductor to emit and receive radio waves, thereby functioning as an extended antenna (Figure 4.20) [16]. The system has a limited range and its operating frequency (typically VHF or UHF [VHF, very high frequency; UHF, ultra high frequency]) cannot pass through dirt and rock. The leaky feeder

Figure 4.19 Radiation from a rectangular aperture.

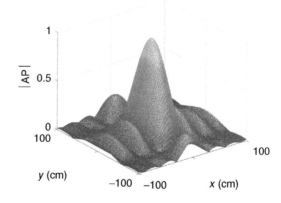

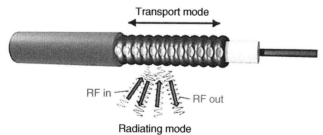

Figure 4.20 Leaky coaxial cable antenna [17]. *Source:* courtesy of Centers for Disease Control and Prevention.

finds applications in mines, underground railways, commercial airplanes, and buildings for communications and Wi-Fi. Line amplifiers inserted at regular intervals (350–500 m) keep the signal strength above a set threshold level [17]. The leaky cable interfaces with portable transceivers for two-way radio communication.

4.2.4 Microstrip Antennas

A microstrip antenna, also known as a patch or printed antenna, is an area of copper etched from or routed on (usually just one side) a printed circuit board (PCB). The metal on the bottom of the PCB serves as a ground plane. The advantages of microstrip antennas include:

- Low fabrication cost
- Conforms to curved surfaces
- Easy to group elements into a large array
- Light weight
- Easily integrates with electronics.

Disadvantages include:

- Low power handling
- Narrow bandwidth
- Low efficiency
- Surface waves in the substrate cause unwanted coupling
- Radiation from feeds and junctions.

Some modifications to the patch counter the disadvantages and make microstrip antennas useful for many applications.

Microstrip patches come in a variety of shapes with the most common being a $L \times w$ rectangle. At 10 GHz, the rectangular microstrip patch has dimensions of about 1 cm × 1 cm, so it easily fits onto a PCB. On the other hand, at 100 MHz, the 1 m × 1 m patch does not fit on a PCB. The PCB size limitation means that patch antennas rarely operate at frequencies below 900 MHz. The dominant patch polarization leaks out of the radiating edges as shown by the arrows in Figure 4.21. The fields at these edges are in phase. In contrast, the field lines along the long edges switch phase at the halfway point. Consequently, the fields cancel rather than radiate. Any radiation coming out of the long edges is orthogonal to the dominant polarization, so it is called cross-polarized.

Electric field lines under the patch resonate with maxima at the radiating edges and zero in the center of the patch. The ratio of the electric to the magnetic field at the point where the transmission line feeds the patch determines the input impedance. The edge-fed microstrip line or bottom-fed coaxial cable attaches to the patch at a point that corresponds to 50 Ω. An alternative to the bottom-fed coax is a stripline or an aperture feed underneath the patch.

Figure 4.21 Diagram of a rectangular microstrip patch antenna.

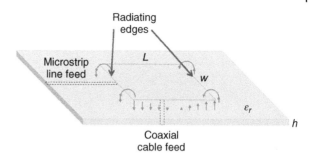

Patch designs typically start by approximating the length and width of the patch using design formulas. Two simple formulas are given by [18]

$$L \simeq 0.49 \frac{\lambda_0}{\sqrt{\varepsilon_r}}, \quad w \simeq \frac{\lambda_0}{\sqrt{2(\varepsilon_r + 1)}} \tag{4.23}$$

where ε_r is the relative permittivity of the PCB substrate. These dimensions serve as seeds for a numerical optimization algorithm in an antenna design software package [20]. Substrate height and permittivity primarily determine patch bandwidth [19].

$$B \propto \frac{h}{\sqrt{\varepsilon_r}} \tag{4.24}$$

A typical patch has a bandwidth of a few percent. The two normalized electric field polarizations associated with rectangular patches are [5]

$$E_\theta = \sin c \left(\frac{w}{\lambda} \sin \theta \sin \phi \right) \cos \left(\frac{L}{\lambda} \sin \theta \cos \phi \right) \cos \phi$$

$$E_\phi = - \sin c \left(\frac{w}{\lambda} \sin \theta \sin \phi \right) \cos \left(\frac{L}{\lambda} \sin \theta \cos \phi \right) \cos \theta \sin \phi \tag{4.25}$$

The total electric field is

$$|E_t(\theta, \phi)| = \sqrt{E_\theta^2 + E_\phi^2} \tag{4.26}$$

Figure 4.22 shows a plot of (4.26) with principal plane cuts. The principal planes pass through the peak gain and are at $\phi = 0°$ and $\phi = 90°$. When $\phi = 0°$ then $E_\phi = 0$, so the field is θ-polarized. When $\phi = 90°$ then $E_\theta = 0$, so the field is ϕ-polarized. For linearly polarized antennas, the E-plane contains the electric field, while the H-plane contains the magnetic field.

Circularly polarized patches have two different modes excited in the cavity under the patch with one phase delayed by 90° relative to the other [21]. A simple feed approach uses a Wilkinson power divider that feeds two different locations on the patch. The power divider splits the signal evenly. One of the signals receives a 90° phase shift (longer path). Other approaches include feeding the patch along the diagonal and mitering the patch corners. As with other antennas, the AR changes with frequency and angle off boresight.

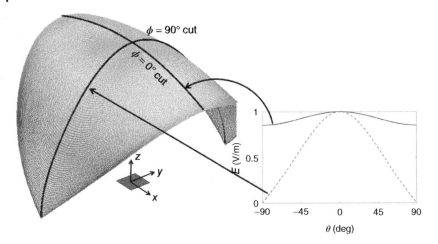

Figure 4.22 Total electric field of patch antenna with orthogonal principal plane cuts.

Example

Estimate the dimensions of a rectangular patch on a substrate with $\varepsilon_r = 2.2$ that operates at 2.5 GHz.

Solution

Use (4.23) to estimate the patch dimensions: $\lambda_0 = \dfrac{3 \times 10^{10}}{2.5 \times 10^9} = 12\,\text{cm}$

$$L \simeq 0.49 \dfrac{12}{\sqrt{2.2}} = 3.96\,\text{cm}, \quad w = \dfrac{12}{\sqrt{2(2.2 + 1)}} = 4.74\,\text{cm}$$

4.3 Antenna Arrays

An antenna array combines many smaller antennas, called elements, into a larger antenna. The array size and hence its gain increases as N (number of elements) increases. Phased arrays have phase shifters that electronically steer the main beam in a desired direction. Electronic beam scanning surpasses slow mechanical antenna steering that tends to frequently breakdown. Appropriately, weighting the signals at the elements before adding them together results in low sidelobes or moving nulls in order to reject unwanted signals. Arrays also gracefully fail. In other words, if one antenna element fails, the array still functions but at a lower performance level. Remote locations like outer space and on top of tall towers where repairs are extremely difficult benefit from an array's robust performance. The many performance advantages of a phased array antenna come at a high cost proportional to the number array elements.

Figure 4.23 Array geometry relative with an incident plane wave.

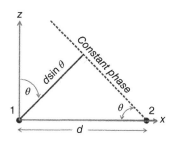

Figure 4.23 shows a plane wave (constant amplitude and phase) incident at an angle of θ on an array with two elements separated by d. The plane wave first arrives at element 2 then at element 1 with a time delay of

$$\Delta = (d/c)\sin\theta \tag{4.27}$$

The transmitted signal $s_t(t)$ arrives at both elements and adds together to get the received signal.

$$s_r(t) = s_t(t) + s_t(t - \Delta) \tag{4.28}$$

At a single frequency, f, the time difference converts to phase by

$$\psi = 2\pi f \Delta = 2\pi f(d/c)\sin\theta = kd\sin\theta \tag{4.29}$$

The array factor equals the sum of the signals at one frequency:

$$\text{AF} = 1 + e^{j\psi} \tag{4.30}$$

AF depends on f, d, and θ. The transmitting array phase has a 180° phase difference from the receive array, so its array factor is given by

$$\text{AF} = 1 + e^{-j\psi} \tag{4.31}$$

The phase sign indicates whether the plane wave travels toward (positive) or away (negative) from the array.

4.3.1 Element Placement

The variables Δ and ψ depend on the element locations. When N signals add together in phase ($\psi = 0°$), then the resulting signal amplitude increases by N. For example if $\theta = 0°$, then $s_r(t) = 2s_t(t)$ and $\text{AF} = 2$. The main beam or array factor maximum points at $\theta = 0°$. For all other angles, $\text{AF}(\theta) \le \text{AF}(0)$.

4.3.1.1 Linear Array

A signal arriving at an N element array with element n located at x_n (Figure 4.24) encounters the closest element first then continues to successive elements until it reaches the last one. Each element receives the signal at a different time. The

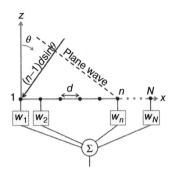

Figure 4.24 Diagram of an N element linear array with a plane wave incident at θ.

weighted (w_n) and time delayed (τ_n) sum of N element signals produces the output signal [22]:

$$s_r(t) = \sum_{n=1}^{N} w_n s_t \left(t - \frac{x_n}{c} \sin \theta + \tau_n \right) \tag{4.32}$$

An array factor represents the weighted and phase shifted sum of one frequency or tone from each element.

$$AF = \sum_{n=1}^{N} w_n e^{j(kx_n \sin \theta + \delta_n)} = \sum_{n=1}^{N} w_n e^{j\psi_n} \tag{4.33}$$

A phased array has phase shifters that change the signal phase at element n by δ_n.

A uniform linear array has $w_n = 1$ and elements spaced d apart along the x-axis: $x_n = (n-1)d$. The main beam peak points at broadside ($\theta = 0°$). Placing the phase center ($\psi = 0°$) at the first element of the array results in an array factor given by

$$AF = 1 + e^{j\psi} + e^{j2\psi} + \cdots + e^{j(n-1)\psi} = \sum_{n=1}^{N} e^{j(n-1)\psi} \tag{4.34}$$

Multiplying both sides of (4.34) by $e^{j\psi}$ and subtracting the resulting product from (4.34) leads to a simpler expression for the array factor

$$AF = \frac{1 - e^{jN\psi}}{1 - e^{j\psi}} = \frac{\sin(N\psi/2)}{\sin(\psi/2)} e^{j\frac{N-1}{2}\psi} \tag{4.35}$$

The maximum of (4.35) occurs when $\psi = 0°$.

$$AF = \sum_{n=1}^{N} e^{j(n-1)0} = N \tag{4.36}$$

Dividing (4.35) by N normalizes the array factor to a main beam peak of 1.0. Moving the phase center to the physical center of the array causes the phase term in (4.35) to disappear and leaves the normalized array factor:

$$\text{AF}_N = \frac{\sin(N\psi/2)}{N\sin(\psi/2)} \tag{4.37}$$

AF_N has a first sidelobe approximately 13 dB below the main beam peak.

Example

Plot the magnitude of the unnormalized array factors for $N = 4, 6,$ and 8 element arrays with elements spaced $d = \lambda/2$ along the x-axis.

Solution

Figure 4.25 has a plot of the array factors. The eight-element array has the highest main beam peak, narrowest main beam, and most sidelobes.

The beamwidth determines the antenna resolution and relates to the directivity of the array. A uniform linear array with half wavelength spacing has an approximate 3 dB beamwidth of [23]

$$\theta_{3\text{dB}} \simeq 101.5°/N \tag{4.38}$$

Increasing the number of elements decreases the beamwidth. A half wavelength spaced linear array has a directivity given by

$$D = \frac{\left|\sum_{n=1}^{N} w_n\right|^2}{\sum_{n=1}^{N} |w_n|^2} \tag{4.39}$$

which for a uniform linear array with $d = \lambda/2$ is

$$D = N \tag{4.40}$$

Adding more elements results in a higher directivity and a skinnier main beam.

Figure 4.25 Array factors for $N = 4, 6,$ and 8 element arrays along the x-axis with $d = \lambda/2$.

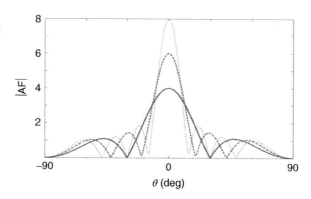

Example

Find the 3 dB beamwidth for a six-element array with $d = \lambda/2$ and compare to (4.38).

Solution

Find the 3-dB beamwidth by setting (4.37) equal to $1/\sqrt{2}$ and solving for θ using the MATLAB command `fzero`:

$$\frac{\sin(3\pi \sin\theta)}{6\sin(0.5\pi \sin\theta)} = \frac{1}{\sqrt{2}} \Rightarrow \theta = 8.6° \Rightarrow \theta_{3dB} = 17.2°$$

which is close to (4.38): $\theta_{3dB} \simeq 101.5°/6 = 16.9°$

The approximation in (4.38) improves as N gets larger.

4.3.1.2 Arbitrary Array Layouts

Elements located anywhere in three-dimensional space (x_n, y_n, z_n) have an array output represented by

$$s_r(t) = \sum_{n=1}^{N} w_n s_t \left[t - \frac{1}{c}(x_n \sin\theta \cos\phi + y_n \sin\theta \sin\phi + z_n \cos\theta) + \tau_n \right]$$

$$(4.41)$$

Converting the time difference between elements to a phase at one frequency results in an array factor given by

$$AF(\theta, \phi) = \sum_{n=1}^{N} w_n e^{j[k(x_n \sin\theta \cos\phi + y_n \sin\theta \sin\phi + z_n \cos\theta) + \delta_n]} = \sum_{n=1}^{N} w_n e^{j\psi_n} \quad (4.42)$$

Planar arrays have all the elements in a plane (e.g. $z_n = 0$). Figure 4.26 shows a 4×4 uniform planar array of microstrip patches. Microstrip lines from any element to the feed point at the center of the array have the same path length, so each element has the same phase. A planar array with uniform phase at the elements has its main beam pointing orthogonal to the plane containing the elements. Note that the microstrip line width changes in order

Figure 4.26 A 4×4 uniform planar array of microstrip patches.

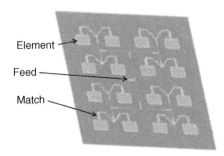

Element

Feed

Match

to match the impedance when the line splits and when it enters the patch. Conformal arrays mold the elements to a structure, like an airplane fuselage. Other configurations, such as spherical and random arrays, find specialized uses in wireless systems.

Example
A planar array has N_x elements spaced d_x in the x-direction and N_y elements spaced d_y in the y-direction arranged in a square grid for a total of $N = N_x \times N_y$ elements. Use (4.37) to write an expression for the normalized array factor.

Solution
From superposition, the array factor is a product of the linear array factor from the elements in the x-direction and the linear array factor in the y-direction.

$$AF = \frac{\sin\left(\dfrac{N_x k d_x \sin\theta}{2}\right)}{N_x \sin\left(\dfrac{k d_x \sin\theta}{2}\right)} \frac{\sin\left(\dfrac{N_y k d_y \sin\theta}{2}\right)}{N_y \sin\left(\dfrac{k d_y \sin\theta}{2}\right)} \tag{4.43}$$

A linear array with N elements has a 3 dB beamwidth of

$$\theta_{3dB} = \sin^{-1}\left(\frac{0.443\lambda}{Nd}\right) \text{ rad} \tag{4.44}$$

The beamwidth of a planar array is defined by principal plane cuts at $\phi = 0°$ and $\phi = 90°$.

At broadside, the directivity is proportional to the projected area of the array (A_p) as long as the element spacing is not much larger than $\lambda/2$

$$D = \frac{4\pi A_p}{\lambda^2}\eta_t \tag{4.45}$$

where η_t is the taper efficiency defined by

$$\eta_t = \frac{\left(\sum_{n=1}^{N} w_n\right)^2}{N \sum_{n=1}^{N} w_n^2} \tag{4.46}$$

This formula assumes that the array only radiates in a hemisphere (e.g. $0 \le \theta \le 90°$, $0 \le \phi < 360°$). If the array has an irregular shape, then assume that each element occupies a unit cell that equals the area of $d_x d_y$ for a rectangular grid. Thus, an N element array has an approximate directivity given by

$$D = \eta_t N d_x d_y / \lambda^2 \tag{4.47}$$

4.4 Electronic Beam Steering

The main beam peak of an array factor occurs when the phase term in (4.33) at each element equals zero ($\psi_n = 0$).

$$\text{AF}_{\text{max}} = \text{AF}(\psi_n = 0°) = \text{AF}\left[(n-1)kd\sin\theta + \delta_n = 0°\right] = \sum_{n=1}^{N} w_n \quad (4.48)$$

The δ_n term in ψ_n corresponds to the phase shifter setting at element n. Phase shifters change the phase of a signal by $0°$ to $360°$ in increments defined by the number of bits controlling them.

By selecting δ_n such that $\psi_n = 0°$, the main beam points in the desired scan direction, θ_s:

$$\delta_n = -kx_n \sin\theta_s \quad (4.49)$$

An array with the main beam pointing at $\theta_s = 90°$ is called an end-fire array. The value of δ_n in (4.49) depends upon frequency ($k = 2\pi f/c$), so if the frequency changes, the pointing direction of the main beam also changes. When the array scans its beam, then the directivity decreases due to the decrease in the projected area of the array.

$$D(\theta_s) = D\cos\theta_s \quad (4.50)$$

Example
An eight-element uniform array with $d = \lambda/2$ scans to $45°$. Plot the magnitude of the broadside and scanned array factors.

Solution
The scanning phase at element n is $\delta_n = -0.707\pi(n-1)$ radians. Figure 4.27 shows the resulting magnitude of the array factors calculated using (4.33).

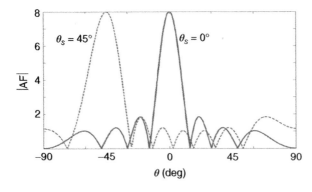

Figure 4.27 The beam of an eight element array steered to $45°$.

4.5 Element Pattern

If a linear or planar array has elements like dipoles or microstrip patches rather than isotropic point sources, then the array antenna pattern equals the element antenna pattern times the array factor.

$$\text{AP} = \text{Element pattern} \times \text{Array factor} \tag{4.51}$$

An array has the same polarization as its elements, so AP has vector components unlike AF.

Example
Assume an eight-element uniform array along the z-axis with $d = \lambda/2$ has a $\sin\theta$ element pattern that is polarized in the z-direction. Plot the element pattern superimposed over the array pattern (in dB) for the array factor scanned to broadside and 45°.

Solution
Substitute the $\sin\theta$ element pattern into (4.42):

$$\text{AP}(\theta) = \sin\theta \sum_{n=1}^{8} e^{jkz_n(\cos\theta - \cos\theta_s)}$$

Figure 4.28 shows the array factor at broadside and steered to 45°. When steered to broadside, the element pattern produces lower relative sidelobe levels than the array factor. When the beam scans, the sidelobes become asymmetric with respect to the main beam. As a result, the main beam peak does not point in the desired direction, and the directivity decreases. Note that the steered beam in Figure 4.27 does not show a decrease in directivity.

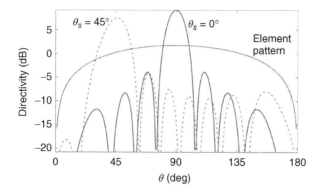

Figure 4.28 Element pattern times the array factor equals the array pattern. As a result, the array directivity decreases with scan.

4.6 Low Sidelobes

Low sidelobes decrease signal amplitudes received through the sidelobes. Transmit antenna low sidelobes attenuate signal propagation in directions outside of the main beam in order to mitigate interference with other wireless systems in the same band. Amplitude weights at the elements produce low sidelobe tapers. Several analytical formulas for amplitude tapers create predictable low sidelobes in the array factor.

The z-transform plays a fundamental role in synthesizing low sidelobe amplitude tapers for arrays. A z-transform converts the array factor into a polynomial by substituting $z = e^{j\psi} = e^{jkd \sin\theta}$ into (4.34) to get [24]

$$\text{AF} = \sum_{n=1}^{N} w_n z^{(n-1)} = w_1 + w_2 z + \cdots + w_N z^{N-1} = (z - z_1)(z - z_2)\cdots(z - z_{N-1})$$

(4.52)

The zeros of this polynomial (z_n) correspond to the nulls in the array factor.

$$z_n = e^{j\psi_n} = e^{jkd \sin\theta_n}$$ (4.53)

Remember that ψ_n represents the phase of the nth zero in AF. The actual null in the array factor occurs at an elevation angle of θ_n. The nth zero relates to null n in the array factor by $\psi_n = kd \sin\theta_n$. Moving one of these zeros results in a new array factor polynomial that has the same order but different coefficients, hence different element weights.

Table 4.1 summarizes the three most well-known low sidelobe amplitude tapers for linear arrays, and the locations of the array polynomial zeros. Use the following steps to find the low sidelobe amplitude weights and the corresponding array factor:

Table 4.1 Low sidelobe amplitude tapers [23].

	Peak sidelobe level (dB)	ψ_n
Binomial [25]	$-\infty$	$180°$
Dolph–Chebyshev [26]	$slldB$	$2\cos^{-1}\left\{ \dfrac{\cos\left(\frac{(n-0.5)\,\pi}{N-1} \right)}{\cosh\left(\frac{\pi A}{N-1} \right)} \right\}$
Taylor [27]	$slldB$	$\dfrac{\pm 2\pi}{N}\begin{cases} \bar{n}\sqrt{\dfrac{A^2 + (n-0.5)^2}{A^2 + (\bar{n}-0.5)^2}} & n < \bar{n} \\ n & n \geq \bar{n} \end{cases}$

$A = \frac{1}{\pi}\cosh^{-1}(10^{slldB/20})$ and $\bar{n} - 1$ is the number of zeros moved

1. Calculate ψ_n using Table 4.1.
2. Let $z_n = e^{j\psi_n}$.
3. Substitute z_n into (4.52).
4. Multiply the factored polynomial to find the w_n.
5. Calculate the array factor using (4.33).

Nulls in the array factor occur at

$$\theta_n = \sin^{-1}\left(\frac{\psi_n}{kd}\right) \tag{4.54}$$

Example
Calculate the amplitude weights for a binomial, a 20 dB Chebyshev, and a 20 dB, $\bar{n} = 2$ Taylor taper, then plot the weights and associated array factors. Assume the array has eight elements spaced $\lambda/2$ apart.

Solution
Follow these steps:

1. Calculate ψ_n using Table 4.1. See Table 4.2 for values.
2. Let $z_n = e^{j\psi_n}$.
3. Substitute z_n into (4.52).
4. Multiply the factored polynomial to get the polynomial coefficients or array weights shown in Table 4.2 and Figure 4.29.

Taking these weights and calculating AF for $\phi = 0°$ and $0 \le \theta \le 90°$ results in the array factor plots in Figure 4.30. In order to emphasize the relationship between θ and ψ, ψ axis appears below the θ axis, where

$$\psi = \left(\frac{2\pi}{\lambda}\right) 0.5\lambda \sin\theta = \pi \sin\theta = 180° \sin\theta$$

Table 4.2 Location of zeros and elements weights for the four amplitude tapers in the example.

n	Uniform ψ_n	Uniform w_n	Binomial ψ_n	Binomial w_n	Chebyshev ψ_n	Chebyshev w_n	Taylor ψ_n	Taylor w_n
1	0.79	1	3.14	0.03	0.94	0.58	0.95	0.55
2	1.57	1	3.14	0.20	1.55	0.66	1.57	0.68
3	2.36	1	3.14	0.60	2.33	0.88	2.36	0.87
4	3.14	1	3.14	1	3.14	1	3.14	1
5	−2.36	1	3.14	1	−2.33	1	−2.36	1
6	−1.57	1	3.14	0.60	−1.55	0.88	−1.57	0.87
7	−0.79	1	3.14	0.20	−0.94	0.66	−0.95	0.68
8		1		0.03		0.58		0.55

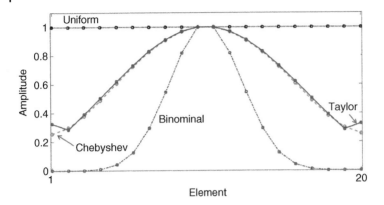

Figure 4.29 Low sidelobe tapers for an eight-element linear array.

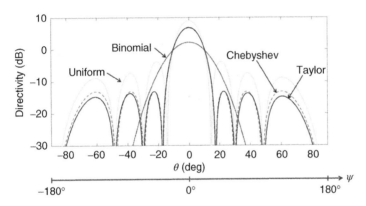

Figure 4.30 Array factors corresponding to the weights in Figure 4.29. There are θ and ψ axes for comparison.

The array factors are normalized to the peak of the uniform array factor ($N = 8$). Note that the array factor directivity decreases as the sidelobe level decreases, so there is a tradeoff between increasing the desired signal entering the main beam and decreasing unwanted signals entering the sidelobes.

4.7 Moving a Null to Reject Interference

An antenna pattern null has zero gain, so any signal arriving in the direction of a null is not received. Adaptive, smart, or null steering antennas protect RF systems from unwanted interference by moving nulls in the directions of the interfering signals. These nulls reduce the received interference and increase the signal to interference plus noise ratio (SINR) [28]. Factoring the array factor polynomial as in (4.52) shows the null locations exist in the array pattern. Moving one null to the direction of an interfering signal changes one of the zeros in the array factor polynomial. Multiplying the factored polynomial produces new w_n that result in a new polynomial. Moving one or more zeros in the array factor polynomial (nulls in the array factor) changes all of the element weights (polynomial coefficients).

Example

A six-element uniform linear array with $d = \lambda/2$ has zeros at

$$\psi_n = \pm 60°, \pm 120°, 180°$$

The factored array factor is given by

$$AF = z^5 + z^4 + z^3 + z^2 + z + 1$$

$$= (z - e^{j60°})(z - e^{-j60°})(z - e^{j120°})(z - e^{-j120°})(z - e^{j180°}) \qquad (4.55)$$

These zeros are shown on the unit circle in Figure 4.31. If a desired signal entering the main beam is 30 dB below a signal at 30°, then the interfering signal dominates the desired signal, because the sidelobe level is only 13 dB below the peak of the main beam (dashed line in Figure 4.32). Move a null to reject the interference.

Solution

Moving a null in the array pattern to 30° mitigates the impact of the interfering signal (see Figure 4.31). Follow these steps:

1. Picking the closest zero ($\psi_n = 120°$) and moving it to $\psi_n = 180° \sin(30°) = 90°$.
2. Put the new zero in (4.57).
3. Multiply the factored polynomial to get the new weights.

$$w_n = \begin{bmatrix} 1\angle 0° & 0.518\angle - 15° & 1\angle - 30° & 1\angle 0° & 0.518\angle - 15° & 1\angle - 30° \end{bmatrix}$$

Figure 4.31 The quiescent array factor for a six-element array with half-wavelength spacing. The null at 41.8° is moved to 30° in order to cancel interference.

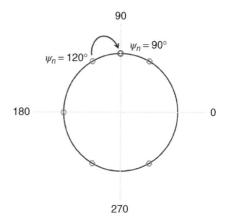

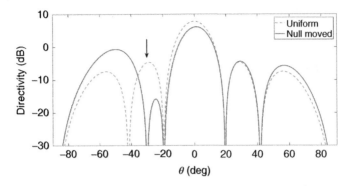

Figure 4.32 Zeros on the unit circle for a six-element array with half-wavelength spacing. The null is moved from $\psi_n = 120°$ to $\psi_n = 90°$ in order to cancel interference.

Some of the element weights have both an amplitude and a nonzero phase. Any of the other zeros could also have been chosen to move to $\psi_n = 120°$. The resulting weights would be different. Old and new array factors appear in Figure 4.32.

4.8 Null Filling

Sector antenna arrays on a tower or building increase antenna gain and coverage on the ground by pointing the main beam at a slight angle toward the ground. Tilting reduces interference between other antennas, since nearby antennas do not point at each other. In addition, tilting creates a larger main beam footprint on the ground that improves power density distribution to users. Mechanically tilting the array or electrically steering the main beam toward the ground (Figure 4.33). Mechanical tilt points a backlobe skyward. Remote electrical tilt (RET) controls the mechanical tilt in real time in order to change coverage. Phase scanning the main beam toward the ground steers

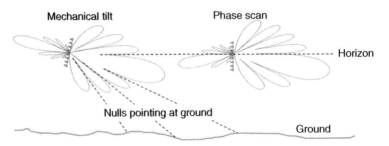

Figure 4.33 Mechanical tilt and phase scan.

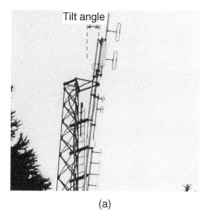
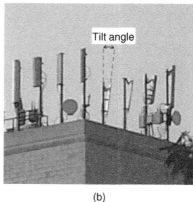

(a)   (b)

Figure 4.34 Tilted arrays: (a) Two 2-element and one 4-element linear array of folded dipoles on Grouse Mt. in British Columbia, Canada and (b) Cell antennas on top of a Penn State building in State College, PA.

the backlobe toward the ground (see right side of Figure 4.33). A high gain backlobe might interfere with other antennas, so its pointing direction matters. The folded dipole arrays in Figure 4.34a tilt downhill. Figure 4.34b shows examples of sector antennas on top of a building that tilt toward the ground.

Areas where nulls in the antenna pattern point at the ground (Figure 4.33) create signal dead zones, so signal dropouts occur. These dropouts disrupt continuous, quality service. Null-filling eliminates the nulls in the antenna pattern pointing toward the ground by moving array polynomial zeros off of the unit circle. Zeros outside of the unit circle have an amplitude greater than one, while zeros inside of the unit circle have an amplitude less than one.

Example

The eight-element phase scanned array in Figure 4.33 has three nulls pointing at the ground. This array has seven zeros that lie on the unit circle (Figure 4.35)

$$\psi_n = \begin{bmatrix} \pm 45° & \pm 90° & \pm 135° & 180° \end{bmatrix}$$

Employ null-filling by moving the appropriate zeros off the unit circle.

Solution

Follow these steps to fill in three nulls:

1. Moving three zeros $\begin{bmatrix} 45° & 90° & 135° \end{bmatrix}$, off of the unit circle (Figure 4.35) by increasing their amplitudes from 1.0 to 1.2:

$$\text{AF} = (z - 1.2e^{j45°})(z - e^{-j45°})(z - 1.2e^{j90°})(z - e^{-j90°})(z - 1.2e^{j135°})$$
$$\times (z - e^{-j135°})(z - e^{j180°})$$

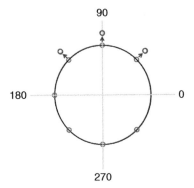

Figure 4.35 Three zeros moved off the unit circle to fill nulls.

Figure 4.36 Null-filled pattern (solid) superimposed on the uniform pattern (dashed).

2. Multiply this factored polynomial to get the coefficients or weights:

$$w = \begin{bmatrix} 1 & 1 - j0.48 & 1.10 - j0.48 & 1.10 - j0.67 & 1.32 - j0.67 & 1.32 - j0.70 & 1.72 - j0.70 & 1.73 \end{bmatrix} \tag{4.56}$$

The array factors associated with both arrays appear in Figure 4.36. Filling the nulls cost the array factor some of the main beam gain.

4.9 Multiple Beams

Some wireless systems need arrays with more than one beam in order to communicate with multiple users or switch between beams. Four multiple beam architectures appear in Figure 4.37 [22]. Figure 4.37a has an array with M different corporate feed networks that produce M beams. Each beam independently scans because each feed network has its own phase shifters. All feed networks share the same elements and perhaps some other electronics like filters and amplifiers. A software approach, called digital beamforming, replaces much of the hardware in the corporate feed with ADCs on receive (Figure 4.37b) and DACs (digital-to-analog converters) on transmit [29]. The ADC converts the RF signal at each element into bits. Software then calculates

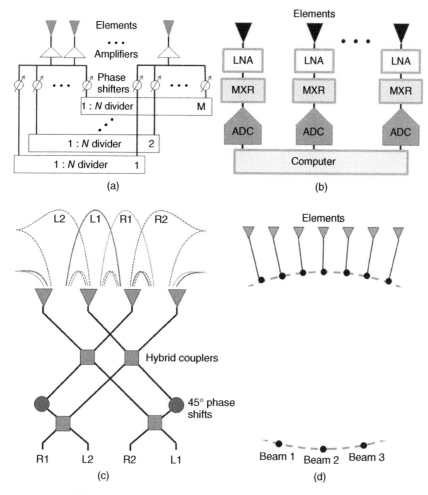

Figure 4.37 Different approaches to multiple beam arrays. (a) Multiple beam corporate feed, (b) digital beamformer, (c) Butler matrix, and (d) Rotman lens.

the beams in the computer and independently steers them. A Butler matrix generates 2^M beams (M is an integer) using hybrid couplers and phase shifters to create a hardware fast Fourier transform (FFT) (Figure 4.37c) [30]. The beams point in fixed relative directions. A Rotman lens has M beams generated from N elements (Figure 4.37d) by having M ports that transmit to/receive from the N elements [31]. These beams also point in fixed directions. The Butler matrix and Rotman lens are useful in beam switching strategies where the beam with the best performance (e.g. SNR) is connected to the receiver or transmitter while the others are not.

4.10 Antennas for Wireless Applications

The application drives the type of antenna used in a wireless system. This section describes some common wireless systems and their antennas.

4.10.1 Handset Antennas

A handset has three functional areas (Figure 4.38): controller, I/O (input/output), and RF. Its brains (the controller) contain the processor and memory. The processor (microprocessors, microcontrollers, and digital signal processing chips) monitors data reception, selects receivers and transmitters, adjusts impedance matching circuits, tunes filters, and selects antennas. A hard disk drive, flash memory, or random access memory (RAM) store applications and data.

A user interfaces with a handset via an IO (input–output) device, such as a touch screen, buttons, keyboard, microphone, speaker, vibrator, camera, sensor, LEDs (light-emitting diodes), biometric sensors, and data ports.

The RF part of the handset generates, transmits, and receives RF signals including:

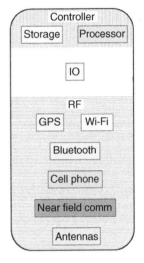

Figure 4.38 Block diagram of a handset with three functional areas.

- Global positioning system (GPS): 1575 MHz
- Wireless local area network (WLAN) or Wi-Fi: 2.4 and 5 GHz
- Bluetooth: 2.4 GHz
- Voice and text: between 700 and 2700 MHz
- Near field communications (NFC): 13.56 MHz.

Each frequency band has its own antenna. Sometimes a wideband or multi-band antenna serves more than one frequency band. More frequency bands, including bands near 27 and 40 GHz, will be in use when 5G (Appendix III) is adopted.

Initially, handsets had either a monopole or normal mode helical antenna, or a combination of the two [32]. These antennas had main beams pointing toward the user's head which caused a high rate of RF energy absorption by the head (Chapter 12). In 1996, the Hagenuk TCP6000 handset [33] introduced an internal slot antenna for GSM (Global System for Mobile) handsets. Engineers wanted the better performing exterior antenna, but consumers thought otherwise and eventually won. Table 4.3 compares the different types of handset antennas. External monopole and helical antennas needed to be as short as possible or to preferably disappear. Moving antennas inside the handset requires

Table 4.3 Comparison of handset antennas [32].

	Antenna	Length	Advantages	Disadvantages
External	Monopole	$\lambda/4$	• Easy to manufacture • Good performance	• Must extend and retract • Not multi-band
	Helix	$\lambda/8 - \lambda/4$	• Easy to manufacture • Short • Good bandwidth	• Does not retract • High SAR
	Monopole + helix	$\lambda/4 - \lambda/2$	See advantages of monopole and helix	See disadvantages of monopole and helix
Internal	Slot	$\lambda/2$	• Low SAR • Not sensitive to hand	• Big • Single band
	ceramic	$\lambda/4$	• Small • Easy to surface mount	• Heavy • Low gain • Narrow bandwidth
	PIFA	$\lambda/4$	• Low SAR • Multi-band	• Low gain • Small bandwidth • Two contacts to PCB
	PMA	$\lambda/4 - \lambda/2$	• Very thin • Similar to helix	• High SAR • Large

them to coexist with all the other handset electronics inside the case. Handset engineers developed clever designs made possible by modern manufacturing technology. Figure 4.39 lists some of the technologies used to manufacture handset antennas.

The first internal handset antennas were manufactured using stamped metal or PCB processing [32]. Metal stamping molds a flat sheet of metal around a die to get the desired shape. Hot stamping makes the antenna by placing a metal foil in a heated die. This thin antenna then fits onto a molded part in the case. Development and tooling costs are low. This approach works when manufacturing large parts with simple geometry but not for thin lines on 2D layouts or for 3D parts. More recently, molded interconnect devices (MID) technology surpassed hot stamping. It integrates the antenna into the handset mechanical housing [34]. This processing enables 3D layouts and reduces component count and cost by integrating connectors, sockets, or other devices. MID antennas, along with other circuits, are chemically or mechanically etched from the thin copper layers that sandwich a dielectric layer in a PCB. These inexpensive antennas have electronics and passive devices co-integrated on the PCB. Unfortunately, this approach is limited to a flat surface which limits the available area in the handset.

Improved manufacturing techniques, like two-shot molding, provide cost-effective production of highly repeatable interconnect devices [35]. It

Type	Theory	Advantages	Disadvantages	Examples
Stamped metal	Stamped steel part integrated together with plastic piece	Easy assembly, versatile, spring clip contacts can be integrated into antenna	Long lead times for patterns, minimum line width, cannot utilize layered antennas or 3D curves	
Flex-film	Copper etched flexible film glued onto plastic piece	Copper has high conductivity, can contain air gap (between antenna and back cover) or mounted on inside of cover to maximum volume	Requires several parts- more logistics, glue and mechanical tolerances	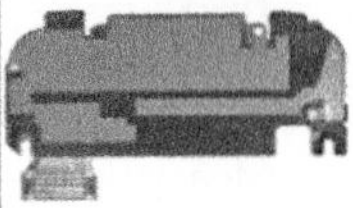
Hot Stamp	Uses heat and pressure to place a metal part to a surface (no 2-D curves to bend over)	Stable pattern, no pealing, similar performance to copper flexfilm	Flat pattern only, no 3D curves	
PCB trace	Antenna trace on PCB	In-expensive, no assembly	Design limited to flat surface, worse performance due to proximity of antenna to PCB components	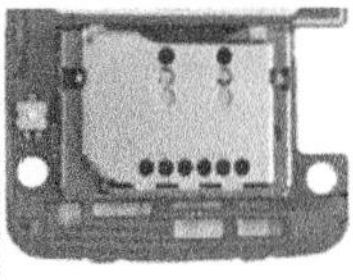
MID (Metal interconnect device)				
One-shot: 3D masking or laser etching	Uses a 3-D mask to print the pattern onto a plastic piece or laser to etch a pattern in electroplated metal	Shorter lead times for antenna pattern changes, stable pattern- no pealing	Expensive, antenna pattern limited to outer surface. not suitable for mass production	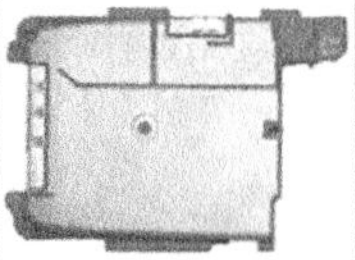
Two-shot	Two different kinds of plastic are molded together-metal adheres to one plastic and not the other	Suitable for mass-production, stable pattern-no pealing, allows complete RF design freedom	Expensive. PCB mount only, long lead times	

Figure 4.39 Comparison of handset antenna manufacturing techniques. *Source:* Reproduced with permission of IEEE [32].

begins with a shape made from a plateable polymer, usually ABS (acrylonitrile, butadiene, and styrene). Next, a second nonplateable material, usually polycarbonate, molds over top of the first shape leaving some areas of the first material exposed. The two parts are then fused together before undergoing electrolysis plating. In this step, the plateable plastic becomes metallized, while the nonplateable plastic remains nonconductive. Two-shot has high initial tooling costs, which limit its use to higher volume applications, because it has limited design flexibility and long development times.

Laser direct structuring (LDS) uses a laser to draw the antenna on the surface of a plastic compounded with a special laser-sensitive metal complex [35]. Exposing the polymer to the laser beam breaks down the metal complex into elemental metal – either copper or palladium – and residual organic compounds. The laser draws the antenna onto the part and leaves behind a roughened surface containing embedded metal particles. These particles act as nuclei for the crystal growth during subsequent plating with copper. Since it uses a single material, LDS offers much lower cost and complexity than two-shot molding. In addition, LDS draws extremely narrow traces compared to hot stamping and two-shot molding. LDS offers more flexibility than other MID processes because circuit redesigns only require reprogramming the laser unit. The laser also performs the plastic surface preparation, unlike two-shot molding, which requires an additional etching step to prepare the plastic for plating.

Multi-band antennas inside the handset serve two or more frequency bands in a smaller space than separate antennas for each band. The tri-band antenna in Figure 4.40 covers NFC at 13.56 MHz as well as the cell bands between 700 and 2700 MHz [36]. Inductors L_1 and L_2 have high impedances at frequencies above 700 MHz and low impedances at frequencies below 100 MHz. Thus, the current in loops ℓ_2 and ℓ_3 only flow at lower frequencies (NFC). Capacitor C_1 has a low impedance at frequencies above 700 MHz and a high impedance at frequencies below 700 MHz, so the loop ℓ_1 only conducts at the higher frequencies in the cell phone bands while L_1 and L_2 act as open circuits, so the antenna has the characteristics of an inverted F antenna.

The Samsung Galaxy S4 antennas showcase MID technology in a handset (Figure 4.41) [37]. All the antennas integrate into the plastic case, so that they are compatible with other functional blocks and difficult to locate. Future handsets will require an increasing number and range of frequencies and wider bandwidths. Figure 4.42 shows the locations of the antennas in the handset of the Samsung Galaxy S9 [38]. The NFC antennas are loops. One also serves as a charging coil. The MST (magnetic secure transmission) antenna enables use of Samsung Pay.

An interesting antenna problem arose when AT&T sold the Apple iPhone 4 in 2010 [39]. Apple's new design had two antennas: (i) Wi-Fi, Bluetooth, and

Figure 4.40 Antenna design that covers NFC as well as the 700–2700 MHz bands.

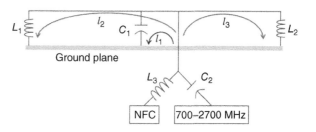

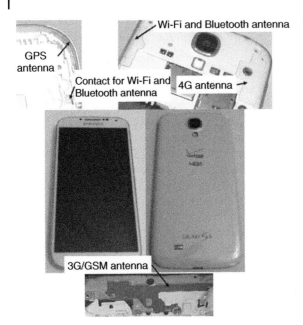

GPS antenna

Wi-Fi and Bluetooth antenna

Contact for Wi-Fi and Bluetooth antenna

4G antenna →

3G/GSM antenna

Figure 4.41 Antennas in a Samsung Galaxy S4.

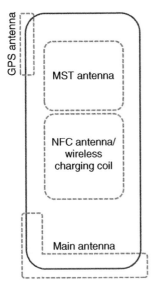

GPS antenna

MST antenna

NFC antenna/ wireless charging coil

Main antenna

Figure 4.42 Antennas on a Samsung Galaxy S9.

GPS: smaller antenna beginning in the bottom left and running to the top left of the handset and (ii) voice and data antenna: much larger antenna running around almost three quarters of the phone. Any conductor (like a hand) that bridged the gap between the two antennas caused them to detune (not resonate at the desired frequency). Naturally holding the bottom left corner of the

Table 4.4 Signal attenuation (dB) for three different cell phones [39].

Attenuation (dB)	Cupping tightly	Holding naturally	On an open palm	Holding naturally inside case
iPhone 4	24.6	19.8	9.2	7.2
iPhone 3GS	14.3	1.9	0.2	3.2
HTC nexus one	17.7	10.7	6.7	7.7

handset makes skin contact between the two antennas resulting in significant signal attenuation. Table 4.4 gives the signal attenuation associated with three different cell phones around the time when the iPhone 4 debuted. Apple fixed the problem by giving owners a nonconductive case that prevented a user's hand from directly touching the antennas. The case significantly improved reception as shown by the last column in Table 4.4.

4.10.2 Cellular Base Station Antennas

A cellular wireless system divides a region into cells with a base station at the center of each cell. The base station communicates with any mobile user entering its cell. Ideally, the power radiated by an omnidirectional base station antenna decreases equally in all directions around the antenna, in which case the cell would be a disk shape. Assuming the cellular base stations are omnidirectional in azimuth and there are no obstacles, a base station transmits and receives from users in the areas designated by circles in Figure 4.43. Each circle corresponds to the outer boundary beyond which the SNR drops too low due to free space loss. The intersection of the circles of the six base stations

Figure 4.43 Coverage area of base stations with omnidirectional antennas.

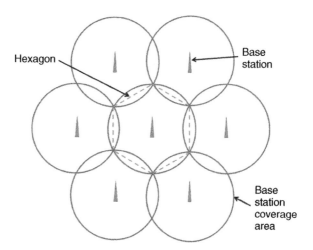

Hexagon

Base station

Base station coverage area

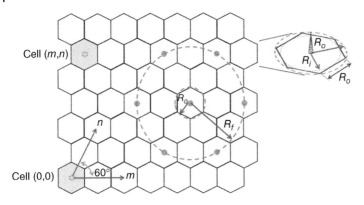

Figure 4.44 Frequency reuse in a cellular system.

around the center base station form the apexes of a regular hexagon designated by the dashed lines. As a result, a hexagon rather than a circle describes a base station coverage area known as a cell (Figure 4.44). An inscribed circle of radius R_i and a circumscribed circle of radius R_o bound a hexagon. The radii of the inscribed and circumscribed circles are related by $R_i = 0.866R_o$. The cells overlap in order that communication occurs seamlessly when a mobile unit passes from one cell to another. A handoff occurs when a user moves from one cell to another causing one base station to transfer the communication to a new base station.

To minimize interference between cells, nearby cells operate in different frequency bands. A cellular system reuses a frequency band in a co-channel cell far enough away to preclude interference. Frequency reuse depends on the size of the cell and the number of available frequency bands. A cluster of cells contains $N_{cluster}$ cells with no cell operating in the same frequency band. Clusters act like puzzle pieces that assemble the coverage area of a cellular system. Increasing $N_{cluster}$ reduces the number of clusters needed to cover a cellular service area and reduces the system capacity [40].

Figure 4.44 divides the cellular region into regular hexagons with a center-to-center spacing of $2R_i = R_o\sqrt{3}$. The cluster size ($N_{cluster}$) is the number of cells in a fixed group in which each cell operates at a unique frequency. This fixed group repeats throughout the cellular region in order to allow frequency reuse but at the same time minimize interference between cells. The ratio R_f/R_o determines the number of cells in a cluster.

$$N_{cluster} = \text{round} \left\{ \frac{1}{3} \left(\frac{R_f}{R_o} \right)^2 \right\} \tag{4.57}$$

where R_f is the distance between two cells operating at the same frequency. Table 4.5 has calculations of (4.57) for various ratios of R_f to R_o. Figure 4.45

Table 4.5 Number of clusters calculated from the reuse distance and cell size.

$N_{cluster}$	3	4	7	9	12	13	16	19	21	25	27
R_f/R_o	3	3.5	4.6	5.2	6	6.2	6.9	7.5	7.9	8.7	9

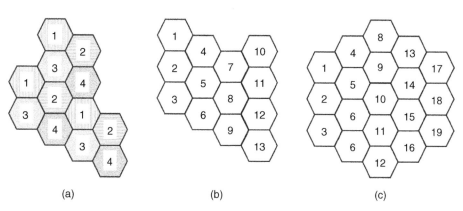

(a) (b) (c)

Figure 4.45 Arrangement of cells for different values of $N_{cluster}$. (a) 4, (b), 13, and (c) 19.

shows possible cell arrangements when $N_{cluster} = 4$, 13, and 19. A number in a cell corresponds to a frequency band that differs from all other cell numbers.

Assume that (m, n) mark the cell centers along the m and n directions shown in Figure 4.44, where m and n are integers. The co-channel distance from a cell at $(0,0)$ to cell (m, n) is given by [40]

$$R_f = R_o \sqrt{3(m^2 + mn + n^2)} \tag{4.58}$$

Cell size depends on the geography and number of mobile users. Cell splitting creates smaller cells from a standard size cell. The layout in Figure 4.46 makes sense in flat terrain where cells extend several kilometers in radius. Buildings, trees, and hills block signals and reduce the size cell as well as create amoeba-like shapes rather than nice circles or hexagons. Cells in metropolitan areas cover a significantly smaller area due to high traffic. In high-use areas, designers divide cells into small cells with names like [41]

- *Macrocell*: Largest cell associated with a base station. Maximum range is 35 km.
- *Microcell*: Smaller cell in high population areas that are low-power. Maximum range is 2 km.
- *Picocells*: Provides very localized coverage (usually inside buildings). Maximum range is 200 m.

- *Femtocells*: Smallest cell that provides coverage inside a room or a small building. Maximum range is 10 m.

The lower right cell in Figure 4.46 shows how small cells fit within a larger cell. A small cell antenna must have an appropriate location that covers the desired area as well as a connection to the network.

Sector antennas are typically 1–2 m long arrays that have 10–20 dB of gain [42]. The top right cell in Figure 4.46 shows a hexagon cell divided into three 120° sectors. Base station antennas have the best coverage when placed high above ground on a tower (Figure 4.47) or a building (Figure 4.48). Figure 4.47 is a tower that has several types of antennas. An omnidirectional antenna on top has a 360° view of the area. Below the omni are three sets of sector antennas.

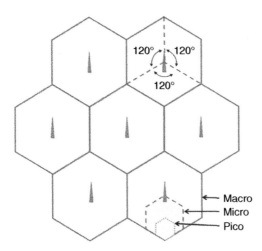

Figure 4.46 Hexagonal cell sectors which are further divided into micro and pico cells.

Macro
Micro
Pico

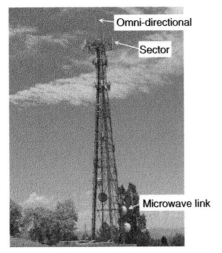

Figure 4.47 Base station antenna tower.

Omni-directional

Sector

Microwave link

Figure 4.48 Sector antennas on top of a building.

Figure 4.49 A six-sector base station antenna.

Several microwave dishes on the tower send and receive signals to/from distant locations. The building below the tower contains transmitters, receivers, control units, cooling, and connections to cables that run underground. Large towers can have several venders renting space for their antennas. Not all base station antennas in the world divide the hexagon into three sectors. The base station in Tasmania, Australia in Figure 4.49 divides the cell into six sectors.

Local ordinances sometimes require camouflaging antenna towers for aesthetic reasons. Approaches to camouflage including putting base station antennas in fake trees, cacti, and rocks as well as inside signs and attics. Figure 4.50 has two examples of base station antennas mounted in fake trees.

(a) Annapolis, MD   (b) Chengdu, China

Figure 4.50 Base stations camouflaged as a trees. (a) Annapolis, MDand (b) Chengdu, China

Companies pay business and home owners to install antennas on or inside structures on their properties [43].

Handset users want to seamlessly move between outdoors and indoors with no disruptions in call quality and service. An in-building cell includes amplifiers to overcome path loss in the building and the internal cable losses. There are two approaches to in-building systems [45]

- Extend an existing cell site by borrowing its signal and bringing it indoors. This approach works when the existing cell site has sufficient capacity to handle the current and projected capacity.
- Locate a cell site or a portion of a cell site in the building. Subscriber traffic must be high enough to justify on-site cellular PCS (personal communications service)/wireless/cellular equipment. The antenna needs a clear view of the closest cell site and a coaxial cable that routes the signal to the in-building system.

Either omnidirectional or leaky coaxial cable antennas distribute the signal inside a building. Omnidirectional antennas work best in large unobstructed areas, such as the interiors of shopping malls and convention centers. A leaky coaxial cable is a better choice for restricted coverage, such as hospitals, elevator shafts, corridors, subway tunnels, and office spaces.

4.10.3 Reflector Antennas

For most applications, the parabolic dish or reflector antenna serves as the workhorse for long distance wireless communications. They convert a small feed antenna into a much bigger aperture antenna. Figure 4.51 shows four methods of illuminating a parabolic reflector surface. The front-fed reflector

Figure 4.51 Four types of reflector antennas.

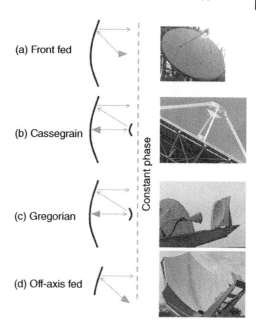

(a) Front fed

(b) Cassegrain

(c) Gregorian

(d) Off-axis fed

Constant phase

places a feed antenna (typically a horn antenna) at the focal point of the paraboloid. The distance from a horn antenna at the focal point to the surface of the reflector then to a plane (constant phase) in front of the reflector is the same for all angles (definition of a parabola). The horn feed blocks the center of the reflector and has difficulty fully illuminating the surface of a large reflector. Two other approaches use a horn that radiates from the center of the main reflector to a small subreflector that then illuminates the main reflector surface. Subreflectors enable feeds to provide a desired illumination for a large reflecting surface. A Cassegrain reflector has a convex subreflector (Figure 4.51b). A Gregorian parabolic reflector has a concave subreflector (Figure 4.51c). The feed or subreflector along with its support structure blocks front-fed, Cassegrain, and Gregorian main reflectors, so an offset design removes the feed from the front of the aperture as shown in Figure 4.51c,d. Front-fed reflectors usually use a Cassegrain design, because the subreflector is closer to the main reflector than in a Gregorian design, making the antenna more compact. Offset-fed reflectors often use Gregorian designs, because the subreflector is closer to the main reflector in the vertical direction, and better control of spillover is possible [46].

Extremely large Cassegrain or Gregorian reflectors use a beam-waveguide [44] that has the feed horn and support equipment in a room far below the antenna (Figure 4.52). The signal from the feed horn travels to the subreflector via several reflecting mirrors. High-power, water-cooled transmitters and low-noise cryogenic amplifiers do not have to tilt as the antenna moves. This

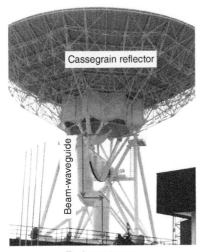
Cassegrain reflector

Beam-waveguide

Equipment room

Figure 4.52 Alfouvar satellite ground station Cassegrain beam-waveguide reflector antenna in Portugal.

design provides easy access to system components. An example of a beam waveguide Cassegrain reflector is the Alfouvar satellite ground station located about 30 km North of Lisbon (Figure 4.52). It provides radio, telephone, and television communications services to Portuguese speaking people around the globe.

The offset feed eliminates feed blockage by using only a portion of the parabolic reflector. In this case, the feed illuminates the reflector while being out of the way of the reflected rays (off-axis feed in Figure 4.51d). This offset reflector design is widely used in dish antennas for satellite television as shown by the house mounted reflector in Figure 4.53.

Figure 4.54 has some examples of typical front-fed satellite communication system reflector antennas. These antennas at the National Center for Atmospheric Research (NCAR) in Boulder, CO receive weather data from satellites.

Figure 4.53 Satellite TV offset reflector antenna.

(a)   (b)   (c)

Figure 4.54 Three types of satellite dishes at NCAR. (a) Wire mesh, (b) solid surface and (c) solid surface with shroud.

The dish in Figure 4.54a has a wire mesh reflector surface. Operating at frequencies where the mesh spacing is less than $\lambda/10$ minimizes the amount of signal that passes through the mesh. The wire mesh reduces the weight as well as the stress due to wind. The reflector in Figure 4.54b has a solid reflector. The satellite dish in Figure 4.54c has a solid surface with a shroud around the edge that suppresses sidelobes due to diffraction from the edges.

4.10.4 Antennas for Microwave Links

Terrestrial microwave antennas relay information from transmitter to receiver over long distances. Microwave links have highly directional antennas for LOS communications at microwave frequencies in order to overcome the large free space loss. A microwave link is cheaper and faster to install than cables, because it does not require laying cable over long distances. Figure 4.55 shows several microwave link towers on top of Colorado Mines Peak (12 497 ft/3809 m). These antennas have an unobstructed view for relaying microwave signals over the Rocky Mountains.

Figure 4.55 Microwave link antennas on top of Mines Peak in Colorado.

Reflector antennas work well in microwave links, because they have very narrow beamwidths for high gain and resolution. The narrow beamwidths of transmit and receive antennas require precise pointing, however. Small pointing deviations cause severe signal deterioration, so the antennas need mounts on very stable, wind-resistant towers. If the frequency of a 244-cm diameter antenna increases from 2 to 6.5 GHz, the maximum allowable tower motion decreases from 3.5° to 1.0° which increases the tower rigidity requirement by 3.5 times [45]. A 2-ft diameter dish at 22 GHz has a 2° beamwidth and requires even more stringent wind load specifications.

Another microwave link antenna, the Hogg or horn-reflector antenna, has a portion of a parabolic antenna mounted at the mouth of a pyramidal horn antenna (Figure 4.56) [47]. The reflector focus lies at the apex of the horn. This design has low sidelobes, because nothing blocks the aperture and the hood discourages radiation out of the sides. This very heavy, wind-resistant horn-reflector requires a large support structure as shown in Figure 4.57a. Its biggest use is for point-to-point communications where it does not have to rotate. Since the 1970s the shrouded parabolic dish antennas have largely replaced the Hogg antenna, because they have equally good sidelobe performance with a lighter more compact construction that can be mounted on simple towers (Figure 4.57b). Flat or cone-shaped radomes over the dishes increase the stability and protect the reflector surfaces (Figure 4.58).

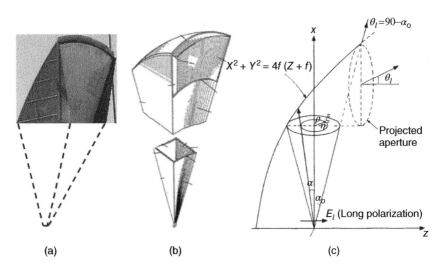

Figure 4.56 Horn reflector antenna. (a) Horn-reflector on Miners Mountain, CO (Figure 4.57) (b) Diagram from patent [47]. (c) Conical horn reflector antenna. *Source:* Reprinted by permission of Ref. [48]; © 2010 IEEE.

Horn-reflector (a)   Shrouded reflector (b)

Figure 4.57 Horn-reflector antennas require hefty mounts compared with shrouded reflector antennas (Miners Mt, CO).

Figure 4.58 Radomes that protect microwave link antennas. The two antennas on the let have conical radomes while the one on the right has a flat radomes over a shroud.

Example

Estimate the loss in dB of a 1 m dish transmitting antenna operating at 5 GHz that is 1 km away from the receive antenna. The transmit antenna has a pointing error of 1° while the receive antenna has no pointing error.

Solution

Start with (4.21) when $\lambda = 3 \times 10^8 / 5 \times 10^9 = 0.06$ m and $z_0 = 1000$ m

$r_0 = 1000 \tan 1° \approx 17.5$ m for a $1°$ pointing error using the small angle approximation for tan

$$AP \simeq \frac{J_1\left(\frac{kr_a r_0}{z_0}\right)}{\frac{kr_a r_0}{z_0}}$$

$$\frac{kr_a r_0}{z_0} = \frac{\frac{2\pi}{0.06}(1°)\frac{\pi}{180°}(17.5)}{1000} = 0.032$$

$$\text{Loss} \simeq -20\log\left\{\frac{J_1(0.032)}{0.032}\right\} = -20\log\left\{\frac{0.016}{0.032}\right\} = -20\log\{0.5\} = 6\,\text{dB}$$

This high loss due to a small pointing error emphasizes the importance of aligning transmit and receive antennas as well as mounting the antennas on solid support structures.

4.11 Diversity

Diversity in the base station mitigates the impact of signal amplitude drops known as signal fading. If a transmitted signal has multiple, independent paths, then the receiver lies in a region of high amplitude variations. The receive antenna either selects the largest signal or combines the signals in a way that enhances reception. Most diversity schemes require multiple antennas. Each additional antenna contributes to the real estate problem on handsets but not as much on base stations. Diversity gain is the increase in SNR due to a diversity scheme. Diversity means receiving a signal at multiple places, frequencies, polarizations, or times and picking the best one. It works best when the signals at the different places, frequencies, polarizations, or times are not correlated.

4.11.1 Spatial Diversity

Spatial diversity requires two or more receive antennas separated in space. Figure 4.59 shows the received power density distributed over one dimension in space. Some regions have low signal power due to fading. An antenna in a region of fading does not receive enough signal for detection. The three antennas in Figure 4.59a all receive low signal power. The antennas are close together, so they lack enough spatial diversity (separation) in order to have one of the antennas in a region of high signal power. Spatial diversity increases by adding more antennas or by putting more space between the existing antennas. Figure 4.59b demonstrates that increasing the antenna separation distance puts at least one of the antennas in a high signal power area. The

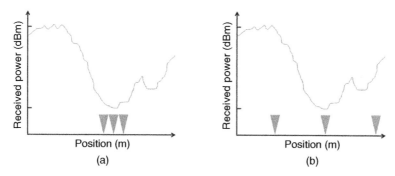

Figure 4.59 Spatial diversity increases the odds of an antenna receiving a strong signal.

Figure 4.60 Wireless routers with multiple monopoles for spatial diversity.

wireless routers in Figure 4.60 have more than one monopole in order to increase spatial diversity. When multipath signals come from all directions, antenna spacing between 0.5λ and 0.8λ results in independent (decorrelated) channels. Spatial diversity requires a significant antenna separation in the VHF and lower UHF bands, because the effectiveness of the antenna separation depends on λ.

Selection diversity picks the strongest signal out of N received signals in an antenna array. For instance, the left most antenna in Figure 4.59b receives the highest signal power, so it is selected as the active antenna while the other two are ignored. The average selection diversity gain for N independent and Rayleigh distributed signals is [49]

$$G_{sd} = \sum_{n=1}^{N} \frac{1}{n}$$
$$\approx \gamma_E + \ln N + 0.5/N \qquad (4.59)$$

where $\gamma_E = 0.5772$ K is Euler's constant. Each additional element contributes a smaller amount to G_{sd}. The average increase to the SNR is given by

$$\overline{\text{SNR}} = (\gamma_E + \ln N + 0.5/N)\overline{\text{SNR}}_n \qquad (4.60)$$

where $\overline{\mathrm{SNR}}_n$ is the average SNR at one antenna element. Selection diversity only requires measuring the SNR at each element but not the amplitude and phase of the signals at the elements (Figure 4.61).

Rather than just selecting the antenna with the highest signal power, maximum ratio combining (MRC) adds the received signals together after equalizing the element weights in the array. It has the optimum statistical fading reduction compared to all other linear diversity combiners. The diversity combiner SNR equals the sum of the SNRs of the signals from each antenna. On average, the diversity gain of MRC is

$$G_{\mathrm{MRC}} = N \tag{4.61}$$

which beats G_{sd}. The average SNR of MRC is

$$\overline{\mathrm{SNR}} = N\overline{\mathrm{SNR}}_n \tag{4.62}$$

MRC requires measuring the amplitude and phase of the signals at the elements. MRC has trouble keeping up with weight changes in fast varying channels.

Equal gain combining (EGC) equalizes the phases but keeps the amplitude weights uniform to improve the SNR. On average, the EGC diversity gain is

$$G_{\mathrm{EGC}} = 1 + (N - 1)\frac{\pi}{4} \tag{4.63}$$

which is very close to G_{MRC} and significantly better than G_{sd}. The average SNR of MRC is

$$\overline{\mathrm{SNR}} = \left[1 + (N - 1)\frac{\pi}{4}\right]\overline{\mathrm{SNR}}_n \tag{4.64}$$

EGC requires a measurement of the signal phase at each element.

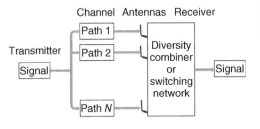

Channel Antennas Receiver

Transmitter
Signal
Path 1
Path 2
Path N
Diversity combiner or switching network
Signal

Figure 4.61 Diversity exploits the different paths a transmitted signal takes to arrive at the receiver.

4.11.2 Frequency Diversity

Frequency diversity switches between multiple frequency bands to until it finds the band that increases the total signal power. For the same transmitting and receiving antenna locations, the peaks and nulls due to multipath occur at different locations for different frequency channels. Figure 4.62 shows the signal level at a receive antenna for three different frequencies. Switching from f_1 to f_2 or f_3 significantly improves the received signal level from a_1 to a_2 to a_3. The frequency separation that yields independent fading on the different channels depends on the path lengths. Large path length differences require small frequency differences in the channels. Microwave line of sight links use frequency diversity. A frequency diversity system that

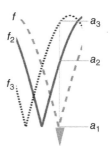

Figure 4.62 Frequency diversity means picking the frequency which results in the highest signal level.

experiences an SNR decrease at one frequency switches to a backup frequency band that has a better SNR.

4.11.3 Polarization Diversity

Since reflections change the polarization of a signal, a transmit antenna with one polarization delivers two orthogonal polarizations to the receiver. Polarization diversity has two antennas with orthogonal polarizations. The receiver selects the antenna with the strongest signal. Cellular base stations use polarization diversity. Figure 4.63 shows two approaches to polarization diversity. The first has an array of vertically polarized antennas next to an

Figure 4.63 Polarization diversity has adjacent antenna arrays with orthogonal polarization.

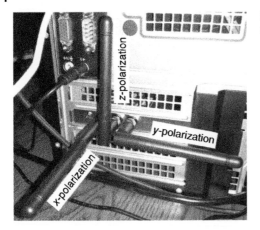

Figure 4.64 Polarization diversity antennas on a computer.

array of horizontally polarized antennas. The more common configuration on the right uses slant 45° polarization diversity, because both polarizations contain horizontal and vertical components (Fresnel reflection coefficients are polarization dependent). One slant polarization duplexes transmit and receive, while the other slant polarization only receives. In this way, one polarization diversity antenna replaces two vertically polarized antennas spaced several feet apart. Decreasing the number and size of the antennas decreases the tower loading as well as the cost of renting space on the tower. Two orthogonal polarizations mean that polarization diversity has only two channels that lack the independence found in spatial and frequency diversity but are independent enough to enhance performance. Polarization diversity is simple and cheap to implement. Figure 4.64 is an example of polarization diversity on a computer wireless card using three orthogonal monopoles.

4.11.4 Time Diversity

Time diversity transmits the same signal at multiple time slots separated by at least the coherence time of the channel [49]. Coherence time is the period in which the channel impulse response does not change. Outside the coherence time, multiple versions of the same signal encounter independent fading conditions on their way to the receiver, because channel conditions change.

Problems

4.1 An antenna has a pattern given by $AP = \text{sinc}(10\sin\theta)$ when $-90° \leq \theta \leq 90°$ for (a) AP rectangular plot, (b) $|AP|$ rectangular plot, (c) $20\log|AP|$ rectangular plot, and (d) $|AP|$ polar plot.

4.2   Calculate the directivity of an antenna when $AP = \cos\theta$ for $0 \le \theta \le \pi/2$ and zero elsewhere.

4.3   Find the directivity of an antenna with $AP(\theta) = \sin\theta$.

4.4   An antenna has θ $AP(\theta) = \sin(5\pi\sin\theta)/(\pi\sin\theta)$ for $0 \le \theta \le \pi/2$ and zero elsewhere. Find the gain if the aperture is 95% efficient.

4.5   Estimate the gain of a 50-m diameter antenna at 1 GHz.

4.6   An antenna operates over all frequencies in X band, what is its bandwidth?

4.7   Find the percent power received by a linearly polarized antenna when a LHP circular signal is incident.

4.8   Plot the PLF vs. rotation angle when a linearly polarized wave is incident on a linearly polarized antenna that rotates in a plane perpendicular to the incident field.

4.9   The IEEE defines the far field of an antenna as $R = 2D_{max}^2/\lambda$ for $\Delta R \le \lambda/16$. Low sidelobe antennas require greater phase accuracy. What should the far field distance be if $\Delta R \le \lambda/48$?

4.10 Find the equation for the electric field of a $\lambda/2$ long dipole and plot the antenna pattern in dB.

4.11 Use MATLAB to calculate and plot the 3D antenna pattern of a $\lambda/2$ dipole.

4.12 Find the radiation resistance of a loop with (a) 1 turn and (b) 100 turns when $r_\ell = \lambda/25$.

4.13 Find the magnitude of the relative electric field of a small loop ($r_\ell = 0.01\lambda$) as a function of θ.

4.14 Plot AR of an axial mode helix in dB vs. N (from 1 to 20).

4.15 Assume a $12\,cm \times 6\,cm$ with a uniform field projects onto a flat screen that is 20 cm away at 10 GHz. Plot the relative antenna pattern over a 2 m by 2 m area that is centered 20 cm from the aperture.

4.16 A rectangular patch on a substrate with $\varepsilon_r = 2.2$ that operates at 2.5 GHz. Plot: (a) E_θ vs. θ $\phi = 0°$ and 90°, (b) E_ϕ vs. θ $\phi = 0°$ and 90°.

4.17 Plot L and W for $1 \leq \varepsilon_r \leq 10$ when the patch resonates at 2.4 GHz.

4.18 Consider an eight-element uniform array along the x-axis with an element spacing of 1.5 cm at 10 GHz. What is the phase shift needed at each element to steer the main beam to 30°.

4.19 Plot the array factors on one graph in rectangular format for an eight-element uniform array when the element spacing is 0.5, 1.0, and 2.0 wavelengths.

4.20 On one graph, plot the binomial array factors when $N = 2$ through nine elements.

4.21 Derive the Chebyshev weights for a six-element array with peak sidelobes that are 20 dB below the main beam. Plot the array factor.

4.22 Derive the Taylor weights for a 20-element array with peak sidelobes that are 20 dB below the main beam and $\bar{n} = 5$. Plot the amplitude weights and array factor.

4.23 Calculate the array weights shown in Figure 4.29 and the corresponding array factors in Figure 4.30.

4.24 Plot the array factor and unit circle representation of a four-element uniform array when the element spacing is 0.5 wavelengths. Now, move the two zeros at $\pm 90°$ to $\pm 120°$. Plot the unit circle representations and array factors before and after moving the nulls. What are the new amplitude weights? Plot the array factors in dB on a rectangular plot.

4.25 Start with a Taylor $\bar{n} = 4$ $sll = 25$ dB taper and place a null at $u = 0.25$ when $d = 0.5\lambda$. Do not allow complex weights. Plot the unit circle representation, the array weights, and the nulled array factor superimposed on the quiescent array factor.

4.26 A four-element array has roots at $z = -1$ and $\pm j$. Move the zeros to (a) $z = -1.2$ and $\pm 1.2j$ and (b) $z = -1.4$ and $\pm 1.4j$.

4.27 Calculate the weights for the array in the previous problem when the zeros are moved to $z = -0.8$ and $\pm 0.8j$.

4.28 Use null-filling for linear array along the x-axis with six elements spaced half a wavelength apart for θ between $0°$ and $90°$. Try to totally eliminate those nulls. Is it possible? Plot the unit circle representation and the array factor.

4.29 Use null filling on an eight-element array with half wavelength spacing along the x-axis for θ between $0°$ and $90°$. Compare and contrast moving the zeros inside and outside the unit circle.

4.30 An eight-element uniform array along the x-axis has interference incident at $\theta = -38°$. Move the array factor null at $\theta = -48.6°$ so that it points at the interference.
(a) Calculate the new array weights.
(b) Plot the unit circle representation.
(c) Plot the adapted array factor.

4.31 Repeat the previous problem but also move the null at $\theta = 48.6°$ to $\theta = 38°$.
(a) Calculate the new array weights.
(b) Plot the unit circle representation.
(c) Plot the adapted array factor.

4.32 Derive (4.58).

4.33 Derive $N_{cluster} = m^2 + mn + n^2$ from (4.57).

4.34 A reflector antenna has an f/D (focal length to diameter ratio) of 1.0. A dipole antenna is placed at the feed facing the bottom of the paraboloid reflector. Calculate the field strength at the bottom and edge of the reflector in the plane containing the focal point and the paraboloid vertex when the dipole lies perpendicular to that plane.

4.35 A reflector antenna has an f/D (focal length to diameter ratio) of 1.0. A small dipole antenna is placed at the feed facing the bottom of the paraboloid reflector. Calculate the field strength at the bottom and edge of the reflector in the plane containing the focal point and the paraboloid vertex when the dipole lies in that plane.

4.36 Two mountain tops are 10 km apart operate at 7 GHz. The transmitter on one mountain is a Hogg antenna that has an aperture of 2 m by 2 m. The receive antenna on the other mountain is a 2 m dish antenna. Estimate the pointing error in degrees that results in a 3 dB loss in power.

4.37 On the same figure, plot the selection diversity gain vs. N using the exact and approximate formula in (4.59).

4.38 On the same figure, plot the average SNR for MRC vs. N for $\overline{SNR}_n = 5$, 10, 15, and 20 dB.

References

1 OSD/DARPA (1990). Assessment of Ultra-Wideband (UWB) Technology, Ultra-Wideband Radar Review Panel, R-6280, Office of the Secretary of Defense, Defense Advanced Research Projects Agency, July 13.

2 FCC 02-48 (2002). Revision of Part 15 of the Commission's Rules Regarding Ultra-Wideband Transmission Systems, Apr 22.

3 IEEE-SA Standards Board (2013). *IEEE Standard for Definitions of Terms for Antennas*. IEEE Std 145.

4 Balanis, C.A. (2005). *Antenna Theory Analysis and Design*, 3e. Hoboken, NJ: Wiley.

5 Stutzman, W.L. and Thiele, G.A. (2013). *Antenna Theory and Design*, 3e. Hoboken, NJ: Wiley.

6 Milligan, T.A. (2005). *Modern Antenna Design*, 2e. Hoboken, NJ: Wiley.

7 Kraus, J.D. and Marhefka, R.J. (2002). *Antennas for All Applications*, 3e. New York: McGraw-Hill.

8 Chattha, H.T., Huang, Y., Ishfaq, M.K., and Boyes, S.J. (2012). A comprehensive parametric study of planar inverted-F antenna. *Wireless Engineering and Technology* 3: 1–11.

9 (2013). Designing with an Inverted-F PCB Antenna for the EM250 and EM260 Platforms, Silicon Labs, Application Note 697, 11 Mar 2013.

10 Chattha, H.T., Huang, Y., Zhu, X., and Lu, Y. (2009). An empirical equation for predicting the resonant frequency of planar inverted-F antennas. *IEEE Antennas and Wireless Propagation Letters* 8: 856–860.

11 Liu, Z.D., Hall, P.S., and Wake, D. (1997). Dual-frequency planar inverted-F antenna. *IEEE Transactions on Antennas and Propagation* 45: 1451–1458.

12 Lehtola, A. (2002). Internal multi-band antenna. US Patent 6, 476,769.

13 Kraus, J.D. (1988). *Antennas*. New York: McGraw-Hill.

14 Saldell, U. (1997). Antenna device for portable equipment. US Patent 5, 661,495.

15 Goodman, J.W. (1968). *Introduction to Fourier Optics*. New York: McGraw-Hill.

16 Updyke, D.T., Muhler, W.C., and Turnage, H.C. (1980). An evaluation of leaky feeder communication in underground mines. Final report for US Dept. of the Interior Bureau of Mines.

17 http://www.cdc.gov/niosh/mining/content/emergencymanagementand response/commtracking/commtrackingtutorial1.html (accessed 28 November 2016).

18 Kara, M. (1996). Formulas for the computation of the physical properties of rectangular microstrip antenna elements with various substrate thicknesses. *Microwave and Optical Technology Letters* 12: 234–239.

19 Jackson, D.R. and Alexopoulos, N.G. (1991). Simple approximate formulas for input resistance, bandwidth, and efficiency of a resonant rectangular patch. *IEEE Transactions on Antennas and Propagation* 39 (3): 407–410.

20 https://www.pasternack.com/t-calculator-microstrip-ant.aspx (accessed 19 June 2019)

21 Bancroft, R. (2004). *Microstrip and Printed Antenna Design*. Atlanta, GA: Noble Publishing Corp.

22 Haupt, R.L. (2015). *Timed Arrays Wideband and Time Varying Antenna Arrays*. Hoboken, NJ: Wiley.

23 Haupt, R.L. (2010). *Antenna Arrays: A Computational Approach*. Hoboken, NJ: Wiley.

24 Schelkunoff, S.A. (1943). A mathematical theory of linear arrays. *Bell System Technical Journal* 22: 80–107.

25 Stone, J.S. (1927). Directive antenna system. US Patent 1, 643,323, 27 September 1927.

26 Dolph, C.L. (1946). A current distribution for broadside arrays which optimizes the relationship between beam width and sidelobe level. *Proceedings of the IRE* 34 (6): 335–348.

27 Taylor, T.T. (1955). Design of line source antennas for narrow beamwidth and low side lobes. *Transactions of the IRE Professional Group on Antennas and Propagation* 3: 16–28.

28 Monzingo, R.A., Haupt, R.L., and Miller, T.W. (2011). *Introduction to Adaptive Antennas*, 2e. SciTech Publishing.

29 Steyskal, H. (1996). Digital beamforming at Rome Laboratory. *Microwave Journal* 39 (2): 100–124.

30 Butler, J. and Lowe, R. (12 Apr 1961). Beam-forming matrix simplifies design of electronically scanned antennas. *Electronic Design* 9: 170–173.

31 Rotman, W. and Turner, R. (1963). Wide-angle microwave lens for line source applications. *IEEE Transactions on Antennas and Propagation* 11 (6): 623–632.

32 Rowell, C. and Lam, E.Y. (2012). Mobile-phone antenna design. *IEEE Antennas and Propagation Magazine* 54: 14–34.

33 http://www.gsmhistory.com/vintage-mobiles/#no_antenna (accessed 17 September 2018).

34 http://www.ptonline.com/articles/mids-make-a-comeback (accessed 28 November 2016).

35 http://www.edn.com/design/pc-board/4427506/What-are-molded-interconnect-devices- (accessed 28 November 2016).

36 Ouyang, Y., Sclub, R.W., Jin, N., and Pascolini, M. (2014). Shared antenna structures for near-field communications and non-near-field communications circuitry. US Patent 0139380 A1, 22 May, 2014.

37 http://www.instructables.com/id/How-to-Disassemble-a-Motorola-Razr (accessed 28 November 2016).

38 https://gadgetguideonline.com/s9/galaxy-s9-layout-and-layout-of-galaxy-s9/ (accessed 28 June 2018).

39 http://www.anandtech.com/show/3794/the-iphone-4-review/2 (accessed 28 November 2016).

40 Ghosh, R.K. (2017). *Wireless Networking and Mobile Data Management*. Singapore: Springer.

41 https://www.repeaterstore.com/pages/femtocell-and-microcell (accessed 28 November 2016).

42 http://www.repeater-builder.com/antenna/andrew/andrew-base-station-antenna-systems-psg-2008.pdf (accessed 28 November 2016).

43 http://www.wirelessdesignmag.com/articles/2013/04/challenges-microcell-deployment-configuration (accessed 28 November 2016).

44 Imbriale, W.A. (2003). *Large Antennas of the Deep Space Network*. Hoboken, NJ: Wiley.

45 https://www.tessco.com/yts/customerservice/techsupport/whitepapers/antennas.html (accessed 28 November 2016).

46 http://www.antennamagus.com/database/antennas/antenna_page.php?id=197 (accessed 6 December 2018).

47 Hogg, D.C. (1965). Horn reflector antenna with concentric conical reflectors at mouth to increase effective aperture. US Patent 3, 224,006, 14 December 1965.

48 Yassin, G., Robson, M., and Duffett-Smith, P.J. (1993). The electrical characteristics of a conical horn-reflector antenna employing a corrugated horn. *IEEE Transactions on Antennas and Propagation* 41 (3, pp. 357, 361).

49 Rappaport, T.S. (2002). *Wireless Communications Principles and Practice*, 2e. Upper Saddle River, NJ: Prentice Hall.

5

Propagation in the Channel

A water channel directs water from a source to a destination. Smooth channels produce easily predictable laminar flow. Obstacles like rocks induce turbulence that makes flow predictions in the channel difficult. A similar situation exists for the flow or channeling of a radio frequency (RF) signal from a transmitter to a receiver. In free space, the signal propagates from the transmitter to the receiver with a loss proportional to the square of the distance traveled. Unfortunately, the RF channel usually has many obstacles like buildings, cars, trees, etc. that attenuate, reflect, refract, and diffract the signal along several different paths. In addition, the transmitter and/or receiver motion induces Doppler shift in the signal. The simple idea of free space loss needs modification or augmentation to get accurate results. Interference and noise corrupt the signal and cause the receiver additional problems.

Geometrical optics (GO), or ray tracing, typically models signal propagation in a channel. Figure 5.1 illustrates some of the mechanisms encountered by a transmitted signal (ray) traveling through a channel to the receiver. In general, a satellite system has a pure line of sight (LOS) channel. The LOS signal travels a long distance through the atmosphere which attenuates, refracts, and depolarizes it. These effects highly depend on the signal frequency. Diffraction, reflection, and shadowing significantly impact signals in the channel. Statistical models approximate complex relationships in a practical setting with relatively simple stochastic models.

This chapter introduces RF propagation modeling in a channel. Approaches to modeling a channel include measured impulse responses, electromagnetic models, and statistical models. Determining the channel impulse response requires experimental measurement of electromagnetic waves propagating in the channel. Channel measurements require (i) expensive equipment, (ii) a significant time investment, and (iii) complex setup. Realistic electromagnetic computer models based on Maxwell's equations help with siting antennas and analyzing specific scenarios. Numerical approximations, like ray tracing,

Wireless Communications Systems: An Introduction, First Edition. Randy L. Haupt.
© 2020 John Wiley & Sons, Inc. Published 2020 by John Wiley & Sons, Inc.

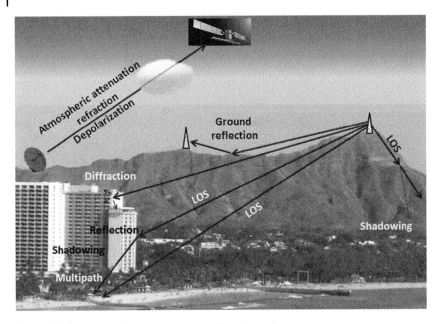

Figure 5.1 Ray tracing in the signal propagation channels.

accommodate large, complex channels. Accurate electromagnetics models are expensive and time-consuming. Several statistical models developed from measurements provide simple estimates in place of more accurate measurements or electromagnetic models.

5.1 Free Space Propagation

Free space propagation assumes a signal propagates in a vacuum with no obstacles. Even the path between a satellite and a ground station has obstacles, including reflections and diffraction from the satellite and ground. The LOS communications link in Figure 5.2 has minimal obstacles. The transmitter generates P_t which the antenna magnifies with gain G_t. Assuming that the transmitted signal travels in all directions from the antenna means that the power density at the receive antenna is $P_t G_t / 4\pi R^2$. If the receiver has an effective aperture A_e, then the received power is

$$P_r = \frac{P_t G_t A_e}{4\pi R^2} \tag{5.1}$$

The Friis transmission formula (5.1) serves as the fundamental model for a communication link [1].

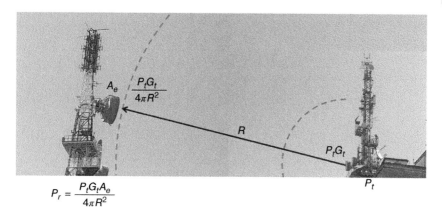

$$P_r = \frac{P_t G_t A_e}{4\pi R^2}$$

Figure 5.2 Friis transmission formula.

A more realistic version of (5.1) includes losses lumped into a constant (L) and a power loss due to distance (γ).

$$P_r = \frac{P_t G_t A_e L}{4\pi R^\gamma} = \frac{P_t G_t G_r c^2 L}{(4\pi f)^2 R^\gamma} \tag{5.2}$$

where $A_e = G\lambda^2/(4\pi)$. This chapter explains how to find appropriate values for L and γ that depend on the propagation channel characteristics. The Friis equation forms the foundation of the link budget of a wireless system. Typically, the link budget analysis converts (5.2) into dB:

$$10\log(P_r) = 10\log(P_t) + 10\log G_t + 10\log G_r + 20\log \lambda + 10\log(L)$$
$$- 10\gamma \log R - 20\log(4\pi) \tag{5.3}$$

The received power must exceed the receiver sensitivity in order for meaningful communications to occur.

5.2 Reflection and Refraction

A signal encounters obstacles as it travels through the channel. These obstacles reflect and refract the signal. By the time the signal reaches the receiver, its amplitude, phase, and duration differ from the transmitted signal. Updating the Friis formula begins with calculating the effects of these reflections and refractions on the signal.

An electromagnetic signal incident on a large perfectly flat medium ($\mathbf{E}_i$, $\mathbf{H}_i$, θ_i) reflects from ($\mathbf{E}_r$, $\mathbf{H}_r$, θ_r) and transmits into ($\mathbf{E}_t$, $\mathbf{H}_t$, θ_t) the medium as shown in Figure 5.3. The plane of incidence contains the incident wave vector and the normal to the boundary (dashed line in Figure 5.3). Any plane wave

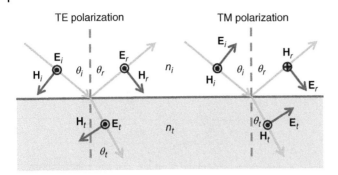

Figure 5.3 Two orthogonal polarizations incident on the boundary between two media. The plane of incidence contains the incident vector and the dashed line that is normal to the surface.

consists of two orthogonal polarizations: transverse magnetic (TM) has the electric field parallel to the plane of incidence and transverse electric (TE) has the electric field perpendicular to the plane of incidence. Boundary conditions determine the directions as well as the amplitude and phase of the reflected and transmitted electric and magnetic fields.

The incident, reflected, and transmitted angles obey Snell's law of reflection

$$\theta_i = \theta_r \tag{5.4}$$

and Snell's law of refraction

$$n_i \sin \theta_i = n_t \sin \theta_t \tag{5.5}$$

where
$$n = \sqrt{\varepsilon_r \mu_r} \quad = \quad \text{index of refraction}$$
$$\varepsilon_r \qquad\qquad = \quad \text{relative permittivity}$$
$$\mu_r \qquad\qquad = \quad \text{relative permeability}$$

and the subscripts i, r, and t refer to the incident, reflected, and transmitted waves. Reflection and transmission depend on the polarization of the signal as well as the material properties and angle of incidence. The Fresnel reflection coefficients for the two polarizations are different:

$$\Gamma_{TM} = \frac{E_{0r}}{E_{0i}} = \frac{n_t \cos(\theta_i) - n_i \cos(\theta_t)}{n_t \cos(\theta_i) + n_i \cos(\theta_t)} \tag{5.6}$$

$$\Gamma_{TE} = \frac{E_{0r}}{E_{0i}} = \frac{n_i \cos(\theta_i) - n_t \cos(\theta_t)}{n_i \cos(\theta_i) + n_t \cos(\theta_t)} \tag{5.7}$$

The Fresnel transmission coefficients likewise depend on polarization.

$$T_{TM} = \frac{E_{0t}}{E_{0i}} = 1 - \Gamma_{TM} = \frac{2n_t \cos(\theta_i)}{n_i \cos(\theta_t) + n_t \cos(\theta_i)} \tag{5.8}$$

$$T_{\mathrm{TE}} = \frac{E_{0t}}{E_{0i}} = 1 - \Gamma_{\mathrm{TE}} = \frac{2n_i \cos(\theta_i)}{n_i \cos(\theta_i) + n_t \cos(\theta_t)} \tag{5.9}$$

These coefficients play an essential role in calculating the amplitude and phase of the signal at the receiver.

Under certain conditions, a plane wave totally reflects from or totally transmits into a flat interface between two media. When $n_i > n_t$, The critical angle, θ_c, occurs when the signal totally reflects from the surface.

$$\theta_c = \sin^{-1}\left(\frac{n_t}{n_i}\right) \tag{5.10}$$

The entire signal incident at $\theta \geq \theta_c$ reflects from the surface, while signals incident at $\theta < \theta_c$ have partial transmission and reflection. The critical angle is polarization independent.

No reflection (total transmission) occurs at the Brewster angle when a TM polarized wave impinges on a boundary between two media

$$\theta_b = \tan^{-1}\left(\frac{n_t}{n_i}\right) \tag{5.11}$$

Unlike θ_c, θ_b exists only for TM polarization. In addition, total transmission only occurs at exactly θ_b. Angles above and below θ_b have some reflection. A Brewster angle exists on both sides of the medium. In fact, the two Brewster's angles on either side of the interface sum to 90°.

$$90° = \tan^{-1}\left(\frac{n_t}{n_i}\right) + \tan^{-1}\left(\frac{n_i}{n_t}\right) \tag{5.12}$$

Table 5.1 summarizes the properties of the critical and Brewster angles.

Oftentimes, a signal passes through several layers of media as shown in Figure 5.4. Buildings, windows, and the atmosphere have multiple layers with different electrical properties. Calculating the reflection and transmission coefficients for each layer requires an iterative process [2]. A single dielectric slab of height h in free space has a relatively simple expression for the reflection coefficient at normal incidence ($\theta_i = 0°$).

$$\Gamma = \frac{\Gamma_d - \Gamma_d e^{-j2k_d h}}{1 - \Gamma_d^2 e^{-j2k_d h}} \tag{5.13}$$

where $\Gamma_d = \frac{1-\sqrt{\varepsilon_r}}{1+\sqrt{\varepsilon_r}}$ and $k_d = \frac{2\pi}{\lambda\sqrt{\varepsilon_r}}$.

Example
What is the minimum height of a dielectric slab in which the reflection at normal incidence is zero?

Table 5.1 Critical and Brewster angles.

	Critical angle	Brewster angle
Characteristic	Total reflection when $\theta \geq \theta_c$	Total transmission when $\theta = \theta_b$
Polarization	TE and TM	TM
Index of refraction	$n_i > n_t$	$n_i \geq n_t$ or $n_i \leq n_t$
Diagram		

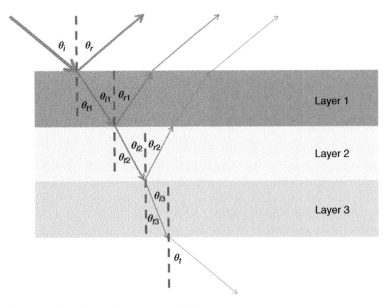

Figure 5.4 Ray tracing through multiple layers.

Solution

No reflection from the slab means that $\Gamma_{\text{slab}} = 0 \Rightarrow \Gamma_d - \Gamma_d e^{-j2k_1 h} = 0$ so $e^{-j2k_1 h} = 1$ which only occurs when $2k_1 h = 2n_1 \pi$ or $h = \frac{n_1}{2}\lambda_d$.

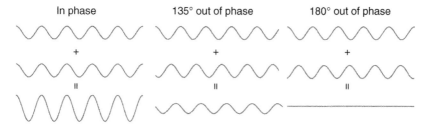

Figure 5.5 Adding one signal to another of the same frequency but different phases.

5.3 Multipath

Reflection changes the path of a signal as noted by the Fresnel equations. When the reflected signal interacts with the LOS signal, interference patterns form. In-phase signals add in amplitude as shown in Figure 5.5, otherwise the resulting sum is less than the sum of the two amplitudes. In fact, when they are out of phase by 180°, they cancel or fade.

Multipath means that a transmitted signal arrives at the receiver by more than one route. Figure 5.6 depicts a communication channel with a direct or LOS signal plus one reflected or multipath signal. Since the signals take different paths to the receiver, they have different amplitudes due to reflections and diffractions as well as additional free space loss. The phase and time of arrival of each signal at the receiver differ, because their paths have different lengths. All the signals sum to create a dispersed signal (longer duration than transmitted

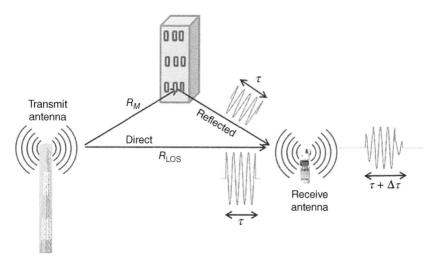

Figure 5.6 Multipath occurs when the signal takes more than one route to get to the receiver. The received signal shows dispersion or broadening of the pulse.

signal) resulting from the time delays of the various paths. If the signal takes two paths of length R_{LOS} and R_M ($R_M > R_{LOS}$), then the multipath signal arrives $\Delta\tau = (R_M - R_{LOS})/c$ after the LOS signal.

Example
Use MATLAB to demonstrate dispersion with an 8.33 ns pulse at a carrier frequency of 2.4 GHz. There are three multipath signals that travel an extra 0.27, 0.97, and 1.29 m. The amplitude of these signals at the receiver is 0.7, 0.5, and 0.2 V compared to 1 V of the LOS signal. Plot the sum of these signals.

Solution
The multipath signals are delayed by 0.90, 3.23, and 4.30 ns. Adding the signals results in the plot in Figure 5.7 which has the received dispersed pulse (solid line) superimposed on the LOS signal (dashed line). The dispersed pulse is 4.3 ns long or 52% longer than the LOS pulse.

When two signals from different directions collide, they add together to form an interference pattern like the one in Figure 5.8. A fade occurs when the signal amplitude drops below the receiver sensitivity level and is not detected. If the sources move, then the interference pattern changes. A receiver moving through the interference region experiences fluctuations in the received signal amplitude.

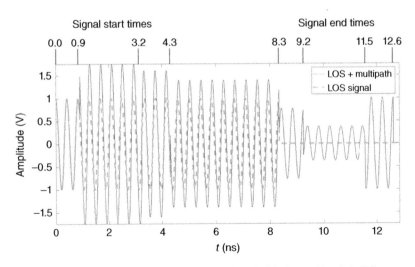

Figure 5.7 Received signal for LOS only (dashed) and LOS plus multipath (solid).

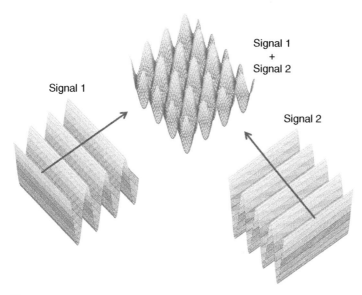

Figure 5.8 Two electromagnetic signals at the same frequency but from different directions meet and add together to produce interference with high- and low-signal amplitudes.

5.4 Antennas over the Earth

Figure 5.9 presents a microwave link between transmit and receive antennas placed on towers. The LOS path (R_{LOS}) starts at the transmitter mounted h_t above the ground and ends at the receiver mounted h_r above the ground. The transmit antenna appears to have an image antenna beneath the ground that radiates the reflected signal. Since the ground is not a perfect reflector, the image antenna radiates a reduced amplitude compared to the actual antenna. Using the Friis transmission formula, the received signal, $s_r(t)$, equals the sum of the electric fields of the LOS and image (reflected) signals.

$$
\begin{aligned}
|E_T| &\propto \left| \underbrace{\sqrt{G_t(\theta_{t1})G_r(\theta_{r1})}\frac{e^{-jkR_{\mathrm{LOS}}}}{4\pi R_{\mathrm{LOS}}}}_{\text{LOS}} + \underbrace{\Gamma_g\sqrt{G_t(\theta_{t2})G_r(\theta_{r2})}\frac{e^{-jkR_i}}{4\pi R_i}}_{\text{Image}} \right| \\
&\approx \left| \sqrt{G_t(\theta_{t1})G_r(\theta_{r1})}\frac{e^{-jkR_{\mathrm{LOS}}}}{4\pi R_{\mathrm{LOS}}} \right| F_{\mathrm{pg}}
\end{aligned}
\tag{5.14}
$$

where

$$G_t, G_r \qquad = \quad \text{gains of the transmit and receive antennas}$$
$$R_{\mathrm{LOS}} \qquad = \quad \sqrt{d^2 + (h_t - h_r)^2}$$

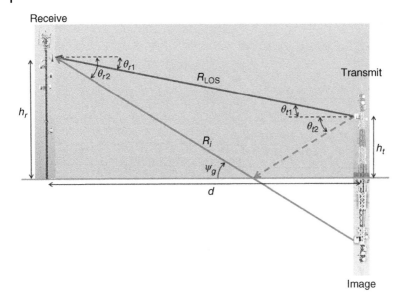

Figure 5.9 Diagram of direct and reflected rays.

$$
\begin{aligned}
R_i &= \sqrt{d^2 + (h_t + h_r)^2} \\
\Gamma_g &= \text{ground reflection coefficient} \\
R_{\text{LOS}} &\approx R_i \text{ in the amplitude terms} \\
F_{\text{pg}} = 1 + \Gamma_g e^{-jk(R_i - R_{\text{LOS}})} &= \text{path gain factor} \\
d &= \text{separation between transmitter and receiver.}
\end{aligned}
$$

The path difference between the LOS and reflected signals is

$$
R_i - R_{\text{LOS}} \approx \frac{2h_t h_r}{d} \tag{5.15}
$$

which delays the reflected signal by

$$
\tau = \frac{2h_r h_t}{cd} \tag{5.16}
$$

As d gets large, ψ_g (angle between the Earth and incident and reflected rays) gets small, so from (5.6) and (5.7), $\Gamma_g \rightarrow \pm 1$. The path gain factor describes the relative signal variation as a function of tower separation and tower heights [3]:

$$
F_{\text{pg}} \approx |1 \pm e^{-jk2h_1 h_2/d}| =
\begin{cases}
2\left|\sin\left(\frac{kh_r h_t}{d}\right)\right| = 2\left|\sin(kh_t \tan \psi_g)\right| & \text{TE polarization} \\[2mm]
2\left|\cos\left(\frac{kh_r h_t}{d}\right)\right| = 2\left|\cos(kh_t \tan \psi_g)\right| & \text{TM polarization}
\end{cases}
\tag{5.17}
$$

when $\psi_g = \theta_{t2} = \theta_{r2}$. The received power is a function of the tower heights and their separation distance.

$$P_r = P_t G_t G_r \frac{h_t^2 h_r^2}{d^4} \tag{5.18}$$

When $d \gg \sqrt{h_t h_r}$, then the power decreases as $1/d^4$ which far exceeds free space loss. Note that the received power and path loss become independent of frequency at large values of d. The path loss in dB is given by

$$L_{dB} = 40 \log d - 10 \log G_t - 10 \log G_r - 20 \log h_t - 20 \log h_r \tag{5.19}$$

Example
Plot the loss for a 2.4 GHz signal when the transmitter and receiver separation distance increases from 100 m to 100 km for

(a) free space
(b) $h_t = 30$ m, $h_r = 2$ m
(c) $h_t = 30$ m, $h_r = 10$ m
(d) $h_t = 30$ m, $h_r = 30$ m

Assume the antennas are isotropic point sources.

Solution
Figure 5.10 shows the resulting plots. Up to a distance called the breakpoint, the loss approximately equals that of free space $(1/d^2)$. After the breakpoint the decay increases to $1/d^4$. This behavior is typical in a multipath environment. In fact, more than one multipath signal produces a $1/d^\gamma$ loss where $1.5 \le \gamma \le 4$ and γ is the path loss exponent.

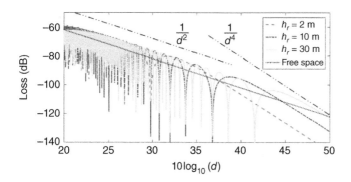

Figure 5.10 Propagation loss due to one bounce between two towers as a function of tower height (h_r) and distance between the towers (d) when $h_t = 30$ m.

The reflection coefficient in (5.14) depends on the type of ground (desert, pavement, farm, etc.) and is given by [3]

$$
\Gamma_g = \begin{cases} \dfrac{-\varepsilon_r(f)\sin\psi_g + \sqrt{\varepsilon_r(f) - \cos^2\psi_g}}{\varepsilon_r(f)\sin\psi_g + \sqrt{\varepsilon_r(f) - \cos^2\psi_g}} & \text{TE polarization} \\[4mm] \dfrac{\sin\psi_g - \sqrt{\varepsilon_r(f) - \cos^2\psi_g}}{\sin\psi_g + \sqrt{\varepsilon_r(f) - \cos^2\psi_g}} & \text{TM polarization} \end{cases} \tag{5.20}
$$

where $\varepsilon_r(f) = \varepsilon_r'(f) - j\varepsilon_r''(f) = $ relative permittivity of the ground at frequency f. Values of ε_r for various surface media as a function of frequency appear in Figure 5.11 [4]. The plots reveal that the soil water content dramatically changes both the real and imaginary parts of the relative dielectric constant. Also, fresh water and salt water have much higher dielectric constants than snow. This observation makes sense, because snow contains more air than water, so the permittivity approximately equals the weighted average of the permittivities of water and air ($\varepsilon_r \approx 1$).

The reflection coefficient for TM polarization is less than or equal to the reflection coefficient for TE polarization at any incident angle. As a result, when the incident wave has electric field components that are parallel and perpendicular to the plane of incidence, the polarization of the reflected wave differs from the polarization of the incident wave. For example a circularly polarized incident wave, becomes elliptically polarized after reflection, because the orthogonal components of the circular polarization have different reflection coefficients.

Example

A 2 GHz right-hand circular polarization (RHCP) wave is incident on flat ground in which the soil has 0.1 moisture content at $\psi_g = 30°$. What is the axial ratio of the reflected wave?

Solution

From Figure 5.11, the $\varepsilon_r' \approx 8$ and $\varepsilon_r'' \approx 0$, so

$$
\Gamma_g = \begin{cases} \dfrac{-8\sin 30° + \sqrt{8 - \cos^2 30°}}{8\sin 30° + \sqrt{8 - \cos^2 30°}} = -0.1954 & \text{TE polarization} \\[4mm] \dfrac{\sin 30° - \sqrt{8 - \cos^2 30°}}{\sin 30° + \sqrt{8 - \cos^2 30°}} = -0.6868 & \text{TM polarization} \end{cases}
$$

$$
\text{AR} = \frac{0.6868}{0.1954} = 3.5155 = 10.9 \text{ dB}
$$

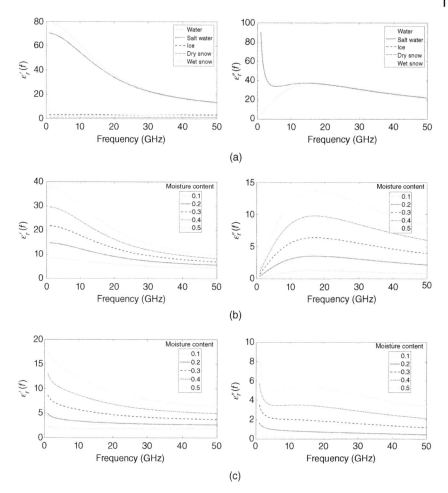

Figure 5.11 Real and imaginary parts of the dielectric constants for various ground conditions as a function of frequency. (a) Forms of water, (b) soil, and (c) vegetation.

A good example that combines multipath and reflections from lossy boundaries occurs when a wireless signal propagates in a tunnel [5]. The signal takes many different paths as it bounces from the walls, ceiling, and floor. Material composition of the tunnel, the tunnel shape and size, obstructions, and tunnel bends create many opportunities for reflection, refraction, and diffraction. The propagation path inside a tunnel may or may not have a LOS component. Consider a communications system in a tunnel with a dipole antenna that transmits 30 dBm of power to an isotropic receive antenna at the center of the tunnel cross section [6]. The square tunnel is 3 m tall and 5 m long with dry granite walls ($\varepsilon_r = 5$, $\sigma = 0.01$ S/m). The received signal as a function of separation distance

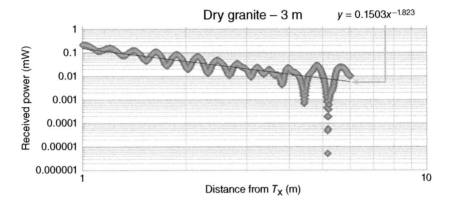

between the transmitter and receiver (d) appears in Figure 5.12. A linear fit to the received power in Figure 5.12 leads to a path loss exponent of 1.823, so the signal decreases according to

$$P_r = 0.1503x^{-1.823} \qquad (5.21)$$

The signal attenuates less than in free space. The tunnel acts like a waveguide with lossy walls, so the signal loses energy in the walls but does not disperse as much as in free space.

5.5 Earth Surface

Earth curvature blocks the LOS signal when the transmitter and receiver are many kilometers apart [5]. Figure 5.13 shows two towers separated by d_{max}, the

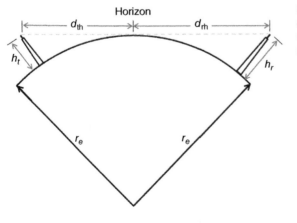

Figure 5.13 The Earth curvature limits the maximum range of LOS signals.

sum of the distances from the transmit (d_{th}) and receive antennas (d_{rh}) to the horizon. A LOS formula that assumes a smooth Earth with no obstructions estimates d_{max} as [7]

$$d_{max} = \sqrt{2h_t r_e} + \sqrt{2h_r r_e} \tag{5.22}$$

where the radius of the Earth is approximately $r_e = 6.378 \times 10^6$ m. In real life, many natural and manmade obstructions between the transmitter and receiver reduce d_{max}, so the LOS is much less than d_{max}.

Example
What is the maximum LOS for two microwave communication towers that are 15 m high over a smooth spherical Earth?

Solution
Using (5.22), it is $d_{max} = \sqrt{2h_t r_e} + \sqrt{2h_r r_e} = 2\sqrt{2 \times 15 \times 6.378 \times 10^6} = $ 27.7 km.

Atmospheric density and the index of refraction decrease with height above the Earth's surface causing electromagnetic waves to bend or refract around the Earth. This phenomenon lets antennas "see" beyond the horizon as shown by the ground station communicating with a satellite in Figure 5.14. Consequently, (5.22) underestimates d_{max}. An atmospheric correction to (5.22) replaces the actual Earth by a larger imaginary Earth where the refracted electromagnetic waves become LOS ($r_e = r_{ec}$) [8]. The new Earth radius is approximately

$$r_{ec} = k_e \times 6.378 \times 10^6 \text{m} \tag{5.23}$$

where the constant k_e is a function of the change in the atmospheric index of refraction (n_a) with height (h) above the Earth:

$$k_e = \frac{1}{1 + r_e \dfrac{dn_a}{dh}} \tag{5.24}$$

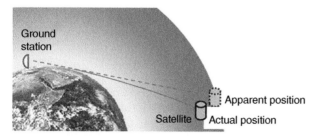

Figure 5.14 The atmospheric index of refraction decreases with altitude, so the signal bends around the Earth.

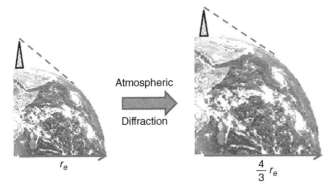

Figure 5.15 Atmospheric refraction bends signals so that the Earth appears to have a larger radius.

Although atmospheric conditions vary tremendously with location, time of day, and season, a reasonable average assumption of $\frac{dn_a}{dh} = 39 \times 10^{-9}\,\text{m}^{-1}$ produces $k_e = 4/3$. Thus, a quick estimate for atmospheric refraction assumes the Earth is 1.33 times larger in radius (Figure 5.15), so the distance to horizon in (5.22) uses (5.23) instead of the actual radius of the Earth.

Example
Repeat the previous example and take atmospheric diffraction into account.

Solution
The revised value is $d_{\max} = \sqrt{2h_t(4/3)r_e} + \sqrt{2h_r(4/3)r_e} = 31.9\,\text{km}$
which is over 4 km longer.

Not many surfaces are perfectly smooth as assumed with the Fresnel equations, so, variations in ground and water height cause reflections in directions other than predicted by Snell's law. The impact of surface variations depends on λ, so surface variations of 1 m appear smooth at high frequency (HF) but very rough at cell phone frequencies. Figure 5.16 shows the effect of various degrees of surface roughness on the reflected and scattered wave. A smooth surface has a surface height standard deviation limited by [9]

$$\sigma_{\text{surf}} < \frac{\lambda}{32\cos\theta_i} \tag{5.25}$$

If (5.25) is true, then the Fresnel equations predict the direction where most of the reflected field travels. As the surface becomes rougher, the scattered field spreads over a wider angle and the Fresnel equations fail to accurately predict the angular scattering of the reflected field. The dominant coherent scattered wave obeys Snell's law. The noncoherent part of the scattered wave radiates isotropically. In this case, coherence implies the same polarization as

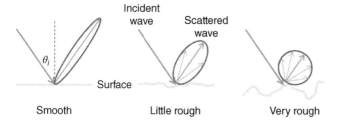

Figure 5.16 Scattering from a surface with various degrees of roughness.

the incident wave at Snell's angle of reflection. Noncoherence refers to the co- and cross-polarized components at angles other than Snell's angle of reflection. The reflection coefficient at Snell's angle of reflection is given by [4]

$$\Gamma = \Gamma_{\text{TE,TM}} e^{-(2k\sigma_{\text{surf}} \cos \theta_i)^2} \tag{5.26}$$

A perfectly smooth surface has $\sigma_{\text{surf}} \to 0$ and $\Gamma = \Gamma_{\text{TE}}$ or $\Gamma = \Gamma_{\text{TM}}$.

Example

A 1 GHz TE wave is incident on a lake at $0 \le \theta_i \le 45°$ with $\sigma_{\text{surf}} < 2.3$ cm. Plot Γ as a function of θ_i and σ_{surf} (Figure 5.17).

Solution

$$\lambda = \frac{3 \times 10^{10}}{1 \times 10^9} = 30 \text{ cm}, n_i = 1, n_t = 1.33$$

$$\Gamma_{\text{TM}} = \frac{\cos(\theta_i) - 1.33 \cos(\theta_t)}{\cos(\theta_i) + 1.33 \cos(\theta_t)}$$

$$\sin \theta_i = 1.33 \sin \theta_t \Rightarrow \theta_t = \sin^{-1}\left(\frac{\sin \theta_i}{1.33}\right)$$

$$\Gamma = \frac{\cos(\theta_i) - 1.33 \cos\left[\sin^{-1}\left(\dfrac{\sin \theta_i}{1.33}\right)\right]}{\cos(\theta_i) + 1.33 \cos\left[\sin^{-1}\left(\dfrac{\sin \theta_i}{1.33}\right)\right]} e^{-4\left(\frac{2\pi}{30}\right)^2 \sigma_{\text{surf}}^2 \cos^2 \theta_i}$$

Figure 5.17 Reflection coefficient for a 1 GHz TE wave is incident on a lake between $0 \le \theta_i \le 45°$ that has a height variation of $\sigma_{\text{surf}} < 2.3$ cm.

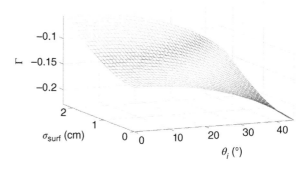

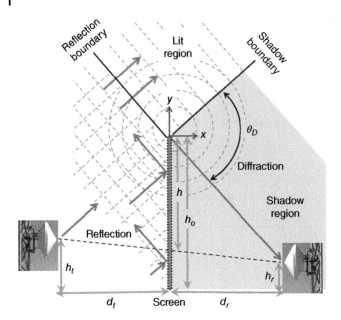

Figure 5.18 Plane wave incident on a single edge. Dashed lines are equal phase fronts.

5.6 Diffraction

Diffraction is electromagnetic scattering from an edge. Figure 5.18 shows a plane wave incident on a thin, perfectly conducting diffraction screen that extends from $y = 0$ to $-\infty$. The lit region contains the incident wave, while the shadow region does not. A shadow boundary divides these two regions. Diffracted waves exist in both the lit and shadow regions. The reflection boundary splits the lit region in two: one with incident, diffracted, and reflected waves and the other with only incident and diffracted waves. Reflected and transmitted waves obey Snell's laws, but diffraction requires substantially more complicated calculations.

5.6.1 Fresnel Diffraction

The diffracted electric field in the shadow region due to the edge of the screen in Figure 5.18 is given by [10]

$$E_{\text{diff}} = E_{\text{LOS}} F(v_F) \tag{5.27}$$

where $F(v_F)$ is the Fresnel integral:

$$F(v_F) = \int_0^v e^{-j\pi\xi^2/2} d\xi = C(v_F) + jS(v_F) \tag{5.28}$$

and

E_{LOS} = LOS electric field at the receiver

$$v_F = \theta_D \sqrt{\frac{2d_t d_r}{\lambda(d_t + d_r)}}$$

d_t = obstacle distance from the transmitter

d_r = obstacle distance from the receiver

$$\theta_D = \tan^{-1}\left(\frac{h_o - h_t}{d_t}\right) + \tan^{-1}\left(\frac{h_o - h_r}{d_r}\right)$$

h_o = obstacle height above the ground

h_t = transmit antenna height above the ground

h_r = receive antenna height above the ground

$C(v) = \int_0^v \cos(0.5\pi\xi^2)d\xi$ = Fresnel cosine integral

$S(v) = \int_0^v \frac{\sin(\xi)}{\xi}d\xi$ = Fresnel sine integral

Note that $S(\infty) = C(\infty) = 0.5$, $S(-\infty) = C(-\infty) = -0.5$, and $S(0) = C(0) = 0.0$. Figure 5.19a has plots of the two Fresnel integrals as a function of v. As v gets large, the oscillations due to the diffracted field attenuate. The Cornu spiral plot in Figure 5.19b has centers at $(-0.5, -0.5)$ and $(0.5, 0.5)$ which correspond to the observation point at an infinite distance from the edge where oscillations due to the diffracted field cease. The diffraction attenuation is [11]

$$L_{\text{diffdB}} = -20 \ \log \left(\frac{\sqrt{[1 - C(v) - S(v)]^2 + [C(v) - S(v)]^2}}{2}\right) \tag{5.29}$$

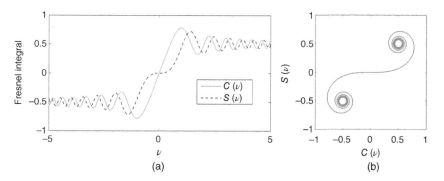

Figure 5.19 Plots of the Fresnel integrals. (a) Cosine and sine integrals and (b) cornu spiral.

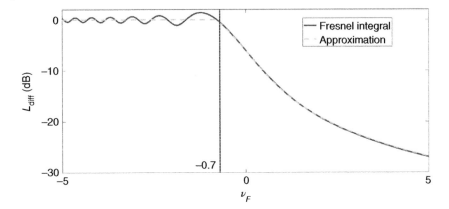

Figure 5.20 Fresnel integral diffraction loss compared with approximation.

The loss in (5.29) can be estimated in dB by [12]

$$L_{\text{diffdb}} \approx \begin{cases} 0 & v_F < -0.78 \\ 6.9 + 20\log\left(\sqrt{(v_F - 0.1)^2 + 1} + v_F - 0.1\right) & v_F \geq -0.78 \end{cases} \quad \text{in dB}$$

(5.30)

Example
Plot (5.29) and (5.30) on the same axis.

Solution
Figure 5.20 shows the comparison.

Since the multipath signal travels a distance greater than R_{LOS}, the LOS signal arrives first, then multipath signals arrive in a time proportional to the distance traveled. In addition, each time the multipath signal reflects from a surface, it attenuates and receives a phase shift. All single bounce multipath signals traveling the same distance have the same phase. The reflected signal has equal phase contours that form a Fresnel ellipsoid in three dimensional space with transmit and receive antennas at the two foci. The sum of the distances from any point on one of the equal phase ellipses to those two foci equals $R_{\text{LOS}} + n\lambda/2$ where n is an integer. In other words, paths A and B in Figure 5.21 equal $R_{\text{LOS}} + \lambda/2$ ($n = 1$) while path C is $\lambda/2$ longer ($n = 2$). The scattered field from any object lying within the first Fresnel zone arrives at the receive antenna with a phase difference from the LOS signal between 180° and 360°. An object in the second Fresnel zone causes a phase shift in the scattered field between 360° and 540°. Thus, objects close to odd numbered Fresnel zone ellipses result in the LOS and multipath signals adding in phase. Objects close to even numbered Fresnel zone ellipses result in the LOS and multipath signals adding out of phase.

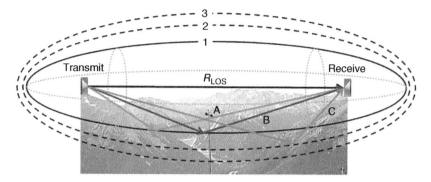

Figure 5.21 Fresnel ellipsoids.

The radius of the nth Fresnel zone ellipse at a point that is d_t from the transmitter and d_r from the receiver is

$$r_n = \sqrt{\frac{n\lambda d_t d_r}{d_t + d_r}} \tag{5.31}$$

Figure 5.22 shows the Fresnel zones mapped onto a diagram of the environment of a microwave link. Odd Fresnel zones experience constructive interference, while even zones have destructive interference. Usually, up to 20% blockage in zone 1 introduces a small loss in the link [13]. More than 40% zone 1 blockage introduces significant signal loss. This analysis breaks down when the separation distance between antennas causes the Earth curvature to enter zone 1.

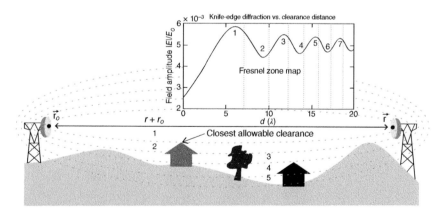

Figure 5.22 The geometry of Fresnel-zone ellipsoids clearing obstacles in a microwave link. The Fresnel zones are labeled moving from the first ellipsoid outward. *Source:* Reprinted by permission of Ref. [14]; ©2009 IEEE.

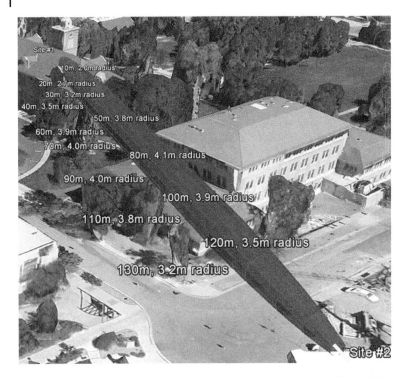

Figure 5.23 A 3D Fresnel zone for a microwave link between two buildings. [15]. Image obtained using Google Earth.

Example

A 715.4 MHz communication link between the tops of two buildings has Site #1 located 15 m above ground at 39°.45′3.3″N, 105°13′22.1″W. Site #2 is located 15 m above ground at 39°44′57.7″N 105°13′21.9″W. These locations were found using Google Maps. Plot the first Fresnel zone ellipsoid for this link using 3-D Fresnel Zone found at https://www.loxcel.com/3d-fresnel-zone.

Solution

The software generates a KML file that viewed in Google Earth as shown in Figure 5.23. The Fresnel zone plots determine whether buildings enter the first Fresnel zone of a communications link. The antennas may have to be raised (or trees trimmed) if obstacles lie inside the first Fresnel zone.

5.6.2 Diffraction from Multiple Obstacles

Signals diffract from edges and the Fresnel integral in (5.28) provides a relatively simple way to calculate diffraction from a single edge. Many times, the

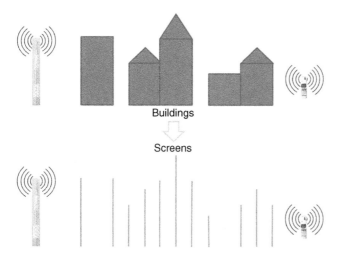

Figure 5.24 Corners, edges, and peaks of buildings are converted to screens.

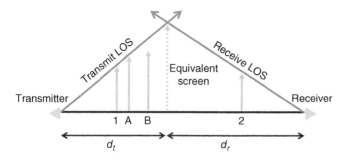

Figure 5.25 Bullington method replaces all of the screens with an equivalent screen.

signal encounters multiple edges that make the diffraction calculations more complicated. Several methods approximate all building edges with diffraction screens as shown in Figure 5.24. The simplest approach, the Bullington method, replaces all of the diffraction screens with one equivalent screen (dashed arrow in Figure 5.25) using the following steps [16]:

1. Draw straight lines from the transmitter through the top of all the screens. The line with the steepest slope passes through the tallest screen in the direction of the receiver, such that all other screens fall below this line.
2. Draw lines from the receiver through the tops of all the screens. The line with the steepest slope passes through the tallest screen in the direction of the transmitter, such that all other screens fall below this line.
3. These two lines intersect at the location and height of the Bullington screen.

4. Calculate the Fresnel attenuation only for the Bullington screen while ignoring all others.

This approach works well for two closely spaced screens with no other significant obstacles.

Example

There are three knife edge obstacles between two antennas in a communications system that operates at 600 MHz:

Object	Height (m)	Distance
Antenna 1	10	0 m
Obstacle A	40	7 km
Obstacle B	60	12 km
Obstacle C	30	22 km
Antenna 2	10	26 km

Find the diffraction loss using Bullington's method.

Solution

Assume x is the distance in km from antenna 1.

$$\frac{30}{7}x = \frac{20}{4}(26 - x)$$

$$x = 14 \text{ km}$$

The height of the Bullington screen is $h = 30\frac{183}{13 \times 7} = 60$ m. Next, the Fresnel parameter, v_F, is found from

$$v_F = h\sqrt{\frac{2}{\lambda}\left(\frac{1}{d_t} + \frac{1}{d_r}\right)} = 60\sqrt{\frac{2}{.5}\left(\frac{1}{14\,000} + \frac{1}{12\,000}\right)} = 1.49$$

where $\lambda = 3 \times 10^8/600 \times 10^6 = 0.5$ m. Now use (5.30) to get the attenuation due to the screen.

$$L_{\text{diffdb}} = 6.9 + 20\log_{10}\left(\sqrt{(1.49 - 0.1)^2 + 1} + 1.49 - 0.1\right) = 16.7 \text{ dB}$$

The screen attenuation is added to the free space attenuation to get the total attenuation.

In general, approaches to multiple screen diffraction loss divide the problem into many one screen diffraction problems. The first step starts by calculating the diffraction loss when the transmitter radiates from the first screen to the second screen. Next, the diffraction loss is calculated from the first screen to

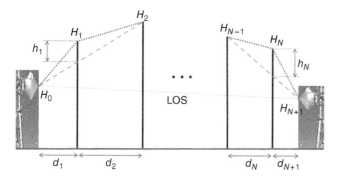

Figure 5.26 Diagram of the Epstein–Peterson multiple screen diffraction model.

the third screen with the second screen in between. This process continues until the last screen diffraction loss is calculated. The final loss due to all the screens is just the sum (in dB) of all the diffraction losses due to the individual screens.

Figure 5.26 diagrams the Epstein–Peterson model [17] with the height and distance variable labeled find the attenuation using the following steps [18]:

1. Calculate the clearance heights from the screen/antenna heights before and after the current diffraction screen:

$$h_n = H_n - H_{n-1} - \frac{d_n(H_{n+1} - H_{n-1})}{d_n + d_{n+1}} \tag{5.32}$$

2. Calculate the Fresnel diffraction parameter for screen n:

$$v_{Fn} = h_n \sqrt{\frac{2(d_n + d_{n+1})}{\lambda d_n d_{n+1}}} \tag{5.33}$$

3. The total diffraction loss in dB due to all the screens is found from (5.30):

$$L_{\text{diffdB}} = \sum_{n=1}^{N} 6.9 + 20 \log \left(\sqrt{(v_{Fn} - 1)^2 + 1} + v_{Fn} - 0.1 \right) \text{ in dB} \tag{5.34}$$

Example
There are three knife edge obstacles between two antennas in a communications system that operates at 600 MHz (Table 5.2):
Find the diffraction loss using the Epstein-Peterson model.

Solution
Substitute into (5.32), (5.33), and (5.34) to get the values in Table 5.2. Adding all the losses from the individual screens results in a total loss of

Table 5.2 Given and calculated values for the Epstein–Peterson example.

Object	Given Height (m)	Distance	Calculated h_n	v_{Fn}	L_{diff}(dB)
Antenna 1	10	0 m			
Obstacle A	40	7 km	11.6	0.3483	9.0
Obstacle B	60	12 km	23.5	0.5340	10.6
Obstacle C	30	22 km	7.1	−0.1298	4.9
Antenna 2	10	26 km			

$9.0 + 10.6 + 4.9 = 24.5$ dB. The Epstein–Peterson model predicts much higher diffraction attenuation than the Bullington model for this case.

Many other multiple screen diffraction methods exist. The Deygout method tends to overestimate loss, especially for multiple closely spaced screens [19]. Its results are best for one dominant screen. This approach beats the Bullington and Epstein–Peterson methods in highly obstructed paths. These approximate methods have been surpassed by much more realistic computer models.

5.6.3 Geometrical Theory of Diffraction

Joseph Keller combined GO and Fresnel diffraction into the geometrical theory of diffraction (GTD) [20].The total electric field equals the GO electric field (incident, refracted, and reflected waves) and the diffracted electric field [21].

$$\mathbf{E}_T = \mathbf{E}_{GO} + \mathbf{E}_{diff} \tag{5.35}$$

Figure 5.27 shows the total, GO, and diffracted amplitude and phase patterns due to a TE plane wave incident on a perfectly conducting screen (at a 60° angle relative to the screen). The discontinuities in the diffracted field exactly compensate for the discontinuities in the GO field to produce a smooth total electric field. The diffracted field has circles of constant phase that are centered on screen's edge. Diffraction occurs at the discontinuity between two materials. The TM version appears in Figure 5.28. These figures show that diffraction depends on polarization. Note the differences and similarities between the two polarizations.

Fresnel diffraction from screens approximates the actual diffraction from obstacles with sharp edges in the channel. The GTD combines with Fresnel diffraction through a more rigorous formulation that includes multiple edge diffractions and obstacle interactions [20]. Shooting and bouncing rays (SBRs) technique practically implements ray tracing using GO and diffraction [22]. It launches rays in directions and at amplitudes proportional to the antenna pattern in those directions. All the LOS, diffracted, refracted, and reflected waves that reach the receiver sum to obtain a realistic received signal.

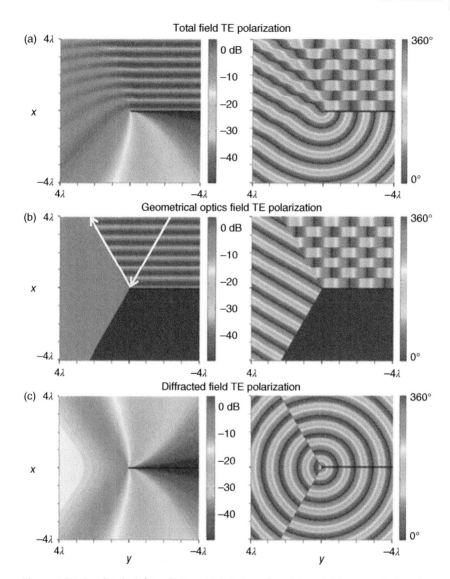

Figure 5.27 Amplitude (left) and phase (right) plots of the (a) total, (b) geometrical optics, and (c) diffracted fields from a screen for TE polarization (electric field perpendicular to the plane of the plots). Source: Reproduced with permission of IEEE.

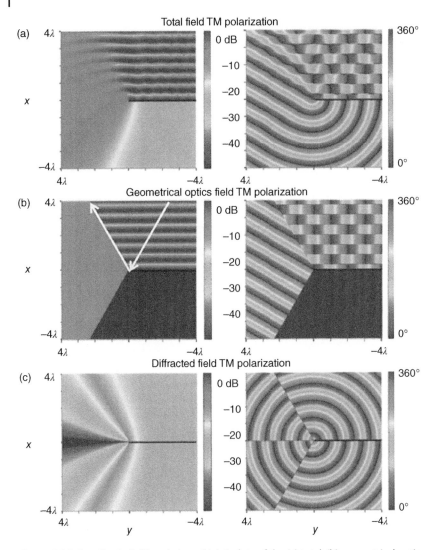

Figure 5.28 Amplitude (left) and phase (right) plots of the (a) total, (b) geometrical optics, and (c) diffracted fields from a screen for TM polarization (magnetic field perpendicular to the plane of the plots). Reprinted by permission of Durgin [14]. © 2009 IEEE.

GTD calculates the polarization-dependent diffracted field from edges and adds the result to the incident and reflected fields to get the total field. Elementary GTD assumes the diffracting edge consists of two infinite half planes that meet to form a wedge as shown in Figure 5.29. The GTD diffracted electric field for an incident plane wave with the electric field either parallel (TM) or

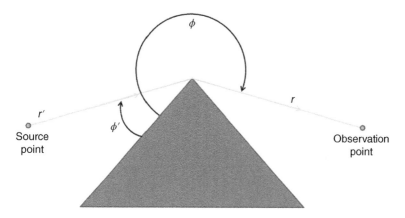

Figure 5.29 Geometry for application of GTD wedge diffraction.

perpendicular (TE) to the edge of a perfectly conducting wedge is [23]

$$\mathbf{E}_{\text{diff}} = \underbrace{\mathbf{E}_i}_{\text{Field at edge}} \times \underbrace{D_{\text{TM}}(\zeta, \phi, \phi')}_{\text{Diffraction coefficient}} \times \underbrace{\frac{1}{\sqrt{r}}}_{\text{Spreading factor}} \times \underbrace{e^{-jkr}}_{\text{Phase}} \qquad (5.36)$$

where

$\mathbf{E}_i$ = incident electric field

r' = distance from source point to diffracting edge

r = distance from diffracting edge to field point

$D_{\text{TM}}(\zeta, \phi, \phi')$ = diffraction coefficient for a finitely conducting wedge
TE

ζ = distance parameter defined by type of wave (planar, spherical, etc.)

ϕ' = incidence angle, measured from incidence face

ϕ = diffraction angle, measured from incidence face

More details and examples of using GTD appear in the literature [20, 23].

Assume that an FM radio station transmits 250 kW at 91.1 MHz. To reach one coverage area, the station must rely on double diffraction over the top of two terrain peaks of equal altitude, which can be modeled as 90° wedges (Figure 5.30). The incident field at A and the two diffraction coefficients at B and C are

At point A: $P_r = \dfrac{P_t G_t G_r \lambda^2}{16\pi^2 r_1^2}$

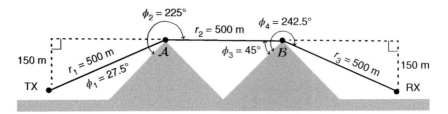

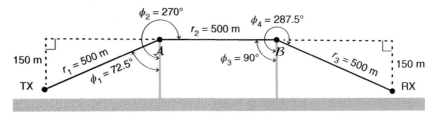

Figure 5.30 An example of FM signal propagation that must diffract over two terrain peaks. Reprinted by permission of Durgin [14]. © 2009 IEEE.

At point B: $P_r = \dfrac{P_t G_t G_r \lambda^3 \left| \underset{TE}{D_{TM}}(L, \phi_2, \phi_1) \right|^2}{32\pi^3 r_1 r_2 (r_1 + r_2)}$

At receiver: $P_r = \dfrac{P_t G_t G_r \lambda^4 \left| \underset{TE}{D_{TM}}(L, \phi_2, \phi_1) \right|^2 \left| \underset{TE}{D_{TM}}(L, \phi_4, \phi_3) \right|^2}{64\pi^4 r_1 r_2 r_3 (r_1 + r_2 + r_3)}$

For this example [14], the LOS signal at the receiver is −11.9 dBm. If the edges are modeled as perfectly conducting screens, then the signal at the receiver for TE is −85.5 dBm and for TM is −80.2 dBm. Modeling the edges as 90° wedges results in a signal level at the receiver of −88.8 dBm for TE and −76.4 dBm for TM.

5.7 Signal Fading

Fading occurs when the receive signal level drops below the minimum detectable level of the receiver due to

- Path loss
- Fluctuations in the received signal power
- Fluctuations in signal phase

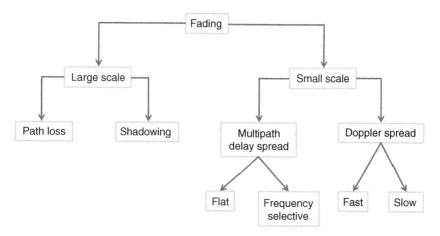

Figure 5.31 Categories of fading.

- Variations in the angle of arrival of the received signal
- Reflection and diffractions from objects
- Frequency shift due to the Doppler effect

The fade margin equals the difference in dB between the received signal strength and the receiver sensitivity. A reliable link has a high fade margin. As an example, a 2.4 GHz, industrial, scientific, and medical (ISM) band radio should have at least a 15 dB fade margin, and a 5 GHz ISM/Unlicensed National Information Infrastructure (U-NII) band radio should have at least a 10-dB fade margin [24]. Figure 5.31 breaks down the different categories of fading that will be discussed in the following paragraphs.

Large-scale fading means that the signal amplitude slowly changes over distances greater than a wavelength. Small-scale fading, on the other hand, implies that signal fades occur quickly over small distances on the order of a wavelength. The differences between these two types of fading are shown by signal power plots at the bottom of Figure 5.32. Small-scale and large-scale fading add together to get a total signal level shown in the top plot of Figure 5.32.

Two types of large-scale fading are path loss and shadowing. Path loss attenuates a signal by $1/R^\gamma$ as previously discussed. Eventually, a signal fades when R becomes large. Shadowing causes large-scale fading when a large object (e.g. building or mountain) blocks the LOS signal, but the diffracted signal exists and slowly decays in the shadow region.

Small-scale fading splits into two categories: multipath delay spread and Doppler spread [25]. Flat fading means that a multipath channel has a bandwidth greater than the symbol period (Figure 5.33). All frequency components in a flat-fading channel have the same amount of fading. Flat fading preserves the spectral characteristics of the transmitted signal. The signal bandwidth is

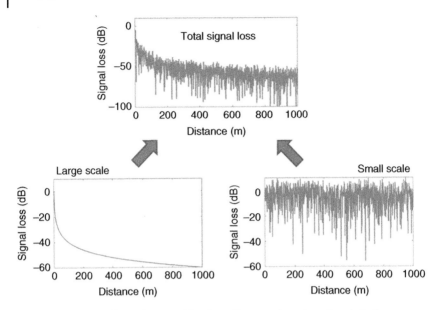

Figure 5.32 The total signal is a combination of large-scale and small-scale fading.

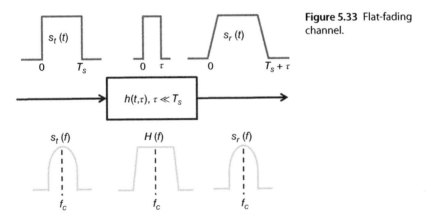

Figure 5.33 Flat-fading channel.

less than the channel bandwidth. Frequency-selective fading (time dispersion) occurs when a multipath channel has a bandwidth that is less than the signal bandwidth (Figure 5.34). Not all signal frequencies experience the same fading. The signal passing through it spreads out in time and has a reduced bandwidth.

Attenuation and phase variations in the channel transfer function produce time-varying distortion in the received signal. Fast and slow fading describe the time rate of change in the channel transfer function and the transmitted signal. Coherence time approximates the time duration when the channel impulse response remains constant. Two received signals have a strong

Figure 5.34 Frequency selective fading.

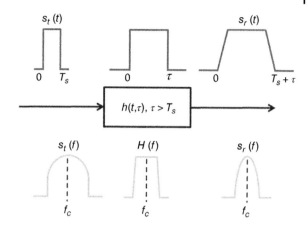

amplitude correlation during the coherence time. Doppler spread refers to the increased bandwidth due to the Doppler spectrum that results from channel/transmitter/receiver motion. Fast fading has a coherence time longer than the symbol period. Signal distortion increases as the Doppler spread increases relative to the transmitted signal bandwidth. Slow fading has a static channel over at least one symbol period. This means that the channel Doppler spread is less than the baseband signal bandwidth.

5.7.1 Small-Scale Fading Models

Small-scale fading has fast variations in the signal over wavelength-size distances. Signal levels appear to be random and are very difficult to precisely predict. Experimental measurements under various practical conditions produced data bases that have resulted in statistical models that have proved extremely useful in system design.

5.7.1.1 Rayleigh Fading

Consider the case of many signals arriving in a region via different paths. Each signal has a different amplitude and phase at the receiver due to the differences in path lengths as well as the diversity of reflections, diffractions, and refractions encountered. If the 2 GHz signals in Table 5.3 converge over a $6\lambda \times 6\lambda$ area, they form an interference pattern. Figure 5.35a,b show the real and imaginary parts of the total electric field due to these multipath signals. Converting the real and imaginary parts of the total field to amplitude and phase produces the plots in Figure 5.35c,d. These plots clearly indicate that the signal over this small area dramatically oscillates. A mobile wireless user passing through this area of interference encounters large signal variations and fading.

Table 5.3 Ten 2 GHz multipath signals at a receiver.

Signals at the receiver
$0.9835\,e^{-jk(x\cos 0°\ +y\sin 0°)}e^{j0°}$
$0.7736\,e^{-jk(x\cos 71.2°\ +y\sin 71.2°)}e^{j341.6°}$
$0.9863e^{-jk(x\cos 310.7°\ +y\sin 310.7°)}e^{j283.1°}$
$0.8574e^{-jk(x\cos 354.0°\ +y\sin 354.0°)}e^{j311.9°}$
$0.8489e^{-jk(x\cos 59.0°\ +y\sin 59.0°)}e^{j62.3°}$
$0.6080e^{-jk(x\cos 215.0°\ +y\sin 215.0°)}e^{j27.0°}$
$0.9881e^{-jk(x\cos 3.2°\ +y\sin 3.2°)}e^{j216.3°}$
$0.5031e^{-jk(x\cos 139.2°\ +y\sin 139.2°)}e^{j60.5°}$
$0.6265e^{-jk(x\cos 15.9°\ +y\sin 15.9°)}e^{j264.0°}$
$0.7174e^{-jk(x\cos 344.4°\ +y\sin 344.4°)}e^{j147.0°}$

Example
Use MATLAB to make histogram plots of the field values in Figure 5.35.

Solution
When these plots were made, the values of the total electric field, E_T, were saved over a $6\lambda \times 6\lambda$ area. The MATLAB command hist(real(ET), 30), where ET contains the sum of all the signals in Table 5.3, results in the plots in Figure 5.36. Gaussian curves nicely fit the real and imaginary histograms but not the amplitude and phase histograms (Table 5.4).

Note that the real and imaginary field values in Figure 5.36a,b resemble Gaussian probability density functions (PDFs). Figure 5.37 has two Gaussian PDF plots: one with $\mu = 0$ and a standard deviation of $\sigma = 1$ (also known as the standard normal distribution) and the other with $\mu = 2$ and $\sigma = 0.5$. The standard deviation describes how close the points cluster about the mean.

The PDFs in Figure 5.36c,d cannot be Gaussian, because

1. The amplitude is always greater than or equal to zero, so the PDF must begin at $x = 0$.
2. One phase is equally likely as another phase, so the PDF must be uniform.

It turns out that when the real and imaginary parts of a random variable have Gaussian PDFs with a standard deviation of σ, then the phase is a uniform PDF while the amplitude is a Rayleigh PDF takes the form of

$$\mathrm{PDF}(x \mid \sigma) = \frac{x}{\sigma^2}e^{-\left(\frac{x^2}{2\sigma^2}\right)} \tag{5.37}$$

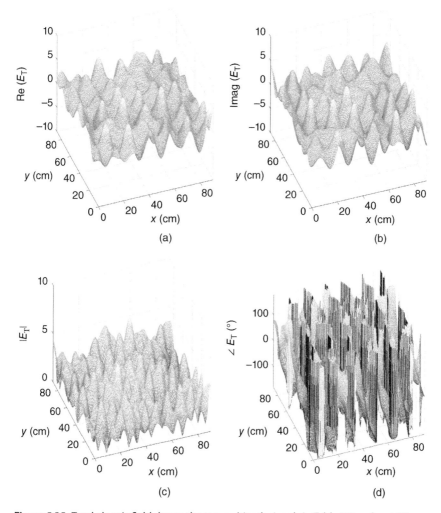

Figure 5.35 Total electric field due to the ten multipath signals in Table 5.3 and no LOS signal. (a) Real, (b) imaginary, (c) amplitude, and (d) phase.

Table 5.4 Statistics of the field data.

	Real	Imaginary	Amplitude	Phase
Mean	0.00	0.00	2.00	−0.08
Variance	2.73	2.73	1.05	3.35

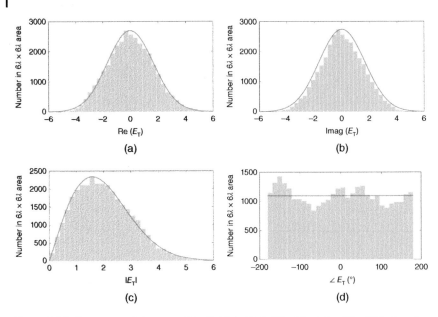

Figure 5.36 Histograms of the sum of the fields in Table 5.3 with plots of the PDFs found from the data statistics. (a) Real, (b) imaginary, (c) amplitude, and (d) phase.

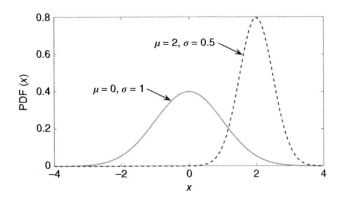

Figure 5.37 Two examples of Gaussian PDFs.

with its cumulative density function (CDF) given by

$$\text{CDF}(x \mid \sigma) = \int_0^x \frac{y}{\sigma^2} e^{-\frac{y^2}{2\sigma^2}} \, dy = 1 - e^{-x^2/(2\sigma^2)} \tag{5.38}$$

where σ is the mode or the location of the maximum value of the PDF. Figure 5.38 contains three examples of the Rayleigh PDF for $\sigma = 0.5$, 1.0, and 2.0. Some statistics of a Rayleigh distribution appear in Table 5.5. Adding

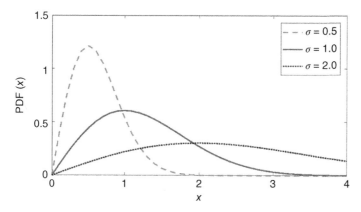

Figure 5.38 Rayleigh PDFs.

Table 5.5 Rayleigh PDF statistics.

Mean	$\sigma\sqrt{\pi/2}$
Mean square value	$2\sigma^2$
Median	$\sigma\sqrt{2\ln(2)}$
Variance	$0.429\sigma^2$

many multipath signals with no LOS signal results in a received signal with an amplitude that is Rayleigh distributed and a phase that is uniformly distributed.

Example
The average power in a Rayleigh faded signal is given by the mean square value in Table 5.5. What is the probability that the signal falls 10 dB below its mean square value?

Solution
Use the CDF in (5.38) to obtain:
$$p\left(\frac{x^2}{2\sigma^2} < \frac{1}{10}\right) = 1 - e^{-1/10} = 0.0952$$

5.7.1.2 Rician Fading
The previous example lacks a dominant or LOS signal. A Rayleigh model works well for modeling a smart phone in a big city where the base station hides behind buildings. Systems like satellite television have a dominant LOS signal with smaller multipath signals, the Rayleigh model does not work. What happens if the first signal in Table 5.3 increases to an amplitude of 6 while the other signals remain the same? Figure 5.39 has plots of the total field consisting

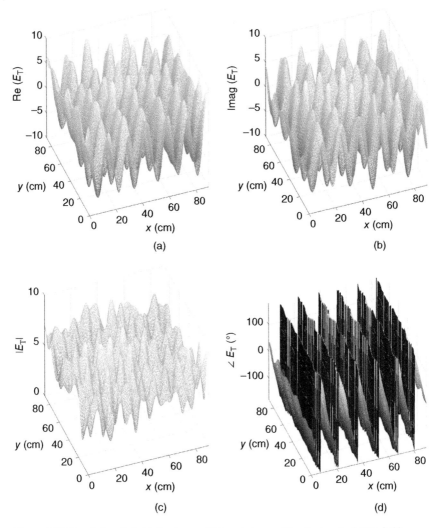

Figure 5.39 Total electric field due to one LOS and nine multipath signals. (a) Real, (b) imaginary, (c) amplitude, and (d) phase

of the LOS signal and 9 multipath signals (the bottom 9 signals in Table 5.3). Figure 5.39a,b show the real and imaginary parts of the total electric field due to these multipath and LOS signals. Converting the total field to amplitude and phase results in the plots in Figure 5.39c,d. These plots of the real, imaginary, and amplitude exhibits fading like those in Figure 5.35. The phase in Figure 5.39d exhibits a periodicity corresponding to the dominant LOS signal.

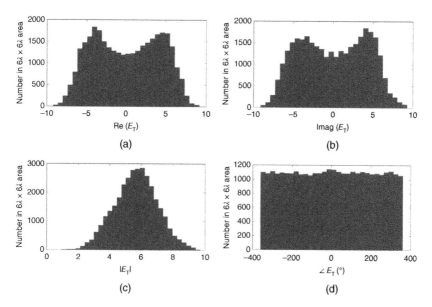

Figure 5.40 Histograms of the sum of nine multipath signals and a LOS signal. (a) Real, (b) imaginary, (c) amplitude, and (d) phase.

Example

Use MATLAB to make histogram plots of the field values in Figure 5.39.

Solution

E_T now contains the sum of the bottom 9 signals in Table 5.3 plus a LOS signal with an amplitude of 6, phase of 0, and arriving at 0°. Figure 5.40 plots the resulting histograms.

The real and imaginary field values in Figure 5.40 do not look like those in Figure 5.36. They have two peaks and cannot be modeled with Gaussian PDFs. The phase histogram looks uniform like the histogram in Figure 5.36d, but the amplitude histogram differs from that in Figure 5.36c. The LOS signal dominates all the other signals, so the mass of the PDF moves closer to the amplitude of the LOS signal. The new PDF in Figure 5.40c takes the form of a Rician PDF [26]

$$\text{PDF}(x \mid \sigma_{\text{MP}}, A_{\text{LOS}}) = \frac{x}{\sigma_{\text{MP}}^2} e^{-\frac{\left(x^2 + A_{\text{LOS}}^2\right)}{2\sigma_{\text{MP}}^2}} I_0\left(\frac{A_{\text{LOS}}x}{\sigma_{\text{MP}}^2}\right) \tag{5.39}$$

where

$$
\begin{aligned}
A_{\text{LOS}} &= \text{amplitude of the LOS signal} \\
2\sigma_{\text{MP}}^2 &= \text{average power of the non-LOS multipath signals} \\
I_0(\xi) = \frac{1}{2\pi}\int_0^{2\pi} e^{\xi\cos\theta} d\theta &= \text{zeroth order modified Bessel function of the first kind}
\end{aligned}
$$

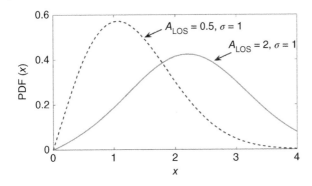

Figure 5.41 Two examples of Rician PDFs.

A graph of two Rician PDFs, both with $\sigma_{MP} = 1$, but one having $A_{LOS} = 0.5$ and the other with $A_{LOS} = 2$ appear in Figure 5.41. As A_{LOS} gets small, the Rician PDF approaches a Rayleigh PDF and as A_{LOS} gets large, the Rician PDF approaches a Gaussian PDF.

The Rice factor equals the ratio of the power in the LOS signal to the average power of the sum of the multipath signals [27]:

$$K_r = \frac{A_{LOS}^2}{2\sigma_{MP}^2} \tag{5.40}$$

If there is no LOS signal, then $A_{LOS} = 0$ and (5.39) becomes a Rayleigh PDF. A strong LOS signal converts (5.39) to an approximate Gaussian distribution with a mean of A_{LOS}. In Figure 5.41, $K_r = 2$ for the solid curve and $K_r = 0.125$ for the dashed curve. More advanced Rician models include two or more dominant signals, such as the LOS and a ground reflection.

5.7.2 Approximate Channel Models

The Rician and Rayleigh models apply to situations that have many multipath signals with and without a LOS signal. A number of refined models applicable to specific types of channels have been developed [28]. Channel models based on experimental measurements have interpolated between measurement points to estimate the signal loss due to commonly encountered wireless environments over specific frequency ranges. Of all the reported measurements, the Okumura report data became a standard [29]. This report presented graphs of the median field strength in a channel as a function of base station height, frequency, and vehicular station antenna height. The report contains data collected from 150 to 1920 MHz, at distances between 1 and 100 km, and with base station antenna heights from 30 to 200 m as well as over various types of urban areas in and around Tokyo.

Table 5.6 Parameters for the Hata model.

Propagation area	$a(h_r)$	C
Metropolitan	$8.29[\log(1.54h_r)]^2 - 1.1 \quad f_c \leq 200\,\text{MHz}$ $3.2[\log(11.75h_r)]^2 - 4.97 \quad f_c \geq 400\,\text{MHz}$	0
Small to medium cities	$(1.1\log f_c - 0.7)h_r - 1.56\log f_c + 0.8$	0
Suburban		$-2[\log(f_c/28)]^2 - 5.4$
Rural		$-4.78(\log f_c)^2 + 18.33\log f_c - 40.94$

Hata derived equations that fit the curves in the Okumura report [30]. The general Hata attenuation model curve fit is

$$L_{\text{Hata}} = A + B\log d + C \tag{5.41}$$

where d is the separation distance in km between the transmitter and receiver. The values of A, B, and C depend upon the propagation environment. Hata's model gives A and B as

$$A = 69.55 + 26.16\log f_c - 13.82\log h_t - a(h_r)$$
$$B = 44.9 - 6.55\log h_t \tag{5.42}$$

where the transmitter height (h_t) and receiver height (h_r) are in meters and the carrier frequency (f_c) is in MHz. The values of a and C given in Table 5.6 depend upon the propagation environment (column 1). Many other propagation models exist for different environments and parameters.

Example

A base station operates at 900 MHz at a height 30 m above the ground. If the receiver is 1 km away at a height of 2 m above ground, then find the path loss using the Hata model for (a) large city, (b) small city, and (c) rural area.

Solution

$$L_{\text{Hata}} = 69.55 + 26.16\log_{10}f_c - 13.82\log h_t - a(h_r) + (44.9 - 6.55\log h_t)$$
$$\log d + C = 126.42 - a(h_r) + C$$

(a) 125.37 dB
(b) 125.13 dB
(c) 96.62 dB

Indoor models predicting signal attenuation based on measurements also exist. For example the Motley–Keenan indoor model attenuation factor due to propagation in and through buildings is given by [31]

$$L_{\text{indoor}} = L_0 + 10\gamma \log\left(\frac{d}{d_0}\right) + L_{\text{wall}} + L_{\text{floor}} \text{ in dB} \qquad (5.43)$$

where

L_0 = path loss at d_0 (dB)
d = separation distance (m) between the transmitter and receiver $(d > 1\,\text{m})$
γ = power loss exponent
L_{wall} = wall attenuation (dB)
L_{floor} = floor attenuation (dB)

If the reference distance is $d_o = 1\,\text{m}$ and assuming free space propagation, then $L_0 = 20 \log 10 f_{\text{MHz}} - 28$ where f_{MHz} is in MHz. Table 5.7 lists measured values of γ for various types of buildings at different frequencies. Table 5.8 lists wall attenuation through commonly used materials in walls. Table 5.9 lists some floor attenuation factors in different scenarios and for different frequencies. Both L_{wall} and L_{floor} in (5.43) include all walls and floors that the signal passes through.

Example
A 900 MHz signal enters a building through a 13-mm thick window. It passes through one concrete wall that is 102 mm thick and one floor. The signal travels a total of 10 m in the building. Estimate the attenuation.

Solution
Let $d_0 = 1\,\text{m}$ and substitute the values from the tables into (5.43).

$$L_{\text{indoor}} = 20 \log(900) - 28 + 10(3.3) \log\left(\frac{10}{1}\right) + 2 + 12 + 9 = 87.1 \text{ dB}$$

5.7.3 Large-Scale Fading

Obstacles in a channel block the LOS signal and create shadows (Figure 5.18). Shadow regions contain the diffracted signal whose amplitude is considerably lower than the LOS signal. Obstacles in a cell change the ideal circular coverage into an amoeba-like coverage area shown in Figure 5.42. Mobile users passing behind an obstacle experience a fade in the obstacle's shadow.

The attenuated signal in a shadow appears random with a log-normal PDF. The PDF is based on the ratio of the received power to the transmit power in dB:

$$\text{PDF}_A = \frac{1}{\sqrt{2\pi}\sigma_{\chi_{\text{dB}}}} e^{-\frac{\left(\chi_{\text{dB}} - \mu_{\chi_{\text{dB}}}\right)}{2\sigma^2_{\chi_{\text{dB}}}}} \qquad (5.44)$$

Table 5.7 Power loss exponents, γ, for indoor transmission loss calculation in dB [32].

Frequency (GHz)	Residential	Office	Commercial	Factory
		γ		
0.8	—	2.25	—	—
0.9	—	3.3	2.0	—
1.25	—	3.2	2.2	—
1.9	2.8	3.0	2.2	—
2.1	—	2.55	2.0	2.11
2.2	—	2.07	—	—
2.4	2.8	3.0	—	—
2.625	—	4.4	—	3.3
3.5	—	2.7	—	—
4	—	2.8	2.2	—
4.7	—	1.98	—	—
5.2	3.0 2.8	3.1	—	—
5.8	—	2.4	—	—
26	—	1.95	—	—
28	—	1.84 2.99	2.76 1.79 2.48	—
37	—	1.56	—	—
38	—	2.03 2.96	1.86 2.59	—
51–57	—	1.5	—	—
60	—	2.2	1.7	—
67–73	—	1.9	—	—
70	—	2.2	—	—
300	—	2.0	—	—

where

$$\chi = P_r/P_t$$
$$\chi_{\mathrm{dB}} = 10\log(P_r/P_t)$$
$$\mu_{\chi\mathrm{dB}} = \text{mean of } \chi_{\mathrm{dB}} \text{ in dB}$$
$$\sigma_{\chi\mathrm{dB}} = \text{standard deviation of } \chi_{\mathrm{dB}} \text{ in dB : typically } 4 \leq \sigma_{\chi\mathrm{dB}} \leq 13 \text{ dB}$$

Figure 5.43 plots χ_{dB} vs. separation distance in dB. Each point on the graph represents the receive signal sample for a constant transmitter power. A linear interpolation of the scatter plot has a slope of $-10\gamma\log d$ where γ is the power loss exponent. Points follow a Gaussian PDF with a mean of $-10\gamma\log d$ and a standard deviation of $\sigma_{\chi_{\mathrm{dB}}}$. The mean of χ_{dB} depends on path loss and obstacle

Table 5.8 Approximate attenuation through a single wall at 900 MHz [32].

Material	Thickness	L_{wall} (dB)
Glass	6	0.8
Glass	13	2
Lumber	76	2.8
Brick	89	3.5
Brick	267	5
Concrete	102	12
Concrete	203	23
Reinforced concrete	203	27
Masonry block	610	28
Concrete	305	35

Table 5.9 Floor penetration loss factors (L_{floor}) for N_{floor} floors [32].

	L_{floor} (dB)		
Frequency (GHz)	Residential	Office	Commercial
0.9	—	9 ($N_{floor} = 1$) 19 ($N_{floor} = 2$) 24 ($N_{floor} = 3$)	—
1.8–2	$4\,N_{floor}$	$15 + 4\,(N_{floor} - 1)$	$6 + 3\,(N_{floor} - 1)$
2.4	$10^{a)}$ (apartment) 5 (house)	14	—
3.5	—	18 ($N_{floor} = 1$) 26 ($N_{floor} = 2$)	—
5.2	$13^{a)}$ (apartment) $7^{b)}$ (house)	16 ($N_{floor} = 1$)	—
5.8	—	22 ($N_{floor} = 1$) 28 ($N_{floor} = 2$)	—

a) Per concrete wall.
b) Wooden mortar.

properties. The total fade margin for a wireless system is the large-scale fade margin plus the small-scale fade margin in dB.

The decorrelation distance in a shadow, X_c, corresponds to the maximum distance in which two signals remain correlated. Two signal samples become decorrelated when their separation distance causes the signal covariance to

Figure 5.42 A free space cell has a circular boundary while a shadowed cell has an irregular cell boundary.

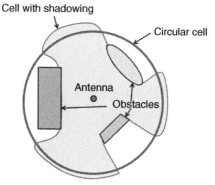

Figure 5.43 Log normal fading with propagation loss.

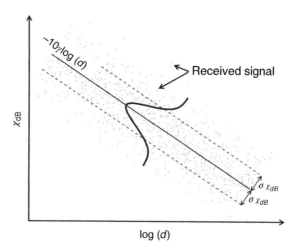

drop $1/e$ below its maximum. A simplified model for the signal covariance is given by [33]

$$\text{cov}(d) = \sigma^2_{\chi_{\text{dB}}} e^{-d/X_c} \tag{5.45}$$

where d is the separation distance between two points in the shadow. The decorrelation distance approximately equals the width of the obstacle.

5.7.4 Channel Ray-Tracing Models

Very accurate computer models use numerical techniques to solve Maxwell's equations. These models have long run times for realistic environments, work best at low frequencies, and only work well for small regions. Consequently, approximations and statistics supplement these approaches. One popular high-frequency approximation, ray tracing, finds practical use in modeling

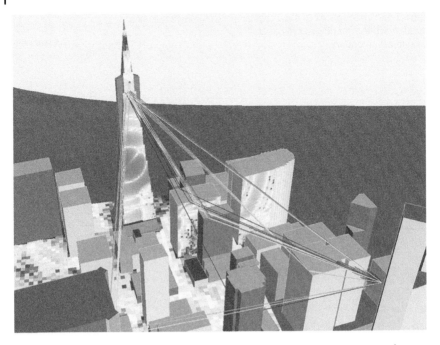

Figure 5.44 Ray tracing in an urban canyon. *Source:* Reproduced with permission of Remcom.

complicated scenarios. A transmitter emits a ray that propagates into a region of interest, such as a city. The ray attenuates, reflects, and diffracts as it takes diverse paths and propagates away from the transmitter. The SBR technique launches a bundle of rays with amplitudes weighted by the transmit antenna pattern [22]. GO and diffraction determine the direction and scattering of a ray. Figure 5.44 illustrates an example of the SBR technique used in Wireless Insight [34] that models an antenna transmitting from a tall building. After positioning transmit and receive antennas, the algorithm only calculates the rays that propagate to the receiver. When the amplitude of a ray drops below a user-defined threshold, it is ignored. SBR in a complex environment is computationally intensive, because a large number of rays must be launched and tracked.

SBR has many applications to wireless system modeling. The example in Figure 5.45 models Wi-Fi coverage in a house due to a transmitter on the first floor. This model helps determine the optimum router location in order to obtain coverage throughout the house. The scenario in Figure 5.46 shows an antenna on top of a building that transmits into another building. Some regions have good coverage while others do not. This type of model helps determine the location of an indoor signal boosting system that provides uniform coverage inside the building.

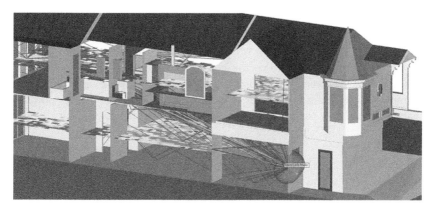

Figure 5.45 Wi-Fi coverage in a house with the router on the first floor. *Source:* Reproduced with permission of Remcom.

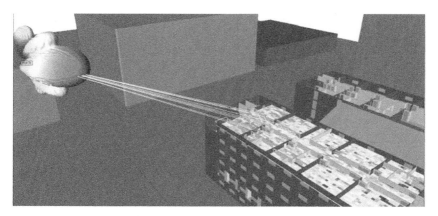

Figure 5.46 Predicting interior coverage from an exterior transmitter. *Source:* Reproduced with permission of Remcom.

5.8 Doppler Effects

When the transmitter and/or receiver moves the transmitted signal experiences a Doppler shift that translates the signal to a lower or higher frequency. Most people associate Doppler shift with sound. A train approaching the listener has a higher pitch than a train going away from a listener. Wireless systems experience Doppler in low-orbit satellite and cellular communication systems due to the velocities of the transmitter and receiver. Doppler causes the carrier frequency to shift higher when the transmitter approaches, and lower when it recedes. The Doppler shifted frequency is calculated using:

$$f_D = \left(\frac{kc - \mathbf{v}_r \cdot \hat{\mathbf{k}}_{tr}}{kc - \mathbf{v}_t \cdot \hat{\mathbf{k}}_{tr}} \right) f_c \tag{5.46}$$

where

$\mathbf{v}_r$	=	receiver velocity vector
$\mathbf{v}_t$	=	transmitter velocity vector
$\hat{\mathbf{k}}_{tr}$	=	unit propagation vector from transmitter to receiver

The difference between the Doppler shifts in signals coming from the same transmitter via two different paths is the Doppler spread. Frequency-shifted signals from multiple paths interfere with each other and cause fading.

Example

A car traveling at 25 m/s receives a 2 GHz signal from a stationary transmitter. Find the received signal frequency when the car travels (a) directly toward the transmitter, (b) directly away from the transmitter, and (c) in a circle with the transmitter as the center.

Solution

The wavelength is $\lambda = \frac{3\times10^8}{2\times10^9} = 0.15$ m

Using (5.46):

(a) $f_D = \left(\dfrac{2 \times 10^9 - 25\hat{\mathbf{x}} \cdot (-\hat{\mathbf{x}})/0.15}{2 \times 10^9} \right) 2 \times 10^9 = 2 \times 10^9 + \dfrac{25}{0.15}$
$= 2.0\,000\,001\,667$ GHz

(b) $f_D = \left(\dfrac{2 \times 10^9 - 25\hat{\mathbf{x}} \cdot (\hat{\mathbf{x}})/0.15}{2 \times 10^9} \right) 2 \times 10^9 = 2 \times 10^9 - \dfrac{25}{0.15}$
$= 1.999\,999\,833$ GHz

(c) $f_D = \left(\dfrac{2 \times 10^9 - 25\hat{\mathbf{x}} \cdot (\hat{\mathbf{y}})/0.15}{2 \times 10^9} \right) 2 \times 10^9 = 0$

This frequency shift looks small, but if the maximum and minimum Doppler shifted signals are added together, the resultant signal experiences fading. Doppler shift occurs when the transmitter and receiver move together or apart. No Doppler shift occurs when the separation distance remains constant.

Fading also takes place when the multipath signals and the LOS signal have different Doppler shifts. For example if the transmitting car in Figure 5.47 moves at a velocity $\mathbf{v}_t$ and the receiving car moves at a velocity $\mathbf{v}_r$, then the LOS signal ($\hat{\mathbf{k}}_{tr}$) and the multipath signal ($\hat{\mathbf{k}}_{mp}$) have different Doppler shifts according to (5.46). These two signals add together to produce fading.

Doppler fading occurs when adding two sinusoids of different frequencies – one due to LOS ($f_c + f_{D1}$) and one due to multipath ($f_c + f_{D2}$). The resulting signal takes the form

$$\cos \left[2\pi(f_c + f_{D1})t \right] + \cos \left[2\pi(f_c + f_{D2})t \right] = 2 \cos \left(2\pi f_{low} t \right) \cos \left(2\pi f_{high} t \right) \tag{5.47}$$

Figure 5.47 Doppler effect due to moving and stationary platforms and obstacles.

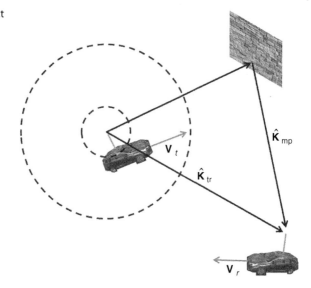

where

$$f_{low} = \frac{f_{D1} - f_{D2}}{2}$$

$$f_{high} = \frac{2f_c + f_{D1} + f_{D2}}{2}$$

The low-frequency signal modulates the high-frequency signal in (5.47). Thus, the envelope of the modulated signal goes to zero or fades every $1/|f_{D1} - f_{D2}|$. A beat frequency is the difference between two closely spaced frequencies, $|f_{D1} - f_{D2}|$. When tuning a guitar, the guitarist adjusts the string tension in order to minimize beat frequency between the plucked string and a tuning fork.

Example

A 1 GHz LOS signal arrives at the receiver with a multipath signal at 1.075 GHz. Plot the resultant signal that shows the beat period in the time domain.

Solution

A time vector was defined between 0 and 20 ns in MATLAB. The LOS and multipath signals were summed and graphed: $\cos(2\pi \times 10^9 t) + \cos(2\pi \times 1.075 \times 10^9 t)$ for $0 \le t \le 20$ ns. The beat period is given by $1/|1.075 \times 10^9 - 1.0 \times 10^9| = 13.3$ ns (Figure 5.48).

The Doppler spectrum extends from $f_c - f_{D_{max}}$ to $f_c + f_{D_{max}}$, where the Doppler spread for a stationary transmitter and moving receiver (v_r) is defined as the

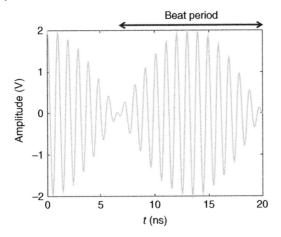

Beat period

Figure 5.48 Beat frequency.

maximum Doppler frequency

$$f_{D_{max}} = \frac{v_r}{\lambda} \tag{5.48}$$

Doppler shift due to any velocity translates into frequencies within the Doppler spectrum. If the baseband signal has a bandwidth greater than $2f_{D_{max}}$, then the Doppler spread has little impact on the received signal, and it is a slow fading channel. A high Doppler spread produces fast fading and requires a receiver that tolerates a fade within one symbol. As long as the symbol rate exceeds $f_{D_{max}}$, then Doppler in the channel can be ignored.

Coherence time, T_c is the inverse of Doppler spread [35].

$$T_c = \frac{1}{f_{D_{max}}} \tag{5.49}$$

A low Doppler spread produces slow fading $(T_c \gg T_s)$, so the impulse response remains invariant over the coherence time. Two received signals have a strong amplitude correlation if they are less than T_c apart. If $T_s > T_c$, then the channel changes during the transmission of the baseband symbol and results in a distorted symbol recovered by the receiver. Two signals arriving more than T_c apart encounter different channel effects. Fading due to the Doppler shift is called time selective fading. $(T_c < T_s)$. A more conservative estimate of T_c is given by

$$T_c = \frac{9}{16\pi f_{D_{max}}} \tag{5.50}$$

The coherence time in (5.49) is longer than in (5.50). A common compromise between the two is given by

$$T_c = \frac{0.423}{f_{D_{max}}} \tag{5.51}$$

Figure 5.49 Doppler spectrum.

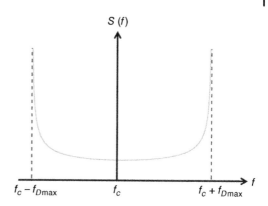

$S(f)$

$f_c - f_{Dmax}$ f_c $f_c + f_{Dmax}$ f

Example

A car is traveling at 25 m/s receives a 2 GHz signal from a stationary transmitter. Find the coherence time using Eqs. (a) (5.49), (b) (5.50), and (c) (5.51).

Solution

$$f_{D_{max}} = \frac{25}{0.15} = 166.67 \text{ Hz}$$

(a) $T_c = \dfrac{1}{f_{D_{max}}} = \dfrac{1}{166.67} = 6 \text{ ms}$

(b) $T_c = \dfrac{9}{16\pi(166.67)} = 1.07 \text{ ms}$

(c) $T_c = \dfrac{0.423}{166.67} = 2.54 \text{ ms}$

If equal amplitude sinusoidal multipath signals at frequency f_c arrive with equal probability from all angles surrounding a receiver that is moving at a velocity $\mathbf{v}_r$, then the Doppler spectrum takes the form [36]

$$S(f) = \frac{1}{\pi \sqrt{f_{D_{max}}^2 - (f - f_c)^2}} \tag{5.52}$$

which is graphed in Figure 5.49. This plot shows the carrier-wave spectrum extending from $f_c - f_{D_{max}}$ to $f_c + f_{D_{max}}$. The low frequency occurs when the receiver moves away from the transmitter, while the high frequency occurs when the receiver moves toward the transmitter. All other motion produces frequencies between these extremes.

5.9 Fade Margin

The level crossing rate (LCR) is the expected number of times that the fading signal amplitude goes above an established threshold (V_{thresh}). Rayleigh fading has an LCR of [10]

$$LCR = \sqrt{2\pi}f_{D_{max}}\rho_T e^{-\rho^2} \tag{5.53}$$

where ρ_T is the threshold voltage normalized to the rms signal level (V_{rms}) written as

$$\rho_T = V_{thresh}/V_{rms} \tag{5.54}$$

The LCR is proportional to the mobile receiver speed.

The average fade duration (AFD) estimates the average length of time that the signal spends below the threshold (signal not received). For Rayleigh fading and isotropic scattering, the AFD below a level ρ_T is [10]

$$AFD = \frac{e^{\rho_T^2} - 1}{\rho_T f_{D_{max}}\sqrt{2\pi}} \tag{5.55}$$

For a particular threshold value, the product of the AFD and the LCR is a constant.

$$AFD \times LCR = 1 - e^{-\rho^2} \tag{5.56}$$

Example
A mobile receiver traveling at 60 mph operates at 900 MHz with $f_{D_{max}} = 88$ Hz. If the threshold is 0 dB, then find the LCR and AFD.

Solution

$$LCR = \sqrt{2\pi}f_{D_{max}}\rho_T e^{-\rho_T^2} = \sqrt{2\pi}(88)(1)e^{-1} = 81 \text{ fades/s}$$

$$AFD = \frac{e^{\rho_T^2} - 1}{\rho_T f_{D_{max}}\sqrt{2\pi}} = \frac{e^1 - 1}{(1)(88)\sqrt{2\pi}} = 7.8 \text{ ms}$$

As a check, use (5.56): $AFD \times LCR = 0.0078 \times 81 = 1 - e^{-\rho_T^2} = 0.632$

5.10 Atmospheric Propagation

The atmosphere has five distinct layers of decreasing density with increasing height above the ground. Layer thickness depends on latitude, season, and time of day. Approximate layer thicknesses at mid-latitude (roughly 30–60°) are [37]

- *Troposphere*: up to 14.5 km
- *Stratosphere*: up to 50 km
- *Mesosphere*: up to 85 km
- *Thermosphere*: up to 600 km
- *Exosphere*: up to 10 000 km

The troposphere and thermosphere substantially impact RF propagation compared to the other layers, because the troposphere contains almost all the water vapor as well as precipitation. This moisture attenuates electromagnetic waves due to the complex permittivity of water in its various forms (Figure 5.11). At a much greater height, the ionosphere (part of the thermosphere) either reflects waves at lower frequencies or attenuates and depolarizes waves at all other frequencies.

Wireless propagation usually occurs in the troposphere where most of the atmospheric moisture resides. Water has a high dielectric constant, so the more water in the troposphere, the more the RF signal attenuates. The rain attenuation coefficient, L_{rain}, is a function of the number and size of the raindrops

$$L_{rain} = \int_0^\infty \mathcal{N}(D)C(D)dD \tag{5.57}$$

Marshall and Palmer [38] proposed a simple exponential rain drop size distribution (DSD):

$$\mathcal{N}(D) = N_{dr}e^{-\rho_r D} \tag{5.58}$$

where $\mathcal{N}(D)$ is the number of drops having an equivalent spherical diameter D (mm) by unit volume, for a diameter interval of 1 mm. They proposed that $N_{dr} = 8 \times 10^3 \, \text{mm}^{-1} \, \text{m}^{-3}$ and $\rho_r = 4.1 R_{ain}^{-0.21} \, \text{mm}^{-1}$ with R_{ain} in mm/h. This distribution is widely used due to its simplicity and good fit for mid-latitude DSDs which have low to moderate rain intensity. $C(D)$ is the attenuation cross-section that corresponds to the attenuation due to a single spherical raindrop. An approximation in the Rayleigh region ($D \ll \lambda$) is

$$C(D) \propto D^3/\lambda \tag{5.59}$$

The attenuation calculated in (5.57) reduces the received power density as shown by the loss term that multiplies the Friis transmission formula in Figure 5.50. Figure 5.51 plots the attenuation vs. frequency for several rainfall rates. The attenuation becomes quite significant at higher frequencies.

Water vapor and atmospheric gases also attenuate signals. Figure 5.52 has plots of the attenuation curves associated with water vapor and oxygen. Tropospheric attenuation due to water and oxygen is negligible at low frequencies.

Figure 5.50 Raindrops attenuate signals passing through.

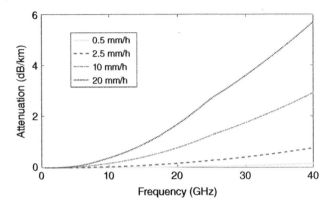

Figure 5.51 Rainfall attenuation as a function of rainfall rate and frequency.

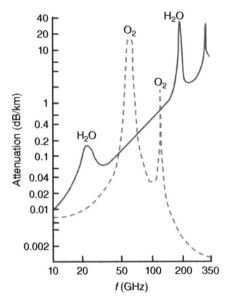

Figure 5.52 Atmospheric gas attenuation: solid line is water vapor and dashed line is oxygen [39].

Attenuation due to water vapor steadily increases with frequency while oxygen drops off above 100 GHz (except for the resonances). Water has resonances at 22 and 183 GHz while oxygen resonates at 60 and 120 GHz. Wireless communications systems avoid the frequency bands around these spikes.

Normally, the warmest air lies next to the Earth. Temperature usually decreases with height above the Earth. Atmospheric waveguides, called ducts, form when

1. An inversion causes cool air to lie next to the Earth with a layer of warm air above it.
2. A layer of cool air gets trapped between two layers of warmer air.

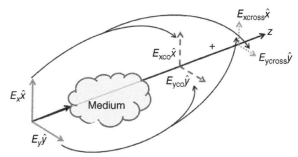

Figure 5.53 Depolarization of an electromagnetic wave.

The layer of cool air has a higher dielectric constant than the warm air, so RF signals at certain frequencies in this atmospheric waveguide encounter low propagation loss over great distances [40]. The signal between a transmitter and receiver in the same duct decays much less than in free space. Ducting mostly occurs over large bodies of water at ultra high frequency (UHF) and microwave frequencies [3]. Elevated ducts occur up to about 1500 m above the surface when the layers of the atmosphere trap the wave from both above and below. Surface ducting is more common and of more practical use, since wireless systems tend to reside there. Ducting conditions occur more frequently in some locations, but their randomness precludes reliable use as a communications channel. Ducting may cause unintentional interference to a distant communication system that is normally too far away to receive the signal.

An electromagnetic wave depolarizes as it passes through a medium. Depolarization means that a plane wave traveling in the z-direction has part of its x-component converted into a y-component and/or vice versa. An electromagnetic wave entering the depolarization medium with an x-component (E_x) and a y-component (E_y) exits the medium with co-polarized (E_{xco}, E_{yco}) and cross polarized (E_{xcross}, E_{ycross}) components (Figure 5.53). Examples of depolarization media include the atmosphere, vegetation, and the ground. Cross polarization discrimination (XPD) measures the amount of depolarization [41]

$$\text{XPD} = 20 \log \frac{E_{co}}{E_{cross}} \tag{5.60}$$

where $E_{co} = \sqrt{E_{xco}^2 + E_{yco}^2}$ is the co-polarized electric field amplitude and $E_{cross} = \sqrt{E_{xcross}^2 + E_{ycross}^2}$ is the cross polarized electric field amplitude.

The other region of the atmosphere that significantly impacts radiowaves is the ionosphere. Radiation from the Sun ionizes atoms and releases free electrons in the ionosphere. These free electrons attenuate, refract, and reflect electromagnetic waves depending upon frequency (Figure 5.54). Electron density starts growing at approximately 30 km above the Earth, but only noticeably impacts radio signals from about 60 to 90 km. Skywaves are radio signals in the low MHz range that reflect from the ionosphere. These HF

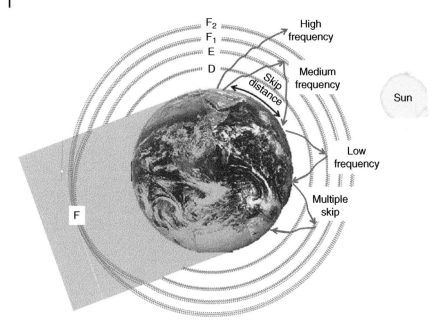

Figure 5.54 The ionosphere affects the propagation distance.

signals propagate around the globe by bouncing between the ionosphere and the Earth multiple times. The distance along the Earth between the transmitter and the point on the ground where the signal reflected from the ionosphere returns to the Earth is called the skip distance and is approximated by [42]

$$d_{\text{skip}} = 2\sqrt{2r_e h'} \tag{5.61}$$

where h' is the virtual height of the ionospheric layer (height where the transmitted and received rays appear to cross). The skip zone is a region transmitter and the skip distance where the signal is too weak to detect.

The ionosphere has several layers as shown in Figure 5.55 (night is on the left half and day on the right half). The D layer lies closest to the Earth at 50–80 km during the day and disappears at night [43]. The electrons quickly recombine at this height, because the relatively high air density provides many opportunities for finding an ion in need of an electron. The Earth blocks solar radiation at night causing the electron levels to fall, and the D layer effectively disappears. Consequently, low frequency signals that reflect from the D layer cannot reach higher layers in the ionosphere except at night when the region disappears. Signals passing through the D region attenuate as the inverse square of the frequency.

The E layer lies above D at 100–125 km above the Earth [43]. Unlike the D layer, it refracts more than attenuates signals passing through. At night, ion

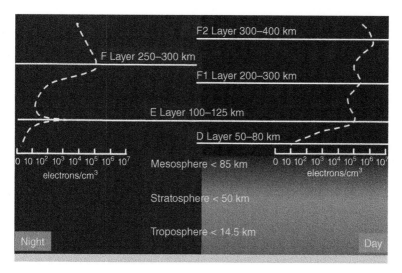

Figure 5.55 Diagram of the ionosphere with nighttime on the left and daytime on the right. The dashed curve represents the typical electron density as a function of height [45].

density in the E layer dramatically decreases as shown by the dashed line plots in Figure 5.55. A typical electron lifetime in the E region is 20 seconds. The residual nighttime ionization at the bottom of the E region causes some attenuation of signals in the lower portions of the HF part of the radio communications spectrum. Signals reflected from the E layer pass back through the D region and get attenuated before returning to Earth. The maximum skip distance for the E region is around 2500 km.

Meteors, electrical storms, auroral activity, and upper atmosphere winds create highly ionized pockets in the E layer that last for a few hours. This sporadic E layer horizontally extends from a few meters to several hundred kilometers while being only a few kilometers thick [44]. When this layer initially forms, it impacts low frequencies; then as time passes, it impacts increasingly higher frequencies until it reaches a peak before decreasing as it disperses. Sporadic E reflects higher-frequency signals from a few minutes to a few hours and lower-frequency signals for much longer. The maximum skip distance is around 2000 km. Sporadic E causes some very high frequency (VHF) signals to propagate further while preventing HF signals from reaching the higher F regions, thus reducing their propagation distance. In temperate regions, sporadic E occurs mainly in the summer. Generally, the higher VHF frequencies are only affected in the middle of the sporadic E season. Auroral sporadic E usually occurs in the morning at the polar regions and have little variation with the seasons. Sporadic E exists primarily in the daytime near the equator with little seasonal variation. In temperate climates, two main peaks occur: one at noon and the other at 19:00.

The F layer lies above the E layer and facilitates long-distance communication at HF frequencies below. Its altitude depends on the time of day, the season, and the amount of solar activity. During the day it splits into two layers called F1 and F2. In the summer, the F1 layer is at 300 km, with the F2 layer at 400 km or more. In the winter, these layers significantly drop in altitude. At night, the F layer is between 250 and 300 km. A typical electron lifetime in the F1 region is 1 minute and the F2 region 20 minutes. The maximum skip distance for the F2 region is around 5000 km.

Spread F contains electron density irregularities caused by ionospheric storms in the F2 region below. It lasts from minutes to hours and horizontally extends up to hundreds of kilometers. HF signals experience fading when passing through the spread F region. It usually occurs in high latitudes at night but can occur during the day as well. Spread F is not common at mid-latitudes.

The number of electrons as a function of height above the Earth determines the reflection and attenuation properties of the ionospheric layers. The Total Electron Content (TEC) is the number of electrons present along a path between a radio transmitter and receiver [46]. One TEC Unit corresponds to 10^{16} electrons/m^2. The TEC depends on the local time, latitude, longitude, season, geomagnetic conditions, solar cycle and activity, and troposphere conditions. A high TEC reduces the position measurement accuracy of satellite navigation systems. The Global Positioning System (GPS), the US part of Global Navigation Satellite System (GNSS), uses an empirical model of the ionosphere, the Klobuchar model, to calculate and remove part of the positioning error caused by the ionosphere when single frequency GPS receivers are used.

The ionospheric electron density depends upon the solar activity. Measuring the ionizing radiation from the sun provides an accurate prediction of ionospheric behavior. The solar radio flux at 10.7 cm or 2800 MHz (F10.7 index) is measured in Solar Flux Units (SFUs). This index indicates the amount of solar activity, because emissions at 2800 MHz correlate well with the sunspot number as well as ultraviolet and extreme ultraviolet emissions that impact the ionosphere. Measurements of the F10.7 index started in Ottawa, Canada in 1947 and still continue at the Penticton Radio Observatory in British Columbia. Figure 5.56 lists the predicted sunspot number and radio flux values with expected ranges on 8 August 2015. The Wolf number (S_{wolf}) [47] measures the number of sunspots and groups of sunspots observed on the surface of the sun. A model that relates S_{wolf} to SFU is given by [48]

$$SFU = 63.7 + 0.728S_{wolf} + 0.00089S_{wolf}^2 \qquad (5.62)$$

As of 2015, the Brussels International Sunspot Number used in Figure 5.56 (S_B) supplanted S_{wolf} due to improved observations [49]. They are related by

$$S_{wolf} = 0.6S_B \qquad (5.63)$$

```
:Predicted_Sunspot_Numbers_and_Radio_Flux: Predict.txt
:Created: 2015 Aug 08 0100 UTC
# Prepared by the U.S. Dept. of Commerce, NOAA, Space Weather Prediction Center (SWPC).
# Please send comments and suggestions to swpc.webmaster@noaa.gov
#
# Sunspot Number: S.I.D.C. Brussels International Sunspot Number.
# 10.7cm Radio Flux value: Penticton, B.C. Canada.
# Predicted values are based on the consensus of the Solar Cycle 24 Prediction Panel.
#
# See the README3 file for further information.
#
# Missing or not applicable data:  -1
#
#           Predicted Sunspot Number And Radio Flux Values
#                     With Expected Ranges
#
#        -----Sunspot Number------  ----10.7 cm Radio Flux----
# YR MO   PREDICTED    HIGH   LOW    PREDICTED    HIGH    LOW
#-----------------------------------------------------------------
2015 02      58.9     59.9   57.9     134.1     135.1   133.1
2015 03      57.1     59.1   55.1     132.3     133.3   131.3
2015 04      55.8     58.8   52.8     129.1     131.1   127.1
2015 05      55.0     60.0   50.0     125.4     128.4   122.4
2015 06      53.4     58.4   48.4     121.4     125.4   117.4
2015 07      51.8     57.8   45.8     117.8     121.8   113.8
2015 08      51.4     58.4   44.4     115.3     120.3   110.3
```

Figure 5.56 Sunspot numbers (S_B) and radio flux [50]. *Source:* Reproduced with permission of NOAA.

Geomagnetic activity also affects the ionosphere. The solar wind (charged particles emanating from the sun) changes the shape of the Earth's magnetic field, which in turn disturbs the charged particles in the ionosphere [51]. The convective motion of charged, molten iron inside the Earth generates the magnetosphere (right side of Figure 5.57). The solar wind constantly bombards the sun-facing side of the Earth's magnetic field and causes it to compress until it extends from 6 to 10 times the radius of the Earth. The magnetosphere on the night side of the Earth has a long tail that fluctuates but extends up to hundreds of Earth radii (beyond the moon).

Magnetic observatories around the globe record the largest magnetic change (maximum minus minimum) that occurred in a three-hour period. Two indices describe the observed level of geomagnetic activity [53]:

1. K_p: The K index lies between 0 and 9 with higher numbers corresponding to higher geomagnetic activity. The estimated planetary K index (K_p) is the mean standardized K-index from 13 geomagnetic observatories between 44° and 60° northern or southern geomagnetic latitude. Figure 5.58 plots the estimated Kp over a three-day period.
2. A_p: The three-hour A_p (equivalent range) index is derived from the K_p index. Since the A is derived from the K index then averaged over the period of a day. Like the K index, values are averaged around the globe to give the planetary A_p index. Table 5.10 equates the K_p and A_p indexes to geomagnetic activity.

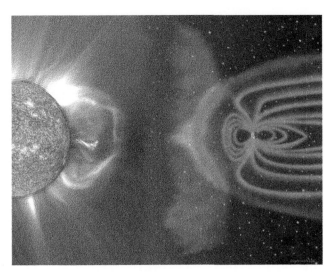

Figure 5.57 Image of Earth's magnetosphere [51]. *Source:* NASA [52] https://images-assets.nasa.gov/image/0201490/0201490~orig.jpg.

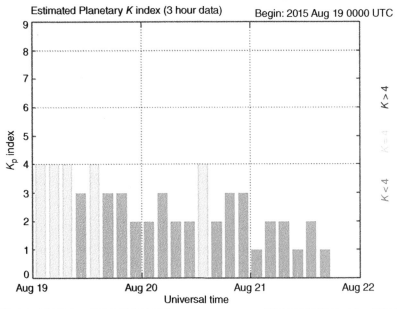

Figure 5.58 Estimated K_p index over a three day period [52]. *Source:* Reproduced with permission of NOAA.

Table 5.10 Measures of geomagnetic activity [53].

A_p index	K_p index	Activity
0	0	Quiet
4	1	Quiet
7	2	Unsettled
15	3	Unsettled
27	4	Active
48	5	Minor storm
80	6	Major storm
132	7	Severe storm
208	8	Very major storm
400	9	Very major storm

The critical frequency (f_{crit}) is the lowest frequency that passes through the ionosphere. An ionosonde measures the critical frequency by transmitting a signal vertically from the ground and recording the return at the same location. The signal time delay indicates the height of the ionospheric layers. The critical frequency for the E layer is from 1 to 4 MHz and for the F2 layer from 2 to 13 MHz [54]. The lowest critical frequency occurs at night during the lowest sunspot years. The highest critical frequency occurs at daytime in high solar activity years. Periods of high solar activity drive the critical frequency to as high as 20 MHz for brief periods. The critical frequency relates to the electron density in an ionospheric layer by

$$f_{crit} = 9\sqrt{N_e} \tag{5.64}$$

where N_e is the electron density in number of electrons per m³.

The maximum usable frequency (MUF) is the highest frequency that reflects from the ionosphere when the signal is incident at an angle θ

$$MUF = \frac{f_{crit}}{\cos\theta} \tag{5.65}$$

where $\theta = 0$ is vertical. The skip distance dramatically increases as the frequency approaches the MUF.

The Lowest Usable Frequency (LUF) is the lowest frequency that results in satisfactory reception. The LUF increases with increasing solar activity. Sometimes the LUF exceeds the MUF between a transmitter and receiver, because the highest possible frequency that propagates through the ionosphere to the receiver gets absorbed [54].

The Frequency of Optimum Traffic (FOT) is the highest frequency that is usable 90% of the days in a month for a specific propagation path [55].

Sometimes FOT is referred to as Optimum Working Frequency (OWF). FOT is estimated to be 85% of the median monthly MUF. To maintain continuous operation, the wireless system operates at a higher frequency during the day and a lower frequency at night.

Near Vertical Incidence Skywave (NVIS) propagation works below the critical frequency (between 3 and 10 MHz) by reflecting a signal from the ionosphere at an angle less than 15° from normal [56]. NVIS covers an area about 200 km in radius that has no intermediate man-made infrastructure. It works well when mountains, buildings, and trees block LOS signals. The desired coverage area and solar activity dictate the optimum operating frequency. It is ideal for disaster relief communication, communication in developing regions, and military applications.

Problems

5.1 Find the received power if 100 W is transmitted 40 km through a 40 dB antenna. The receive antenna has a gain of 23 dB and the frequency is 5 GHz.

5.2 A handset radiates 1 W through a 3 dB linear polarized antenna at 2.4 GHz. Assume that the base station is polarized matched to the antenna 1 km away. Estimate the size of the base station aperture that is needed to receive the signal for a receiver with a sensitivity of -100 dBm.

5.3 A plane wave is incident from medium 1 with $n_1 - 1.0$ onto two layers: medium 2 with $n_2 = 1.5$ and medium 3 with $n_1 = 2.0$. Find the transmission angle into medium 3.

5.4 A wave is incident from a region with $n_i = 1.0$ to a region with $n_t = 1.5$. For $0 \leq \theta_i \leq 90°$ plot the TE and TM (a) reflection coefficients and (b) transmission coefficients.

5.5 A wave is incident from a region with $n_i = 1.5$ to a region with $n_t = 1.0$. For $0 \leq \theta_i \leq 90°$ plot the magnitude of the TE and TM reflection coefficients.

5.6 On the same graph, plot the reflection coefficient vs. angle for $n_t = 1.0$ and a transmission region with (a) $n_1 = 1.5$ and (b) $n_2 = 2.5$.

5.7 Glass has an $\varepsilon_r = 2.25$. Calculate the percentages of reflected and transmitted powers for a 2.4 GHz signal at normal incident on a large planar sheet of glass.

5.8 A 100 ns pulse at $f_c = 2.4$ GHz takes two paths. One path is 100 m and the other is 105 m. How long is the received pulse?

5.9 A 100 MHz LOS communications system has transmitting and receiving antennas separated by 15 km over a flat Earth. If the transmitting antenna is 20 m above ground, what is the optimal height of the receive antenna?

5.10 An LOS link at 50 MHz has a transmit antenna 20 m above ground. The receive antenna is 15 km away. Find the height of the receive antenna that will maximize signal reception for TE polarization.

5.11 A 1 GHz RHCP wave is incident on flat ground covered in deep wet snow at $\psi_g = 20°$. What is the axial ratio of the reflected wave?

5.12 If antennas are placed on 20 m towers what should their separation be on a spherical Earth when (a) no atmospheric refraction and (b) atmospheric refraction.

5.13 A 1 GHz wave is incident on a lake that has a height variation of $\sigma_{surf} = 2$ cm. Find the reflection coefficient as a function of angle.

5.14 Plot a three-dimensional cornu spiral $S(v)$ vs. $C(v)$ vs. v.

5.15 Estimate the diffraction loss at 900 MHz for
(a) $d_t = 100$ m, $d_r = 100$ m, $h_t = 5$ m, $h_r = 5$ m, $h_o = 10$ m
(b) $d_t = 100$ m, $d_r = 100$ m, $h_t = 5$ m, $h_r = 5$ m, $h_o = 5.1$ m
(c) $d_t = 10$ m, $d_r = 190$ m, $h_t = 5$ m, $h_r = 5$ m, $h_o = 5.1$ m
(d) $d_t = 10$ m, $d_r = 190$ m, $h_t = 5$ m, $h_r = 1$ m, $h_o = 10$ m
(e) $d_t = 10$ m, $d_r = 10$ m, $h_t = 1$ m, $h_r = 1$ m, $h_o = 10$ m

5.16 Find the radii of the first three Fresnel zones when
(a) $d_t = 100$ m, $d_r = 100$ m, $f = 500$ MHz
(b) $d_t = 100$ m, $d_r = 100$ m, $f = 5$ GHz
(c) $d_t = 150$ m, $d_r = 50$ m, $f = 500$ MHz
(d) $d_t = 150$ m, $d_r = 50$ m, $f = 5$ GHz

5.17 There are two knife-edge obstacles between two antennas in a communications system:

Object	Height (m)	Distance (m)
Antenna 1	20	0
Obstacle A	30	100
Obstacle B	20.75	225
Antenna 2	1.5	250

Find the diffraction loss using Bullington's method at 2.4 GHz.

5.18 There are four knife-edge obstacles between two antennas in a communications system:

Object	Height (m)	Distance (km)
Antenna 1	20	0
Obstacle A	40	1
Obstacle B	50	1.5
Obstacle C	60	2
Obstacle D	10	2.2
Antenna 2	10	3

Find the diffraction loss using Bullington's method at 800 MHz.

5.19 There are three knife-edge obstacles between two antennas in a communications system:

Object	Height (m)	Distance (m)
Antenna 1	10	0
Obstacle A	20	100
Obstacle B	25	200
Obstacle C	30	250
Antenna 2	2	300

Find the diffraction loss using Bullington's method at 5 GHz.

5.20 Repeat Problem 17 using the Epstein–Peterson method.

5.21 Repeat Problem 19 using the Epstein–Peterson method.

5.22 For Rayleigh fading, what is the probability that the received signal power is at least (a) 20, (b) 6, and (c) 3 dB below the mean power?

5.23 Determine the fade margin in a Rayleigh channel if the received power falls below the receiver sensitivity 1% of the time.

5.24 Show that for $K_r = 0$ the Rician PDF becomes a Rayleigh PDF.

5.25 A base station operates at 2.4 GHz at a height 10 m above the ground. If the receiver is 100 m away at a height of 1.5 m above ground, then find the path loss using the Hata model for (a) large city, (b) small city, (c) suburban, and (d) rural area.

5.26 A base station operates at 800 MHz at a height 20 m above the ground. If the receiver is 500 m away at a height of 10 m above ground, then find the path loss using the Hata model for (a) large city, (b) small city, (c) suburban, and (d) rural area.

5.27 A 2400 MHz signal enters an office building through a brick wall that is 267 mm thick. The signal travels a total of 20 m in the building. Estimate the attenuation.

5.28 A 2.4 GHz signal enters a house through a brick wall that is 89 mm thick and passes through one floor. The signal travels a total of 5 m in the house. Estimate the attenuation.

5.29 A wireless communication system operates at two frequencies: (a) 900 MHz and (b) 1800 MHz. Find the maximum Doppler spread if the transmitter is stationary and you are on a train traveling at 200 km/h. Give an estimate of the coherence time.

5.30 Find the LCR for $\rho_T = 0.1$ and a maximum Doppler frequency of 30 Hz.

5.31 Find the AFD for a maximum Doppler frequency of 200 Hz and (a) $\rho_T = 1$, (b) $\rho_T = 0.1$, and (c) $\rho_T = 0.001$.

5.32 Compute LCR for an $f_{Dmax} = 20$ Hz, $f_c = 900$ MHz, and $\rho_T = 1$. What is the maximum velocity of the vehicle?

5.33 The U.S. Geological Survey (USGS) categorizes rain fall as follows: violent shower $R_{ain} > 50$ mm/h, heavy shower $50 \geq R_{ain} > 10$ mm/h, moderate shower $10 \geq R_{ain} > 2$ mm/h, slight shower $R_{ain} < 2$ mm/h. Plot attenuation vs. frequency for 1, 5, 25, and 50 mm/h.

5.34 Plot rainfall attenuation vs. rainfall rate $50 \geq R_{ain} > 1$ for $f = 10, 20, 30,$ and 40 GHz.

5.35 Find the sunspot number, 10.7-cm radio flux, and the estimated planetary K index for the day

References

1 Friis, H.T. (May 1946). A note on a simple transmission formula. *IRE Proc.* 34 (5): 254–256.

2 Yahalom, A., Pinhasi, Y., Shifman, E., and Petnev, S. (2010). Transmission through single and multiple layers in the 3–10 GHz band. *WSEAS Trans. Commun.* 9 (12): 759–772.

3 Collin, R.E. (1985). *Antennas and Radiowave Propagation.* New York, NY: McGraw-Hill.

4 Ulaby, F.T. and Long, D.G. (2014). *Microwave Radar and Radiometric Remote Sensing.* University of Michigan Press.

5 Dudley, D.G., Lienard, M., Mahmoud, S.F., and Degauque, P. (2007). Wireless propagation in tunnels. *IEEE AP-S Mag.* 49 (2): 11–26.

6 Infantolino, J.K., Kuhlman, A.J., Weiss, M.D., and Haupt, R.L. (2013). Modeling RF attenuation in a mine due to tunnel diameter and shape. IEEE AP-S Symposium, Orlando, FL (July 2013).

7 Burrows, C.R. (1935). Radio propagation over spherical earth. *Proc. Inst. Radio Eng.* 23 (5): 470–480.

8 Doerry, A.W. (2013). Earth curvature and atmospheric refraction effects on radar signal propagation. Sandia National Laboratories Albuquerque, New Mexico, Sandia Report SAND2012-10690 Unlimited Release Printed, January 2013.

9 Ulaby, F.T., Moore, R.K., and Fung, A.K. (1982). *Microwave Remote Sensing Active and Passive* Vol. II. Addison-Wesley.

10 Rapaport, T.S. (2002). *Wireless Communications Principles and Practice.* Upper Saddle River, NJ: Prentice Hall.

11 Molisch, A.F. (2011). *Wireless Communications*, 2e. West Sussex: Wiley/IEEE.

12 (Nov 2013). Propagation by diffraction. In: *Recommendation ITU-R P.526-13.*

13 http://www.proxim.com/products/knowledge-center/calculations/calculations-fresnel-clearance-zone (accessed 30 July 2019.)

14 Durgin, G.D. (2009). The practical behavior of various edge-diffraction formulas. *IEEE Antennas Propagat. Mag.* 51: 24–35.

15 https://www.loxcel.com/3d-fresnel-zone (accessed 17 July 2018).

16 Bullington, K. (1947). Radio propagation at frequencies above 30 megacycles. *Proc. IEEE* 35: 1122–1136.

17 Epstein, J. and Peterson, D.W. (1953). An experimental study of wave propagation at 850 Mc. *Proc. IEEE* 41: 595–611.

18 Samuel, W., Oguichen, T.C., and Worgu, S. (2017). Computation of 10 knife edge diffraction loss using Epstein–Peterson method. *Am. J. Software Eng. Appl.* 6 (1): 1–4.

19 Deygout, J. (1966). Multiple knife-edge diffraction of microwaves. *IEEE Trans. Antennas Propagat.* AP-14: 480–489.

20 Keller, J.B. (1962). Geometrical theory of diffraction. *J. Opt. Soc. Am.* 52 (2): 116–130.

21 Stutzman, W.L. and Thiele, G.A. (2013). *Antenna Theory and Design*, 3e. Wiley.

22 Ling, H., Chou, R.-C., and Lee, S.-W. (1989). Shooting and bouncing rays: calculating the RCS of an arbitrarily shaped cavity. *IEEE Trans. Antennas Propag.* 37 (2): 194–205.

23 ITU (2013). Propagation by diffraction. In: *P Series Radiowave Propagation, Recommendation ITU-R P*, 526–513.

24 http://www.invictusnetworks.com/faq/RF%20Technical%20Info%20and %20FCC%20Regs/Fade%20Margin%20Calculator%20-%20Basic.htm (accessed 17 July 2018).

25 National Instruments White Paper (2018). Understanding RF signal fading types.

26 Rice, S.O. (1945). Mathematical analysis of random noise part III. *Bell Syst. Tech. J.* 24: 46–108.

27 Abdi, A., Tepedelenlioglu, C., Kaveh, M., and Giannakis, G. (2001). On the estimation of the K parameter for the rice fading distribution. *IEEE Commun. Lett.* 5 (3): 92–94.

28 Andersen, J.B., Rappaport, T.S., and Yoshida, S. (1995). Propagation measurements and models for wireless communications channels. *IEEE Commun. Mag.* 33 (1): 42–49.

29 Okumura, Y., Ohmori, E., Kawano, T., and Fukuda, K. (1968). Field strength and its variability in VHF and UHF land-mobile radio service. *Rev. Elec. Commun. Lab.* 16 (9–10): 825–873.

30 Hata, M. (1980). Empirical formula for propagation loss in land mobile radio services. *IEEE Trans. Vehicular Technol.* 29 (3): 317–325.

31 Motley, A.J. and Keenan, J.P. (1988). Personal communication radio coverage in buildings at 900 MHz and 1700 MHz. *Electron. Lett.* 24: 763–764.

32 ITU (2017). Propagation data and prediction methods for the planning of indoor radiocommunication systems and radio local area networks in the frequency range 300 MHz to 100 GHz. Recommendation ITU-R P.1238-9.

33 Gudmundson, M. (1991). Correlation model for shadow fading in mobile radio systems. *Electron. Lett.*: 2145–2146.

34 Wireless InSite, Remcom.

35 National Instruments White Paper(2013).Doppler spread and coherence time. http://www.ni.com/white-paper/14911/en/ (accessed 17 July 2018).

36 Clarke, R.H. (1968). A statistical theory of mobile radio reception. *Bell Syst. Tech. J.* 47 (6): 957–1000.

37 https://www.nasa.gov/mission_pages/sunearth/science/atmosphere-layers2 .html (accessed 31 January 2019).

38 Marshall, J. and Palmer, W. (1948). The distribution of raindrop with size. *J. Meteorol.* 5: 165–166.

39 https://www.itu.int/dms_pubrec/itu-r/rec/p/R-REC-P.676-3-199708-S!!PDF-E.pdf Rec. ITU-R P.676-3 1 (accessed 30 July 2019).

40 Seybold, J.S. (2005). *Introduction to RF Propagation.* Hoboken, NJ: Wiley.

41 http://www.mike-willis.com/Tutorial/PF10.htm (accessed 25 February 2019).

42 http://ionolab.org/index.php?page=ionosphere&language=en, (accessed 29 May 2018).

43 https://www.electronics-notes.com/articles/antennas-propagation/ionospheric/ionospheric-layers-regions-d-e-f1-f2.php (accessed 25 February 2019).

44 https://www.electronics-notes.com/articles/antennas-propagation/ionospheric/sporadic-e-es.php (accessed 25 February 2019).

45 http://www.sws.bom.gov.au/Educational/1/2/5 (accessed 9 December 2018).

46 http://www.swpc.noaa.gov/phenomena/total-electron-content (accessed 8 August 2015).

47 https://science.nasa.gov/heliophysics/focus-areas/magnetosphere-ionosphere (accessed 17 July 2018).

48 https://www.electronics-notes.com/articles/antennas-propagation/ionospheric/solar-indices-flux-a-ap-k-kp.php (accessed 29 May 2018).

49 http://www.sidc.be/silso/datafiles-old (accessed 30 May 2018).

50 http://www.swpc.noaa.gov/products/predicted-sunspot-number-and-radio-flux (accessed 17 July 2018).

51 https://images-assets.nasa.gov/image/0201490/0201490~orig.jpg (accessed 29 May 2018).

52 https://www.swpc.noaa.gov/products/planetary-k-index (accessed 21 August 2015).

53 Poole, I. (2002). Understanding solar indices. *QST Magazine*, pp. 38–40.

54 American Radio Relay League (2015). *ARRL Handbook*, 93e.

55 Kishore, K. (2009). *Antenna and Wave Propagation.* New Delhi: I.K. International Publishing House.

56 Witvliet, B.A. and Alsina-Pagès, R.M. (2017). Radio communication via near vertical incidence skywave propagation: an overview. *Telecommun. Syst.* 66: 295. https://doi.org/10.1007/s11235-017-0287-2.

6

Satellite Communications

Satellite communications refers to a communication link that involves a satellite. Most early satellite communication operated at C band with relatively low power and low antenna gain. The Earth stations for these satellites transmitted several kilowatts of power through large reflector antennas that were many meters in diameter. Future applications require higher frequencies and smaller ground stations. Table 6.1 lists the satellite frequency bands currently in use. Direct Broadcast Satellite (DBS) provide consumers with a direct-to-home Satellite TV link. Fixed Satellite Service (FSS) connects two ground stations via a satellite.

6.1 Early Development of Satellite Communications

Arthur C. Clarke initiated the idea of providing worldwide communications using satellites in 1945 [2]. His idea did not materialize until the first rocket launches in the late 1950s, though. On 4 October 1957, the Soviet Union started the space race by launching a 58-cm diameter sphere weighing 84 kg called Sputnik into orbit [3]. For three weeks, it transmitted rapid beeps at 20.007 and 40.002 MHz. Sputnik circled the Earth 1440 times before burning up in the atmosphere on 4 January 1958. In 1960, the United States responded with Echo, NASA's first communications satellite. It was a 30-m diameter balloon made of Mylar (Figure 6.1) [4]. A ground station transmitted circularly polarized signals at 960 MHz and 2.39 GHz to the Echo satellite that reflected the signal to a different ground station. One of the ground stations, Bell Laboratory in Holmdel, NJ, used the newly invented Hogg antenna (Chapter 4) shown in Figure 6.2. The goals of Echo were to demonstrate long-distance voice communications, study propagation effects, try different satellite tracking techniques, and determine whether a passive satellite fits into telecommunications [5].

Wireless Communications Systems: An Introduction, First Edition. Randy L. Haupt.
© 2020 John Wiley & Sons, Inc. Published 2020 by John Wiley & Sons, Inc.

Table 6.1 Satellite frequencies (GHz) [1].

Frequency band	Downlink (DL)	Uplink (UL)
C	3.700–4.200	5.925–6.425
X (Military)	7.250–7.745	7.900–8.395
Ku (Europe)	FSS: 10.700–11.700 DBS: 11.700–12.500 Telecom: 12.500–12.750	FSS & Telecom: 14.000–14.800 DBS: 17.300–18.100
Ku (USA)	FSS: 11.700–12.200 DBS: 12.200–12.700	FSS: 14.000–14.500 DBS: 17.300–17.800
Ka	18–31	
EHF	30–300	
V	36–51.4	

Source: www.nasa.gov.

Figure 6.1 Echo satellite [4]. *Source:* Courtesy of NASA. https://www.nasa.gov/multimedia/imagegallery/image_feature_559.html.

Figure 6.2 Ground station Hogg horn antenna in Holmdel, NJ [4]. *Source:* Courtesy of NASA. www.nasa.gov.

In 1961, OSCAR 1 (Orbiting Satellites Carrying Amateur Radio), became the world's first nongovernment satellite [6]. Some amateur radio operators built the satellite for less than US$100. It transmitted the Morse code message "hi–hi" at 144.983 MHz for nearly 20 days, using a 60 cm monopole. The small power supply with no solar cell for recharging limited the satellite's lifespan. Thousands of radio operators in 28 different countries listened to the signal. Today, the Radio Amateur Satellite Corporation (AMSAT) operates several satellites for use by amateur radio operators worldwide.

In 1962, the American Telephone and Telegraph Company's (AT&T) Telstar 1 [7, 8] became the first active communications satellite as well as the first commercial payload in space. It was 0.9 m in diameter and weighed 77 kg. Telstar made possible the first transatlantic television transmission (Andover Earth Station, Maine to the Pleumeur-Bodou Telecom Center, Brittany, France). The Andover Earth Station used an even larger version of the Hogg horn antenna shown in Figure 6.2. A solar array with a 15-W battery back-up powered the spin-stabilized satellite. The uplink (UL) had eight channels at 6 GHz, while the downlink (DL) had eight channels at 4 GHz [9]. It received telemetry commands through a VHF quadrafilar helical antenna. Telstar failed when passing through the Van Allen Belt, because the electronics were not sufficiently radiation hardened. As a result, the satellite was deactivated in 1963. During its short life, Telstar demonstrated the feasibility of practical satellite communications, made advances in satellite tracking, and provided critical satellite design information on the Van Allen radiation belts. Telstar instigated a debate in the United States on whether communications satellites should be operated and controlled by the private sector or government.

Figure 6.3 Syncom IV-3 Satellite. *Source:* Courtesy of NASA. www.nasa.gov.

Syncom (synchronous communication satellite) launched into a geosynchronous orbit in 1963 [10, 11]. A satellite in geosynchronous orbit circles the Earth once every sidereal day (23 hours, 56 minutes). An observer on Earth sees a geosynchronous satellite at the same time and in the same place in the sky every day. Syncom had a frequency-translation transponder capable of one two-way telephone or 16 one-way teletype channels. A slotted dipole with 2 dB of gain that operated at two frequencies near 7.36 GHz received ground signals. The satellite retransmitted the received signals through another slotted dipole at 1.815 GHz. Four monopole antennas extended from the cylindrical satellite for telemetry and command. The satellite reached the desired orbit but never functioned.

Later in 1963, Syncom 2 became the first successful geosynchronous communications satellite [12]. Syncom 2 hosted the first telephone conversation between heads of government via satellite (USA President Kennedy and Nigerian Prime Minister Balewa). In addition, the satellite enabled several test television transmissions from Fort Dix, New Jersey to a ground station in Andover, Maine. Other Syncoms followed eventually leading to the INTELSAT satellites. Figure 6.3 is a photograph of a later model, Syncom IV-3, taken from the Space Shuttle.

NASA launched the first commercial communications satellite in 1965, INTELSAT I or Early Bird, into geosynchronous orbit [10]. It enabled 240

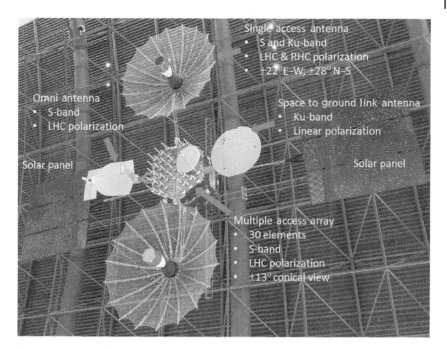

Figure 6.4 Model of the TDRS satellite at the Smithsonian National Air and Space Museum.

two-way transatlantic telephone channels compared to Syncom I's single channel [13]. INTELSAT I through IV were spin-stabilized. Technological advances allowed INTELSAT V to use the more advanced three-axis stabilization. The early INTELSATs operated at C band. Ku band was added later. The C band link had a 500 MHz bandwidth that was divided into 12 subbands of 40 MHz each. The Ku subbands were at least 80 MHz wide.

In the 1980s, NASA deployed the Tracking and Data Relay Satellite (TDRS) System in geostationary orbit (Figure 6.4) [14]. These satellites relay data from user satellites in lower orbits to ground stations that are not in the LOS. Each TDRS satellite has two 5-m dish antennas (S and K bands), and a 30-element S-band phased array to communicate with the user satellites. A Ku-band antenna communicates with the ground station. The multi-beam phased array has 20 beams that link up to 20 different satellites.

6.2 Satellite Orbits

The mission defines the type of orbit for a satellite. An orbit determines a satellite's speed as well as its distance from and position above the Earth. These factors play an important role in determining the link budget.

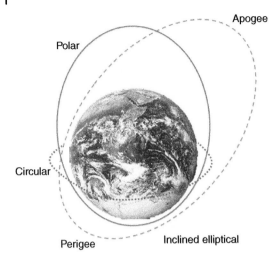

Figure 6.5 Satellite orbits and their inclination.

Inclination describes the orbit tilt angle relative to the equatorial plane. Figure 6.5 shows a circular equatorial orbit (0°) as well as an inclined elliptical orbit and the extreme polar orbit (90°). A polar orbit means the satellite orbits Earth in a plane containing both poles. The point farthest from the Earth on an elliptical orbit is called apogee, while the closest point is perigee.

Orbits are also classified by their distance from Earth (Figure 6.6): low Earth orbit (LEO), medium Earth orbit (MEO), and geosynchronous Earth orbit (GEO). Any satellite orbiting the Earth beyond GEO is in a high Earth orbit (HEO). A satellite's speed and orbital period determine its height above the

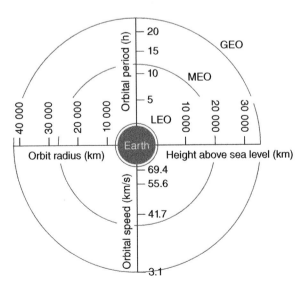

Figure 6.6 Orbit height, speed, and period [15].

Earth. Some space missions require satellites to reach an escape velocity of at least 11.2 km/s to leave Earth orbit.

In 1945, Arthur C. Clarke talked of launching a satellite into geostationary orbit [16], "A rocket which can reach a speed of 8 km/s parallel to the earth's surface would continue to circle it forever in a closed orbit; it would become an 'artificial satellite'." He went on to say that [16] "An 'artificial satellite' at the correct distance from the earth would make one revolution every 24 hours; i.e. it would remain stationary above the same spot and would be within optical range of nearly half the earth's surface." In honor of his brilliant ideas, a geostationary orbit is called the Clarke orbit.

Geosynchronous satellites travel at about 3 km/s in an orbit at 35 800 km above the Earth and circle the Earth once a day. A geostationary orbit is a geosynchronous orbit in the equatorial plane. A single satellite in geostationary orbit "sees" approximately 1/3 of the Earth surface between ±77° latitude, so three equally spaced satellites in geostationary orbit cover the entire Earth outside of the polar regions. The advantages of geostationary satellites are:

- Ground tracking not needed
- No communications handoff needed between satellites
- Three satellites provide complete Earth coverage below 77° latitude
- Very little Doppler shift

The disadvantages include

- Long transmission latencies (travel time from sender to receiver)
- Received signal very weak
- Poor coverage above 77° latitude
- Expensive launch

A geostationary satellite with an antenna that has a 17.3° beamwidth covers approximately 1/3 of the Earth surface outside of the polar regions [17]. Multiple satellites with narrow beams are more typically used for Earth coverage. Ground stations on the equator at the same longitude as the satellite experience a 239 ms minimum time delay – a long time in our nanosecond world. Most ground stations are not directly below the satellite, so the cross range distance to the satellite increases the free space loss which means the signal is even weaker and the time delay greater. In addition, the signal passes through more of the atmosphere which further absorbs and depolarizes the signal. Atmospheric effects limit the elevation angle for C band to 5° above the horizon and for Ku band to 10° above the horizon.

Typically a satellite lasts about 13 years in orbit. The gravity of the sun and moon push a satellite north or south of its geostationary orbit. East and west deviations result from orbital velocity and altitude errors as well as the Earth not being a perfect sphere. Without an orbital correction, a satellite moves about 0.85° per year in the north–south direction [17]. Orbital station-keeping moves a satellite back to its intended orbit by firing thrusters. Orbital corrections use precious fuel, and a satellite's life expectancy depends on the amount of fuel stored. Making corrections only when the satellite shifts $\pm 3°$ extends a satellite's life by about three years. The satellite needs to save enough fuel at the end of its life in order to safely deorbit.

The sun acts like an RF noise source that moves along a trajectory across the sky. Twice a year during the equinoxes, the sun crosses a line between the geostationary satellite and ground station and causes fading at the ground station for several minutes a day over a period of several days. During this time, the increased noise degrades the carrier-to-noise ratio (C/N) of the weak signals from the satellite. The solar fading angle is the angle (measured from the ground station antenna) between the satellite and the Sun at the time when signal degradation begins or ends. The sun outage start and end dates depend on the geographical location of the ground station. Small antennas with very wide beamwidths are out of commission for up to half an hour. High gain antennas (most satellite antennas) typically have the link disrupted for only a few minutes. Complicated algorithms exist to precisely calculate when and how long the sun fade lasts.

Simple geometry leads to an approximation for the sun fade time [18]. From the Earth, the sun has an apparent diameter of $0.53°$. If the ground antenna has a 3 dB beamwidth of θ_{3dB} in degrees, and the sun moves across the sky at a rate of $15°$ per hour, then the maximum sun fade time in minutes is given by

$$T_{sunfade} = 4(\theta_{3dB} + 0.48) \text{ minutes} \tag{6.1}$$

The sun's declination changes by $0.4°$ per day, so the maximum number of sun fade days is given by

$$N_{sunfade} = \frac{\theta_{3dB} + 0.48°}{0.4°} \tag{6.2}$$

Example
What is the diameter of a parabolic reflector ground station operating at 4 GHz that has $N_{sunfade} = 2$.

Solution

$$N_{sunfade} = 2 = \frac{\theta_{3dB} + 0.48°}{0.4°} \Rightarrow \theta_{3dB} = 2(0.4) - 0.48 = 0.32°$$

$$\lambda = \frac{3 \times 10^8}{4 \times 10^9} = 0.075 \text{ m}$$

$$\theta_{3dB} = 0.32° = 29.2°/(r_a/\lambda) = 29.2°/(r_a/0.075) \Rightarrow r_a = 6.84 \text{ m}$$

diameter = 13.69 m

Figure 6.7 Front and side views of the Airlink antenna. *Source:* Reprinted by permission of Haupt 2010 [19]. © 2010, IEEE.

Geostationary satellites are important for weather and communications, because ground satellite dishes do not have to track the satellites. Inmarsat, a British satellite telecommunications company, offers worldwide communications services via 12 geostationary satellites. The AIRLINK® conformal array on airplanes provides in-flight communications via the Inmarsat geostationary satellites [19]. Figure 6.7 shows the antenna array that operates between 1530 and 1559 MHz on receive and 1626.5 and 1660.5 MHz on transmit with a gain in excess of 12 dB. The array of rectangular microstrip elements arranged in a triangular grid scans ±60° in azimuth and elevation.

LEO satellites travel at 7.8 km/s or less depending upon the distance from the Earth. At 160–2000 km above the Earth, they avoid the atmospheric drag and lie below the high radiation levels of the inner Van Allen radiation belt. Space debris resides in low orbits, so satellite survival depends on keeping track of space debris. NASA estimates that 500 000 objects between 1 and 10 cm orbit Earth with average impact speeds greater than 22 000 mph [20]. LEO satellites have orbital inclinations anywhere from equatorial to polar. They are relatively cheap to launch, have low latency, and low free space attenuation.

The Iridium constellation has 66 active LEO satellites plus several spares in low-Earth polar orbit at 781 km above ground and an inclination of 86.4° in order to provide voice and data communications [21]. One orbit takes 100.5 minutes. The satellite constellation has six orbital planes spaced 30° apart, with 11 active satellites in each plane (Figure 6.8). Satellites control their orbital altitudes to within 10 m and position errors to within 15 km.

The Iridium-NEXT satellite has an L-Band phased array with 168 transmit and receive modules that produce 48 transmit/receive beams [23]. It employs a Time-Division Duplex architecture that allocates different time slots for UL

Figure 6.8 Iridium constellation. *Source:* Reprinted by permission of Leopold and Miller [22]. © 1993, IEEE.

and DL in the 1616–1626.5 MHz band. The array has a 4700 km footprint on Earth. Iridium has a 2.4-kbs voice communications data rate, a 64 kbs data rate for L-Band Handset Data Services and Short Burst Data, and a capability of 512 kbit/s to 1.5 Mbps link for high data rate applications. There is also a Ka-Band 8 Mbps link.

Iridium-NEXT satellites also have two 20 GHz ULs and 30 GHz DLs connecting to a terrestrial gateway [23]. Four 23.18–23.38 GHz crosslinks allow adjacent satellites in the same orbital plane and in adjacent planes to route data in order to provide worldwide coverage. Crosslink communications occurs at 12.5 Mbps, in half duplex mode. Two fixed antennas enable in-plane communications and steerable antennas lock onto satellites in neighboring planes. Telemetry and command occur over the 20/30 GHz links with omni-directional antennas on the satellite.

MEO satellites lie between 2000 and 35 800 km and orbit the Earth between 2 and 24 hours. The Global Positioning System (GPS) in a MEO orbits the Earth in 12 hours at 20 200 km [24]. Figure 6.9 shows several versions of the GPS satellites. A constellation of 24 GPS III satellites appears in Figure 6.10. The satellites in the GPS constellation surround the Earth in six equally spaced orbital planes. Each plane contains four "slots" occupied by baseline satellites. This 24-slot arrangement ensures users see at least four satellites from virtually any point on the planet. GPS consists of three segments: the space segment, the control

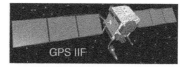

GPS IIA launched 1990–1997
GPS IIR launched 1997–2004
GPS IIR-M launched 2005–2009
GPS IIF launched 2010–2016
GPS III launched 2017-

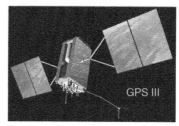

Figure 6.9 GPS satellites. *Source:* Courtesy of United States Government [25]. www.usa.gov/government-works.

Figure 6.10 GPS satellite orbits. One GPS III satellite shown in orbit. *Source:* Courtesy of United States Government [25]. www.usa.gov/ government-works.

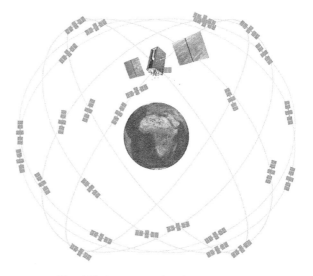

segment, and the user segment. The US Air Force develops, maintains, and operates the space and control segments. Figure 6.11 displays the locations of the Earth stations for ground control and tracking of GPS satellites. Glonass (navigation satellites for Russian defense) and Galileo (navigation system used by European Union) satellite constellations have MEO orbits as well.

Figure 6.11 Earth stations for ground control and tracking of GPS satellites [25]. *Source:* Courtesy of United States Government. www.usa.gov/government-works.

In June 2011, the Air Force successfully completed a GPS constellation expansion called the "Expandable 24" configuration [24]. Three of the 24 slots were expanded, and six satellites were repositioned, so that three of the extra satellites became part of the constellation baseline. As a result, GPS now effectively operates as a 27-slot constellation with improved coverage in most parts of the world. As of 30 June 2017, there were a total of 31 operational satellites in the GPS constellation, not including the decommissioned, on-orbit spares. The GPS constellation contains old and new satellites.

Satellites send timing and position information to a receiver that calculates the distance to the satellite. The receiver accurately calculates latitude, longitude, and altitude when it detects signals from four or more satellites. Determining the latitude and longitude of the receiver only requires three satellites.

Figure 6.12 has a block diagram of the signal generation in a GPS satellite. The GPS signal contains three types of data [26]:

1. Pseudo-random code identifies the transmitting satellite.
2. Ephemeris data contains satellite status as well as the current date and time. This data updates every two hours and is valid for four hours.
3. Almanac data contains the position of the satellite and is updated every 24 hours.

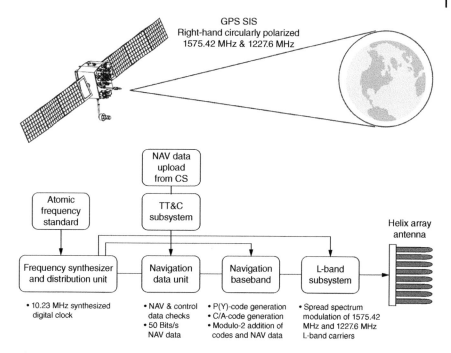

Figure 6.12 GPS signal generation. *Source:* Courtesy of United States Government [26]. www.usa.gov/government-works.

The GPS satellite phased array of helical antennas transmits RHCP signals at two primary GPS frequencies at L band [27]:

- L1 at 1575.42 MHz provides the course-acquisition (C/A) and encrypted precision (P) codes. It is also used for the L1 civilian (L1C) and military (M) codes.
- L2 at 1227.60 MHz carries the P code, as well as the L2C and military codes.

They transmit sufficient power to guarantee a minimum signal power at the Earth surface of −166 dBW.

The GPS signals use CDMA with a unique high-rate PRN Gold code for each satellite [28]. A GPS receiver must have the Gold codes from all the satellites in order to decode the signal. Anyone has access to the C/A code transmitted at 10.23 million chips per second (Mcps). Only the US military has access to the P (precision) code that has a rate of 10.23 Mcps. Both codes send the exact time to the receiver. Only the L1 signal carries the C/A code, whereas the P code appears in both L1 and L2.

Significant subframe contents

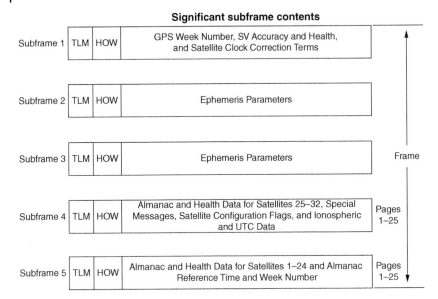

Figure 6.13 Navigation message content and format [26]. *Source:* Courtesy of United States Government. www.usa.gov/government-works.

GPS signals transmit frames containing 15 000 bits at 50 bps over 30 seconds [26]. Frame transmission starts on the minute or half minute according to the atomic clock on each satellite. A frame consists of five 300-bit subframes that each take six seconds to transmit. Subframes contain ten 30-bit words that are 0.6 seconds long. The first two words in each subframe carry telemetry (TLM) and handover (HOW) information. Figure 6.13 lists the data contained within a frame.

6.3 Satellite Link Budget

Space exploration satellites communicate with the Earth ground stations for control and data transfer. Figure 6.14 has a graph of the $1/R^2$ loss between a ground station and satellites at planets in our solar system. Higher frequencies suffer more atmospheric loss, so the transmitter must have more power than at low frequencies. A high power transmitter weighs more, takes up more space, and consumes more operating power than a low power transmitter, so the satellite transmits at a lower frequency than the ground station.

Size and weight constraints on satellites force the ground station to have the large antenna, the high power transmitter, and the sensitive receiver to make up for the huge propagation loss. Reflector antennas or phased arrays provide the high gain needed by the antennas in a satellite link. Parabolic dish antennas

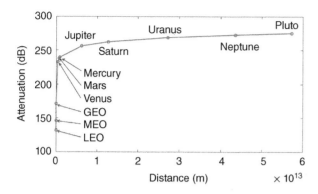

Figure 6.14 Attenuation as a function of distance from the Earth.

are significantly cheaper than phased arrays, so they serve as the workhorses of satellite communications [29]. Phased arrays play an important role in situations that demand high performance like multiple beams or adaptive nulling [30].

Example
NASA launched the Cassini spacecraft in 1997 in order to explore Saturn. The large 4-m Cassegrain dish antenna at the end of the Cassini spacecraft (Figure 6.16) sends data to a dish antenna in the NASA Deep Space Network (DSN) at 8.425 GHz. It has dual high power amplifiers (HPAs) that deliver up to 40 W of power to the antenna. When Cassini was 1.5 billion km from the Earth, then what is the power received by the (i) 70 m (Figure 6.15a) and (ii) 34 m (Figure 6.15b) reflector antennas in the DSN.

Solution
Friis formula: $P_r = \dfrac{P_t G_t A_e}{4\pi R^2} = \dfrac{P_t G_t G_r \lambda^2}{(4\pi R)^2} = \dfrac{P_t \left(\frac{4\pi A_t}{\lambda^2}\right)\left(\frac{4\pi A_r}{\lambda^2}\right)\lambda^2}{(4\pi R)^2} = \dfrac{P_t A_t A_r}{(\lambda R)^2}$

The DL has a wavelength of $\lambda = \dfrac{3 \times 10^{10}}{8.425 \times 10^9} = 3.56 \text{ cm}$

Cassini antenna area: $A_t = \pi 2^2 = 12.6 \text{ m}^2$

The area of the receive antenna is $A_r = \pi r = 3848.5$ or 907.9 m^2

$P_r = \dfrac{P_t A_t A_r}{(\lambda R)^2} = \dfrac{40(12.6)(3848.5)}{(0.0356 \times 1500 \times 10^9)^2} = 6.8 \times 10^{-16} = -151.7 \text{ dBW}$

$P_r = \dfrac{P_t A_t A_r}{(\lambda R)^2} = \dfrac{40(12.6)(907.9)}{(0.0356 \times 1500 \times 10^9)^2} = 1.6 \times 10^{-16} = -157.9 \text{ dBW}$

A combination of the large distance to a satellite from the ground and the low transmit power available on the satellite leads to a very weak signal arriving at

Figure 6.15 Cassini spacecraft. *Source:* Courtesy of NASA. www.nasa.gov.

(a) 70 m   (b) 34 m

Figure 6.16 Reflector antennas in the NASA DSN at Goldstone. (a) 70 m and (b) 34 m. *Source:* Courtesy of NASA. www.nasa.gov

the ground station. The satellite has a fixed signal frequency, transmitter power, and transmit antenna gain due to its limited size and high cost. The ground station on the other hand, has more latitude in size and budget. If the satellite is a distance R from the receiver on the ground, then the SNR at the ground receiver is the ratio of the signal power from the Friis formula to the thermal noise power.

$$\text{SNR} = \frac{P_s}{P_n} = \frac{\dfrac{P_t G_t G_r \lambda^2}{(4\pi R)^2}}{kTB} = \underbrace{\left(\frac{P_t G_t \lambda^2}{kB(4\pi R)^2} \right)}_{\substack{\text{Satellite} \\ \text{transmitter}}} \underbrace{\left(\frac{G_r}{T} \right)}_{\substack{\text{Ground} \\ \text{receiver}}} \tag{6.3}$$

Note that the receiver has no control over the satellite transmitter factor in (6.3). The second factor, G/T, depends on the antenna and LNA design at the Earth station. As a result, satellite ground stations determine the antenna G/T. Increasing G increases the received signal power, while decreasing T decreases the noise power. Thus, increasing G/T enhances a wireless system's probability of detecting the received signal and is an important system specification.

The noise power collected by an antenna comprises [31]:

- Atmospheric attenuation noise caused by absorption and re-radiation of signal energy by water and oxygen molecules in the atmosphere. This noise rapidly increases with decreasing antenna elevation angle and precipitation, because the signal has to travel further through the atmosphere.
- Noise radiated by the Earth that enters through the antenna sidelobes.
- Cosmic (galactic) noise.
- Ohmic losses or losses due to the resistance of the feed system and the antenna reflectors.
- Attenuation between the antenna and the LNA – essential to put the LNA very close to the antenna.

Figure 6.17 presents some of the contributors to the antenna temperature.

The antenna system temperature, T_{ant}, is the system noise temperature in the denominator of G/T and consists of the temperature of external sources (T_{aext}), temperature of antenna losses (T_{aloss}), and temperature of antenna feed line (T_{afeed}) [32].

$$T_{\text{ant}} = T_{\text{aext}} + T_{\text{aloss}} + T_{\text{afeed}} \tag{6.4}$$

The antenna temperature due to external sources comes from integrating the external temperature weighted by the antenna gain over a spherical surface [33].

$$T_{\text{aext}} = \frac{1}{4\pi} \int_0^{2\pi} \int_0^{\pi} G(\theta, \phi) T_0(\theta, \phi) \sin(\theta) d\theta d\phi \tag{6.5}$$

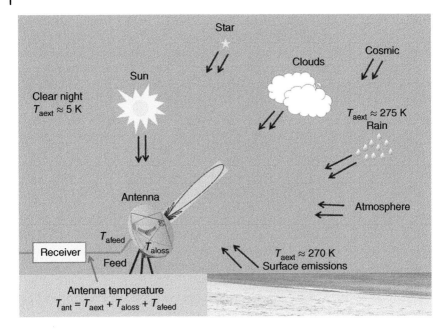

Figure 6.17 Contributions to antenna noise temperature.

where T_0 is the blackbody equivalent temperature. Temperatures range from a few degrees Kelvin to over 290°K depending upon where the antenna main beam points. Typically, the antenna loss and feed temperatures are small.

Example

An antenna has a gain of 40 dB and is 60% efficient. Calculate the G/T when the lossless feed cable is connected to the receiver that has NF = 3 dB for (a) clear night ($T_{\text{aext}} = 5\,\text{K}$) and (b) rain ($T_{\text{aext}} = 275\,\text{K}$).

Solution

$$T_{\text{aloss}} = \left(\frac{1}{0.6} - 1\right) 290 = 193.3\,\text{K}$$

$$T_{\text{rec}} = (10^{3/10} - 1)290 = 288.6\,\text{K}$$

(a) $G/T = \dfrac{10^{40/10}}{193.3 + 288.6 + 5} = 15.2 \Rightarrow 13.1\,\text{dB}$

(b) $G/T = \dfrac{10^{40/10}}{193.3 + 288.6 + 275} = 13.2 \Rightarrow 11.2\,\text{dB}$

6.4   Bent Pipe Architecture

A communications satellite transponder goes by the name of bent pipe architecture. One ground station sends a signal at a high frequency to the bent pipe satellite, and the satellite changes the signal to a lower frequency and retransmits through a HPA to another ground station. The satellite serves to redirect the signal from one ground station to another. Figure 6.18 diagrams the operation of a typical bent pipe satellite.

6.5   Multiple Beams

Rather than using one broad antenna beam (low gain) to cover a large area, a satellite often has multiple high-gain narrow beamwidth beams from one antenna array to cover the same area. The multi-beam approach handles higher data rates and lower powered UL transmitters like those on handsets. Figure 6.19 shows the beam routing for 4 UL and 4 DL beams [34]. Figure 6.20 illustrates a hypothetical example of a satellite having 14 beams covering the United States.

The Ka-band SPACEWAY® satellite system consists of multiple geo-synchronous satellites operating over a 500 MHz Ka-band bandwidth sub-divided into

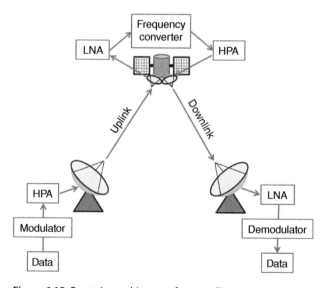

Figure 6.18 Bent pipe architecture for a satellite communications system.

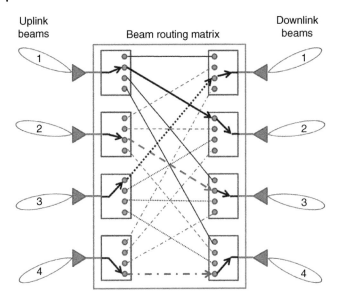

Figure 6.19 Beam routing matrix for 4 UL and 4 DL beams.

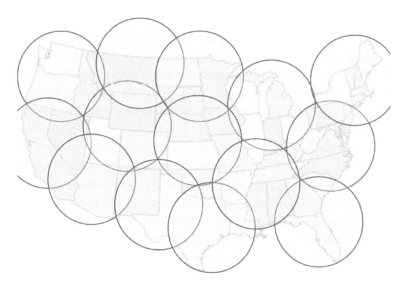

Figure 6.20 US coverage with 14 beams.

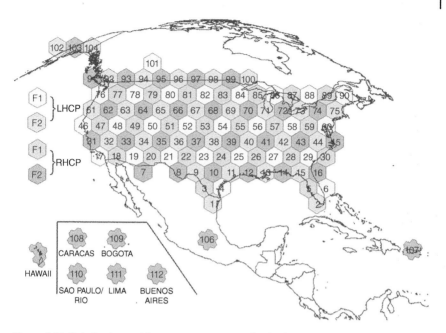

Figure 6.21 Polarization and frequency assignment for the Spaceway beams. *Source:* Reprinted with permission of Whitefield et al. 2006 [35]. © 2018, IEEE.

16, 62.5 MHz subbands with 8 LHCP and 8 RHCP using FDMA and TDMA [35, 36]. The satellites use 112 UL cells positioned over the US, Puerto Rico, and several major cities in Central and Latin America as shown in Figure 6.21. The DL operates from 19.7 to 20.2 GHz and the UL from 29.5 to 30.0 GHz. Each UL cell has a 0.5° fixed beam with the opposite polarization of the DL beam. Seven DL microcells with 0.189° beam widths lie within a UL cell and have a polarization opposite to that of the UL cell. These microcells improve approximately 5 dB DL beam gain loss at the cell edge to less than 1 dB. Adjacent cells have either different polarizations (RHCP or LHCP) and/or different frequencies (F1 or F2) to allow cell reuse.

6.6 Stabilization

Spin and three-axis body stabilization keep the antennas and solar panels pointing in the desired directions. Spin stabilization works for satellites that have symmetry about one axis like the model of the Canadian Alouette satellite in Figure 6.22a, while three-axis body stabilization works for any satellite – no symmetry needed (TDRS in Figure 6.22b).

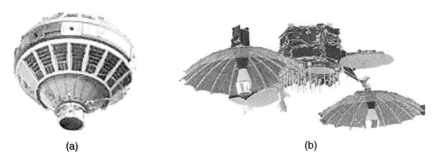

(a)   (b)

Figure 6.22 Satellite stabilization (a) spin and (b) three-axis body.

Spin stabilized satellites fire jet thrusters to start the spinning once in orbit. The satellite spins between 30 and 120 rpm about its axis of symmetry [17]. Spinning induces a gyroscopic effect that keeps the spin axis pointing in a desired direction with little wobble. Spin stabilization requires a design that has antennas and solar panels symmetrically placed about the satellite. Traditionally, the solar panels wrap around the satellite body. Placing multiple extended flat panels around the satellite body works as well.

Three-axis stabilization keeps the satellite in a stationary position relative to its orbit in order to maintain constant antenna and solar panel pointing accuracy. Satellites have three axes that need stabilization: pitch, yaw, and roll. The yaw axis points toward the Earths center, the pitch axis is normal to the orbital plane, and the roll axis is tangent to the orbit. Each axis either has a large flywheel or reaction wheel for stabilization. Reaction wheels keep the satellite much more steady than flywheels and thrusters, but their weight shortens the satellite's lifespan [17]. Sensors monitor external references, such as the sun or stars, then a controller uses the information to adjust the satellite orientation by controlling the reaction wheel spin or firing the thrusters on the appropriate axis.

Problems

6.1 Using trigonometry, show that a geostationary satellite with an antenna that has a 17.3° beamwidth covers approximately 1/3 of the Earth surface outside of the polar regions.

6.2 Find the minimum one-way signal latency (time delay) for an Earth station on the equator communicating with a geostationary satellite.

6.3 Calculate the number of days that significant levels of Sun interference will be experienced at each equinox for an 11-m diameter antenna at 11 GHz.

6.4 Approximate the maximum duration of a Sun transit at each equinox for an 11-m diameter antenna at 11 GHz.

6.5 An Earth station has an antenna temperature of 45 K that feeds to an LNA with $T = 100$ K and $G = 50$ dB, followed by a mixer with $T = 1000$ K. Find the system temperature.

6.6 Find the CNR for a geosynchronous satellite operating at 6.1 GHz having an antenna with a gain of 26 dB. Its receiver has $T = 500$ K, $B = 36$ MHz, and gain of 110 dB. The Earth station transmits 100 W via a 54 dB antenna gain.

6.7 Calculate the gain of a satellite antenna having orthogonal beamwidths of 3° and 6° at the edge of its coverage zone.

6.8 Find the CNR at an Earth station with an antenna gain of 53 dB receiving a signal at 3.875 GHz from a satellite at an orbit of 39 000 km, transmitting 10 W through an antenna with a gain of 30 dB.

6.9 Find the CNR at a ground station with $G_r = 1$ dB, $T = 260$ K, and $B = 20$ kHz. The satellite is 2000 km away and transmits a 2.5 GHz signal at 0.5 W through an antenna with $G_t = 18$ dB.

6.10 A satellite system operating at 4.15 GHz has an Earth station antenna 30 m in diameter and an aperture efficiency of 68%. The system noise temperature varies between 79 and 88 K depending on weather conditions. Find the variation in G/T at the Earth station.

6.11 A satellite at a distance of 40 000 km communicates with a ground station at 4.0 GHz with a 2 W transmitter and a 17-dB antenna gain. Find the power received by a ground station antenna having an effective area of $10\,\text{m}^2$.

References

1 http://www.inetdaemon.com/tutorials/satellite/communications/frequency-bands (accessed 26 August 2017).

2 Clarke, A.C. (1945). V2 for ionosphere research? *Wireless World* L1 (10): 305–308, Letters to the Editor.

3 Smil, V. (2017). Sputnik at 60. *IEEE Spectrum*: 20.

4 https://www.nasa.gov/multimedia/imagegallery/image_feature_559.html (accessed 24 August 2017).

5 Jakes, W.C. (1961). Participation of bell telephone laboratories in project echo and experimental results. *The Bell System Technical Journal* 40 (4): 975–1028.

6 http://www.arrl.org/news/oscar-i-and-amateur-radio-satellites-celebrating-50-years (accessed 24 August 2017).

7 https://airandspace.si.edu/collection-objects/communications-satellite-telstar (accessed 24 August 2017).

8 https://www.nasa.gov/topics/technology/features/telstar.html (accessed 24 August 2017).

9 Shennum, R.H. and Haury, P.T. (1963). A general description of the Telstar spacecraft. *The Bell System Technical Journal* 42 (4): 801–830.

10 https://appel.nasa.gov/2010/02/25/ao_1-7_sf_history-html (accessed 24 August 2017).

11 https://nssdc.gsfc.nasa.gov/nmc/spacecraftDisplay.do?id=1963-004A (accessed 24 August 2017).

12 https://nssdc.gsfc.nasa.gov/nmc/spacecraftDisplay.do?id=1963-031A (accessed 24 August 2017).

13 Williamson, M. (2006). *Spacecraft Technology: The Early Years*. London: IET.

14 Yuan, J., Yang, D. and Sun, X. (2006). Single access antenna pointing control system design of TDRS (pp. 5–1097). 2006 1st International Symposium on Systems and Control in Aerospace and Astronautics, Harbin.

15 https://en.wikipedia.org/wiki/Low_Earth_orbit (accessed 24 October 2017).

16 Clarke, A.C. (1945). V2 for ionosphere research? *Wireless World* L1 (2): 58, Letters to the Editor.

17 H. Hausman, "*Fundamentals of Satellite Communications*, part 1," https://www.ieee.li/pdf/viewgraphs/fundamentals_satellite_communication_part 1 .pdf (accessed 25 October 2017).

18 https://www.itu.int/dms_pubrec/itu-r/rec/s/R-REC-S.1525-1-200209-I!!PDF-E.pdf (accessed 3 January 2018).

19 Haupt, R.L. (2010). *Antenna Arrays: A Computational Approach*. Hoboken, NJ: Wiley.

20 https://www.airspacemag.com/space/how-things-work-space-fence-180957776 (accessed 25 February 2019).

21 Sekiguchi, K. (2016). Iridium contributes to "maritime safety" (pp. 90–92). 2016 Techno-Ocean, Kobe.

22 Leopold, R.J. and Miller, A. (1993). The IRIDIUM communications system (pp. 575–578). 1993 IEEE MTT-S International Microwave Symposium Digest, Atlanta, GA.

23 http://spaceflight101.com/spacecraft/iridium-next (accessed 31 May 2018).

24 https://www.gps.gov/systems/gps (accessed 4 February 2019).

25 https://www.gps.gov/multimedia/images (accessed 31 May 2018).

26 US Department of Defenses and GPS NAVSTAR (2008). *Global positioning system standard positioning service signal specification*, 4e. GPS NAVSTAR.

27 https://www.navtechgps.com/gnss_facts (accessed 26 February 2019).

28 Holmes, J.K. and Raghavan, S. (2004). A summary of the new GPS IIR-M and IIF modernization signals (Volume 6, pp. 4116–4126). IEEE 60th Vehicular Technology Conference – VTC2004-Fall.

29 Rahmat-Samii, Y. and Haupt, R.L. (2015). Reflector antenna developments: a perspective on the past, present and future. *IEEE AP-S Mag* 57 (2): 85–95.

30 Haupt, R.L. and Rahmat-Samii, Y. (2015). Antenna array developments: a perspective on the past, present and future. *IEEE AP-S Mag* 57 (1): 86–96.

31 Ho, C., Kantak, A., Slobin, S., and Morabito, D. (2007). Link analysis of a telecommunication system on earth, in geostationary orbit, and at the moon: atmospheric attenuation and noise temperature effects. IPN Progress Report 42-168, Jet Propulsion Laboratory, 15.

32 Cakaj, S., Kamo, B., Enesi, I., and Shurdi, O. (2011). Antenna noise temperature for low earth orbiting satellite ground stations at L and S band (pp. 1–6, 17–22). Third International Conference on Advances in Satellite and Space Communications, Budapest, Hungary (April 2011).

33 Ulaby, F.T., Moore, R.K., and Fung, A.K. (1981). *Microwave Remote Sensing Active and Passive*, vol. 1. Reading, MA: Addison-Wesley Publishing Co.

34 Elbert, B.R. (2008). *Introduction to Satellite Communication*. Artech House.

35 Whitefield, D., Gopal, R. and Arnold, S. (2006). Spaceway now and in the Future: on-board IP packet switching satellite communication network (pp. 1–7). MILCOM 2006–2006 IEEE Military Communications Conference, Washington, DC.

36 Fang, R.J.F. (2011). Broadband IP transmission over SPACEWAY® satellite with on-board processing and switching (pp. 1–5). 2011 IEEE Global Telecommunications Conference – GLOBECOM 2011, Houston, TX.

7

RFID

Automatic identification (auto ID) technologies identify, track, and collect data about objects of interest. Auto ID includes bar codes, optical character readers, radio frequency identification (RFID), and some biometric technologies, such as retinal and finger print scans. They improve data accuracy and reduce the amount of time and labor needed to manually input data [1]. An RFID system has a reader that communicates with a tag. The tag either backscatters a signal (passive system) or transmits a signal (active system) back to the reader where it is detected. RFID is useful for inventory, tracking packages, identifying lost dogs, verifying identification cards, measuring temperature, etc. This chapter introduces RFID system technology.

7.1   Historical Development

The first passive RFID system originated in WWII when German pilots rolled their planes as they returned to base in order to modulate the backscattered signal from German radar [2]. This modulated return distinguished them from Allied aircraft, so that they did not attract German anti-aircraft fire. In 1939, the British developed the first active identify friend or foe (IFF) system. An IFF transponder on a British plane emitted a unique modulated signal when illuminated by British radar that identified the friendly plane to the radar operators.

   In 1945, the Soviet Union gave a hand-carved replica of the Great Seal of the United States to the US Ambassador. Inside the carving was a clever passive listening device called "The Thing" or "The Great Seal Bug" [3]. This bug had a high-Q resonant cavity with a very thin 75 μm membrane on one end (Figure 7.1). A monopole antenna sticking out of the bottom couples a strong RF signal into the cavity at the resonant frequency. Anyone talking near the carving causes the membrane to vibrate and change the size of the cavity which in turn changes the resonant frequency and modulates the signal retransmitted

Wireless Communications Systems: An Introduction, First Edition. Randy L. Haupt.
© 2020 John Wiley & Sons, Inc. Published 2020 by John Wiley & Sons, Inc.

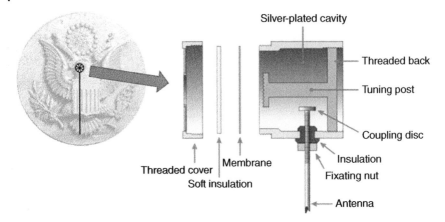

Figure 7.1 Great seal bug [3]. *Source:* Courtesy of NASA. www.nasa.gov.

from the monopole. The Soviet Union demodulated the AM signal, enabling them to listen to the conversations. This bug had no internal power source and functioned like a passive RFID tag.

In 1948, Harry Stockman came one step closer to modern RFID by proposing a communication system in which a transmitter sends a carrier signal to a vibrating reflector. The reflector modulates the backscattered signal that a receiver detects and demodulates [4]. His ideas for modulating the backscattered signal relied on mechanics similar to "The Thing." Stockman noted that modulation with mechanical vibrations limited this approach to low frequencies.

RFID outperforms bar codes for inventory and tracking applications. Joe Woodland based his barcode idea on Morse code [5]. He and his colleagues received a patent in 1952, but his idea did not blossom until the laser and desktop computers were invented [6]. The bar code contains a universal product code (UPC) that is a 12-digit number unique to a manufacturer (first six numbers) and type of product (next five numbers) as shown in Figure 7.2. The data contained within the barcode depends on the relative width of the black and white bars and not on the physical size of the barcode. Dark bars are 1–3 units wide while white bars are 1–4 units wide. A UPC cannot distinguish one can of Coke from another can of Coke, but it can distinguish a liter bottle of Pepsi from a can of Coke. A check digit for error detection at the end of the barcode comes from the following algorithm [7]:

1. Add the numbers in the odd positions then multiply the sum by 3.

$$3(0 + 8 + 4 + 0 + 2 + 6) = 60$$

2. Add the numbers in the even positions.

$$7 + 7 + 2 + 8 + 9 = 33$$

Figure 7.2 UPC number and associated bar code.

Manufacturer identification number	Item number	Check digit
78742	08296	7

0   7 8 7 4 2   0 8 2 9 6   7

3. Add the numbers found in steps 1 and 2 to get the number q. The check digit plus q equals a number that is a factor of 10.

$60 + 33 = 93$ so the check digit is 7

If the scanner does not calculate 7 for the check digit, then it demands a rescan.

The first modern far-field RFID patent envisioned a base station corresponding with a transponder that has memory and data processing power [8]. This patent states in the abstract: "In the preferred inventive embodiment, the transponder generates its own operating power from the transmitted interrogation signal, such that the transponder apparatus is self-contained." Around the same time, Los Alamos National Laboratory developed an RFID system with a transponder on a truck carrying nuclear materials and readers at the security gates [9]. The transponder responded to the reader with data that identified the truck and driver. This system formed the basis for automated toll payment systems for roads, bridges, and tunnels in the mid-1980s. Around the same time, Los Alamos also developed a passive UHF RFID tag to track cows.

A 1983 patent outlined an RFID card system [10]: "An automatic identification system wherein a portable identifier, preferably shaped like a credit card, incorporates an oscillator and encoder so as to generate a programmable pulse position-modulated signal in the radio frequency range for identification of the user. The identifier can be made to generate the identification signal constantly or can be made for stimulated transmission responsive to an interrogation signal. The identification signal can be preset or can be programmable by use of a programmable memory." This type of card is now in wide use throughout the world for public transportation, motel keys, etc.

In the mid-1980s, Fairchild Semiconductor, Motorola, Texas Instruments, IBM, and DEC developed an RFID monetary system that implanted an RFID tag in a person's hand [11]. The tag enabled transactions by connecting to the appropriate financial institution. Implanting tags into humans killed this idea. The resulting patents spurred the very successful pet tagging industry for identifying and recovering lost pets.

IBM developed and patented a UHF RFID system that had a long range and high data rate in the early 1990s [2]. Wal-Mart partnered with IBM to create a commercial system that never came to fruition. In the end, a barcode company called Intermec bought all of the IBM RFID patents. The Intermec RFID systems were used in applications from warehouse tracking to farming. Low volume and high cost limited sales, though. The lack of international standards discouraged wide-spread use.

In 1999, RFID technology exploded after MIT created the Auto-ID Center [12]. This center stimulated the wide acceptance of RFID through the introduction of the Electronic Product Code (EPC). This innovation tracked objects using the Internet. In 2003, EPCglobal replaced the Auto-ID Center in order to promote the EPC standard. It manages the EPC network and standards, while its sister organization, Auto-ID Labs, manages and funds research on EPC technology.

7.2 RFID System Overview

RFID systems have a reader and a tag [13]. The reader and tag communicate through a sequence of commands (called the inventory round) that leads to the tag identification and sometimes an exchange of data. Readers communicate with tags in an area called the interrogation zone (IZ) where the reader signal level exceeds the tag sensitivity. An RFID reader transmits an encoded radio signal that interrogates the tag. The RFID tag in the IZ responds with a carrier modulated by data stored in memory. The data might be a serial number, time stamp, or configuration instructions which aids in taking inventory or directing a package to be moved to a certain location.

Figure 7.3 shows block diagrams of monostatic and bistatic RFID systems. A monostatic reader's transmitter and receiver share an antenna. The circulator in the reader directs the transmitted signal to the antenna and the received signal to the receiver while isolating the transmit and receive circuits. Circulators protect the sensitive low power receive circuitry from the much higher powered transmit signal as well as allows the transmitter and receiver to operate at the same time. Bistatic RFID eliminates the circulator by using separate antennas for transmit and receive. Bistatic systems have high isolation between the transmit and receive circuits.

An inlay is the tag's guts: antenna and circuit on a substrate [14]. A smart label has an inlay inside of a barcode label with adhesive for easy attachment. The label surface contains information such as sender's address, destination address, and product information. Smart labels are cheap because they are

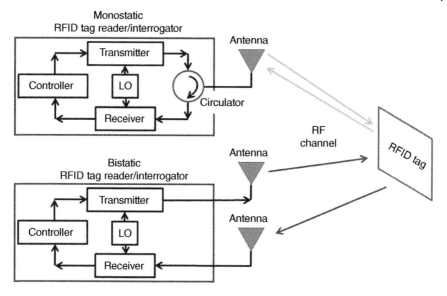

Figure 7.3 RFID system diagram for monostatic and bistatic RFID readers.

printed in large quantities. Hard tags made from polycarbonate, ceramic, steel, polystyrene, and polypropylene are rigid and thicker than labels. They cost much more than smart labels and are attached to an object by adhesive, shrink wrap, stitching, straps, screws, or other means.

RFID tags fall into two broad categories: near field and far field (Table 7.1). Near-field tags operate close to the reader at low frequencies and low data rates. They use loop antennas to couple LF or HF magnetic field between the reader and tag. Low frequencies penetrate materials well, so the environment has little impact on the signal. In contrast, far-field tags use dipole antennas and communicate at much higher frequencies and data rates. They communicate with the reader over much larger distances. Their fields have small penetration depths into materials, so the environment significantly impacts the signal.

Tags are either active, passive, or semi-passive. Active tags are transponders with their own power source that boosts the signal power sent to the reader and increases the communication range. Passive tags harvest energy from the reader's signal in order to power the electronics that retrieves data from memory, then modulates the signal scattered back (backscatter) to the reader. The cheap passive tags have limited range. Semi-passive or battery-assisted tags (BATs) use battery power for data collection and processing but not for boosting the signal strength sent to the reader. The best type of tag to use in a system depends on the physical environment, required read range, and material properties of the tagged object.

Table 7.1 RFID frequency allocations [15–17].

	Frequency band	Range	Tag type	Data rate	Applications
Near field	120–150 kHz (LF)	10 cm	Passive	Low	• Access control • Livestock tracking
	13.56 MHz (HF)	10 cm–1 m	Passive	Low to moderate	• Ticketing • Payment • Data transfer • Airline baggage • Libraries
Far field	433 MHz (VHF)	1–300 m	Active	High	• Tracking • Locating
	Europe: 865–868 MHz North America: 902–928 MHz (UHF)	1–12 m	Passive Semi-passive Active	Moderate to high	• Parking lot access • Toll collection • Supply chain
	2450–5800 MHz (microwave)	1–40 m	Passive Semi-passive Active	High	• Toll roads • Vehicle identification • Supply chain

RFID systems follow international and regional standards. The ISO (International Organization for Standardization) has a series of standards defining the interface between the reader and tag [18]:

1. 18000-1: Generic Parameters for the Air Interface for Globally Accepted Frequencies
2. 18000-2: Parameters for Air Interface Communications below 135 kHz
3. 18000-3: Parameters for Air Interface Communications at 13.56 MHz
4. 18000-4: Parameters for Air Interface Communications at 2.45 GHz
5. 18000-5: Parameters for Air Interface Communications at 5.8 GHz (Withdrawn)
6. 18000-6: Parameters for Air Interface Communications at 860–960 MHz
7. 18000-7: Parameters for Air Interface Communications at 433 MHz

Governments throughout the world already assigned UHF frequencies around 900 MHz long before RFID, so no internationally recognized frequency band exists for RFID. As a result, the newest Gen 2 (second generation) protocol works at a narrow frequency band between 860 and 960 MHz.

Communication between reader and tag is either full duplex (FDX) or half duplex (HDX) [19]. In an FDX system, the reader and tag transmit and receive data at the same time by using different frequency bands. In an HDX system, either the reader or tag transmits data, but not both at the same time. An HDX reader charges a capacitor in the tag in the initial wakeup. After the reader stops transmitting, the tag uses power from the charged capacitor to transmit the requested data to the reader. An HDX reader uses simpler decoding techniques than FDX. A kill command from the reader permanently stops all tag functions. An inoperable tag is called a quiet tag [14].

7.3 Tag Data

A UPC code enables a grocery store to count bottles of Pepsi sold but cannot help Amazon track the purchase and delivery of a customer's item. Tracking individual items requires a much more sophisticated code. The EPC solves this problem by incorporating information about an individual item with the UPC.

The EPC is a unique universal identifier assigned to every product and all categories of products in the world [20]. It contains the information indicated in Figure 7.4 [21]. The header identifies the information structure encoded on the tag, including the type of EPC and the encoding length (64, 96, or 128 bits). Additional instructions for the reader follow the header. The filter value tells the Reader to select or disregard certain tags, while the partition value tells the reader where the company prefix ends and the item reference begins. The EPC manager number identifies the entity that maintains the remaining partitions of the EPC. An EPC Manager assigns a variable-length object class and a serial number to a specific instance of that product. Figure 7.5 is an example of the last three parts of the EPC binary encoding shown in Figure 7.4.

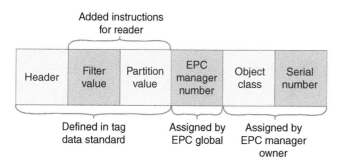

Figure 7.4 EPC binary encoding on RFID tag.

Figure 7.5 EPC code example.

EPC = 34584116155211234567890

Some tags use electrically erasable, programmable, read-only memory (EEPROM) that does not require continuous power to store data [22]. A tag with EEPROM stores the data for a long period (several years), even without any power. The type of data stored in the tag falls into one of the four categories:

1. *EPC*: Electronic Product Code.
2. *User*: user defined data about the item, such as item type, last service date, or serial number (32 to over 64 000 bits).
3. *Reserved*: access and lock passwords that limit viewing and editing data.
4. *TID, tag identifier*: unique random number provided by the manufacturer and cannot be changed. The reader requires special settings in order to read this number (instead of the EPC).

Tag read-only (RO) memory resembles a bar code, because it contains a small amount of static data that cannot change. A write-once-read-many (WORM) tag is programmed only once. A RO or WORM tag contains information like manufacturing date or location. Data in read–write (RW) or smart tags can be modified many times. A pallet may have both RO and RW memory in which the RO memory contains the pallet serial number and the RW indicates the contents at a particular time.

7.4 Tag Classes

A tag needs power for its electronics and signal generation and transmission. Passive tags derive their power from the reader signal and fall under Class 0–2. The battery in a semi-passive tag (Class 3) powers electronics but does not help amplify the signal returned to the reader. Active tags get their power from a battery and fall under Class 4 or 5.

7.4.1 Passive Tags

A passive tag uses energy harvesting to extract operating power from the reader signal. This small amount of power runs the integrated circuit (IC) and powers the backscattered signal. A passive tag outside the IZ cannot harvest sufficient power to properly function. The lack of a continuous power source limits the amount of the data transmitted by a passive tag. Passive RFID tags fall into three of the six standard classifications [23]:

- *Class 0*: A passive tag with RO memory used to detect the tag's presence. A 0+ tag is a write once/read-only memory (WORM) tag.
- *Class 1*: A passive, WORM, backscatter tag with a one-time, field-programmable, nonvolatile memory. Tag data supplied by the manufacturer or user.
- *Class 2*: A passive backscatter tag that has identification as well as other information in memory.

Passive tags are ideal for nonreusable applications. Some advantages of passive tags include: [24]

- Small size
- Lightweight
- Inexpensive (depends on quantity)
- Generates no RF noise
- Longer life (20-plus years)
- Resists harsh environment

Disadvantages of passive tags include:

- Requires a reader to supply the power
- Has limited data storage
- Sends a weak signal to reader
- Has a short read range (several cm to several m)

RF-DC power conversion efficiency (η_{PCE}) equals the power arriving at the load divided by the power received by the antenna. The rectifier, voltage multiplier, and storage elements contribute to η_{PCE}. For a passive tag, $\eta_{PCE}P_r$ must exceed the tag sensitivity in order to enable a response. The energy harvesting circuit charging time depends on distance from reader, antenna size, inductance, Q factor, and charging station field strength.

Passive tags balance the power needs of the IC in the tag with the backscattered power needed to exceed the sensitivity of the reader. The tag antenna passes the reader signal to a matching circuit then through a circulator to the demodulator and energy harvesting circuit (Figure 7.6). Next, the DC output from the harvesting circuit provides power to the memory, modulator, and demodulator. A harvesting circuit or charge pump converts the AC signal into DC then amplifies it using a voltage multiplier. Schottky diodes with low series resistance or metal–oxide–semiconductor field effect transistors (MOSFETs) produce a high η_{PCE} [25]. A large energy-storage capacitor bank smooths the rectifier output waveform and converts it into a nearly DC signal. In addition,

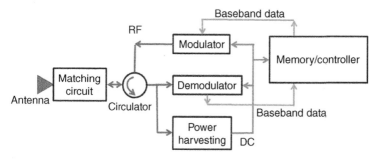

Figure 7.6 Passive RFID tag.

it stores enough energy to operate the tag when the reader signal is low or not present. The rectifier's power conversion efficiency is defined for a voltage, V_{load}, across a load resistance, R_{load} [26]:

$$\eta_{PCE} = \frac{V_{load}^2}{R_{load}P_{in}} \qquad (7.1)$$

where P_{in} is the received rectifier power. This power enables the demodulator to pass a baseband signal to the memory/controller and the memory/controller to send its data to the modulator that passes an RF signal to the antenna for transmission back to the reader.

The DC voltage (V_{DC}) of the Dickson multiplier or Dickson charge pump in Figure 7.7 depends on the number of stages [27]. Adding more stages increases the output voltage but decreases the efficiency [28]

$$\eta_{PCE} = \frac{V_{DC}}{V_{DC} + 2NV_D} \qquad (7.2)$$

where N = number of stages and V_D = forward voltage drop. For a Schottky diode, $0.15 \leq V_D \leq 0.45$ V. The Dickson multiplier has the disadvantages of requiring multiple stages for a usable supply voltage and may also require large passive components (e.g. large capacitors) to reduce the output ripple.

Other circuits similar to the Dickson multiplier are referred to as voltage doublers or multipliers. Another approach uses bridge rectification circuits. Major considerations in choosing the right rectifier circuit include impedance matching between the rectifier and antenna, the input power level to the

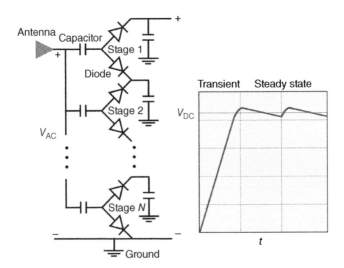

Figure 7.7 Diagram of a Dickson charge pump and its output voltage.

rectifier, the DC power and DC voltage required, and the RF frequency. An RF energy-harvesting architecture has the advantage of fitting on the IC [15].

The Federal Communications Commission (FCC) limits transmit power in the band 902–928 MHz to less than 1 W. Consequently, link analysis concludes that a reader must deliver between 10 and 30 μW to the tag IC in order to activate the tag. If the charge pump is about 30% efficient, then tags with perfectly matched antennas need their antenna to deliver 30–100 μW in order to power the chip.

Example

Assume the 915 MHz tag IC has a threshold power level of −10 dBm. The reader transmits 30 dBm through an antenna with 6 dB gain. Assume the tag antenna is isotropic and $\eta_{\text{PCE}} = 0.3$. What is the maximum extent of the IZ?

Solution

Start with the Friis transmission formula then solve for r:

$$P_r = \frac{P_t G_t c^2}{(4\pi r f)^2} \eta_{\text{PCE}}$$

$$r = \frac{c}{4\pi f} \sqrt{\frac{P_t G_t \eta_{\text{PCE}}}{P_r}} = \frac{3 \times 10^8}{4\pi(915 \times 10^6)} \sqrt{\frac{0.3 \times 1 \times 4}{0.0001}} = 2.86 \text{ m}$$

7.4.2 Tags with Batteries or Supercapacitors

Either a battery or supercapacitor provides power to active and semi-active tags [26]. A battery has a high energy storage capacity, while supercapacitors have a large capacitance that stores a significant amount of electrical charge. Usually, a supercapacitor has less energy-storage capacity than a battery, but smaller parasitics makes it attractive. Larger tags with big IZs need large batteries that require replacement or recharging. A typical active or semi-passive tag uses a lithium-thionyl chloride battery [29]. These cells offer high energy density and a long shelf life, and withstand extreme temperatures better than lithium-ion batteries. They are not rechargeable, though.

7.4.2.1 Semi-Passive Tags

A Class 3 tag is a semi-passive or BAT that uses a battery instead of an energy harvesting circuit to power the IC [23]. A semi-passive tag has a longer read range than a passive tag, because the backscattered signal uses 100% of the harvested power, since the battery only powers the IC. Tags with sensors need batteries to continuously operate in the absence of a reader signal. A tag sensor might measure temperature, pressure, relative humidity, acceleration, vibration, motion, altitude, chemicals, or other physical attributes.

Reusable plastic containers in manufacturing and food-processing justify the expense of semi-passive tags. Some advantages of semi-passive tags include [24]

- Increased read range
- Reduced reader power
- Increased memory storage
- Supports environmental sensors
- Reduced radio noise

while the disadvantages are

- Sensitivity to harsh environment
- Limited battery life
- Costs more than passive tags
- Larger size and weight than passive tags

The link budget calculated from the Friis transmission formula determines the amount of battery power needed to operate an active tag. The tag power received at the reader is given by

$$P_r = \frac{P_t G_r G_t c^2}{4\pi f_c^2 r^2} L \tag{7.3}$$

The generic loss constant, L, includes circuit and channel losses. A tag battery generates P_t that radiates from an antenna with gain, G_t and is received by a reader antenna with gain, G_r. The maximum range between the reader and tag is found by solving (7.3) for r when the receiver sensitivity is given by P_r.

7.4.2.2 Active Tags

Active RFID systems primarily operate at 433 MHz (VHF) but also inhabit the UHF and microwave bands as well. There are two classes of active tags [23]:

- *Class 4*: Active tags that have a battery to power onboard circuits and sensors as well as the transmitter.
- *Class 5*: Same as class 4 tag but also communicates with other classes of tags and devices. It is also called a reader tag.

Active tags have a range of up to 100 m, and their higher power penetrates materials that have low conductivity. Their lifespan depends on the battery life. In general, large objects, such as rail cars, big reusable containers, and other assets that need to be tracked over long distances need active tags. One application of an active tag monitors temperature within a refrigerated truck [24]. Another common application monitors security and integrity of a shipping container.

Active tags are either transponders or beacons [30]. Transponders wake up when they receive a signal from a reader and then respond with their own signal.

They have a long battery life, since they power off most of the time. Beacons have higher power requirements, on the other hand, because they continuously emit a location signal at predetermined intervals in order for the receiver to precisely locate a tag. In a real time locating system, a beacon emits a signal with its unique identifier at preset intervals.

7.5 Data Encoding and Modulation

The reader transmits a signal that wakes up and powers a passive tag as well as sends data. Most small, low cost tags use a simple modulation scheme that minimizes electronics on the tag, such as ASK.

A reader using OOK encoding will generate a long sequence of zeros at some point in the data stream. A string of zeros means no RF signal exists, so a passive tag turns off, because it cannot harvest any power. To overcome the OOK problem, readers encode binary data using pulse-interval encoding (PIE). PIE is reminiscent of Morse Code, because it uses a long pulse (dash) to represent a "1" and a short pulse to represent a "0." A Type A Reference Interval (TARI) measures the width of the data "0" symbol. TARI values range from 6.25 to 25 μs (25 μs gives a data rate of 160 kbps) [31]. PIE dedicates more power to each bit than OOK [29]. Since the one and zero symbols in PIE have a carrier present as shown in Figure 7.8, the tag harvests power from both zeros and ones.

PIE is less efficient than OOK, because a PIE TARI is 1/3 as long as an OOK pulse for the same data rate [31]. As a result, the PIE bandwidth is three times wider than the OOK bandwidth. In order to fit the PIE bandwidth inside a 500 kHz channel, the data rate cannot exceed 85 kbps, which corresponds to reader data rates in the United States when using unfiltered PIE [31]. OOK has equal low and high pulses, so it contains 50% of the maximum power delivered to the tag. If PIE has a high pulse that is three times as long as the low pulse then it delivers about 63% of peak power. The data rate depends on the data: a message with many zeros transmits faster than a message with many ones. Increasing the power delivered to the tag comes at the price of a lower data rate.

Figure 7.8 PIE encoding with ASK modulation.

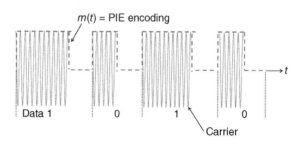

$m(t)$ = PIE encoding

Data 1   0   1   0

Carrier

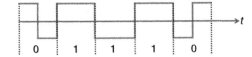

Figure 7.9 Example of FM0 encoding.

Tags encode data using either FM0 or Miller encoding. ASK, FSK, or PSK then modulates the encoded data.

FM0 encoding has the following rules:

1. Phase inversion at the beginning of each new symbol.
2. "1" is constant over the bit period.
3. "0" has one phase change during the bit period.

Figure 7.9 shows an example of FM0 encoding applied to a binary data stream. Its data rate ranges from 40 to 640 kbps. For RFID using FM0, an SNR of around 10 dB or more is usually sufficient [32].

Miller coding starts with Manchester coding of the signal in which a NRZ data bit stream is XORed with a clock at a power of two times the bit rate [26]. This signal is not spectrally efficient for transmission, though, because it has many transitions. Miller encoding increases the spectral efficiency by using these coding rules [26]:

1. "1" has a phase transition during the symbol period
2. "0" is constant over the symbol period
3. No phase transition between symbols unless consecutive zeros
4. Subcarrier modulation: The baseband Miller encoded waveform is multiplied by a square wave in order to move the information frequency further from the carrier frequency. The advantage is better interference rejection. The disadvantage is a reduced data rate.
5. M square wave transitions per symbol: $M = 2, 4, 8$

Miller coding hardware implementation starts with an XOR gate that generates a Manchester code followed by a falling-edge-triggered flip-flop divider that eliminates every other transition. The output drives the backscatter transistor to modulate the carrier.

Example

Given the data stream: 0 1 1 0 0 1 0 1, plot the baseband signal voltage for NRZ, Manchester, and Miller coding.

Solution

Figure 7.10 shows the resulting plots. Note that the Miller coding has much fewer transitions than the Manchester coding.

Figure 7.10 NRZ data in terms of Manchester and Miller coding. *Source:* Reprinted with permission of Shan et al. 2016 [26]. © 2018, IEEE.

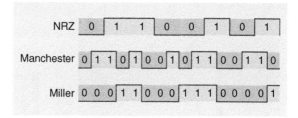

7.6 Reader-Tag Communication

Reader-tag communication either takes place in the near field or far field. Near-field communications (NFCs) uses a loop antenna operating at LF or HF. Typically, the reader and tag are less than 10 cm apart for LF and 1 m apart for HF. LF or HF inductive coupling between loop antennas results in slow data rates. Far-field communication uses dipoles operating at UHF and above to transfer data through capacitive coupling in the far field.

7.6.1 Near Field

Typical components of an NFC system appear in Figure 7.11. At close distances, the magnetic field dominates and decreases as $1/r^3$ in the near field. Figure 7.12 is a plot of the total power radiated by a loop antenna along with its near and far field components. The demarcation between the near field and far field is approximately $1/k = \lambda/2\pi$. An NFC reader radiates a magnetic field that does not have nulls in its IZ. Power transfer between two loops in NFC depends on [33]:

- operating frequency
- number of turns in the coils
- area of the coils
- *angle between the coils*: maximum coupling occurs when the loops are in the same plane
- distance between the coils

Advantages of NFC include:

- the short read range protects sensitive data from intrusion. For instance, credit card tags need a short read range to deny access to other readers.
- the environment has little impact on the antenna performance.
- LF and HF signals penetrate most materials.
- very reliable

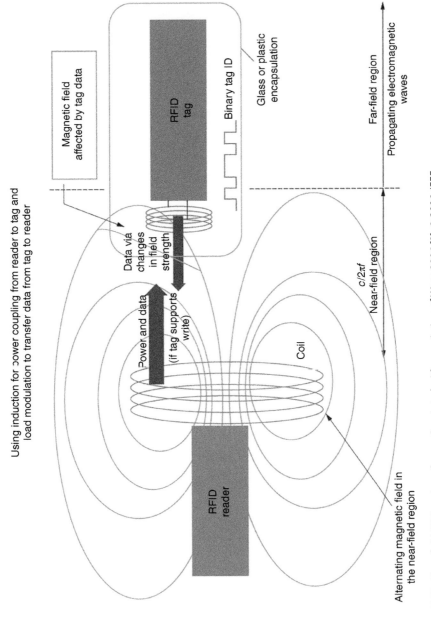

Using induction for power coupling from reader to tag and load modulation to transfer data from tag to reader

Magnetic field affected by tag data

RFID tag

Binary tag ID

Glass or plastic encapsulation

Far-field region

Propagating electromagnetic waves

Data via changes in field strength

Power and data (if tag supports write)

$c/2\pi f$

Near-field region

Coil

RFID reader

Alternating magnetic field in the near-field region

Figure 7.11 Near-field RFID system. *Source:* Reprinted with permission of Want 2006 [2]. © 2006, IEEE.

Figure 7.12 Power decreases as a function of distance from the loop antenna.

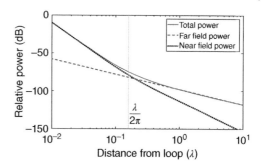

While disadvantages are:

- the short read range requires tags and readers be in close proximity.
- a large antenna
- low data rate
- the inability to discriminate between multiple tags.

LF tags are expensive and thicker than tags at higher frequencies. Figure 7.13 displays two types of LF tags. The multi-turn loop antennas connect to an IC or to a capacitor for power storage. The LF tag in Figure 7.13a is a flat multi-turn loop antenna and has a large area. The one in Figure 7.13b is thicker and has an iron core to minimize the number of turns in the loop [36].

An LF reader takes between 25 and 50 ms to charge a tag [34]. When charging finishes, the reader encodes the data bits using FM0 which modulates a 134.2 kHz carrier for transmission to the tag. The tag responds by switching its load impedance between open and matched in order to modulate the magnetic field with its stored data. LF tags often use FSK with a 129.3 µs one bit at 123.7 kHz and a 128.3 µs zero bit at 134.7 kHz. A 12.2 ms 96-bit FSK modulated end of burst (EOB) signal from the tag lets the reader know that the

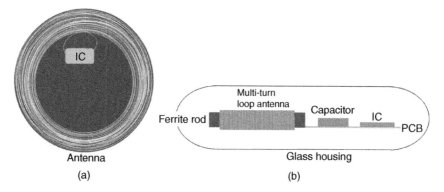

Figure 7.13 Examples of LF RFID tags.

data was received. A typical tag uses Manchester encoding of a 32-bit unique ID followed by a 64-bit data stream (Header + ID + Data + Parity). If the tag receives an invalid command, it sends back 16 bits to tell the reader to retransmit the data. All the stored power dissipates by the end of the tag transmission until the reader transmits another signal.

Example
If $f_c = 125$ kHz and the tag bit rate is $f_c/32$, then find the data rate and the time needed to receive 64 bits.

Solution
Data rate $= 125$ kHz/32 $= 3.9062$ kbps.   Receiving   64 bits: $8\,\mu s \times 32 \times 64 =$ 16.384 ms.

LF tags function well near water, animal tissues, metal, wood, and liquids. The automotive industry (largest LF tag user) embeds an LF tag inside the ignition circuit of an automobile vehicle immobilizer system. Placing the key in the ignition causes an RFID reader to check the tag ID. The car only starts when the ID is verified.

Passive HF tags have a low data rate and a read range less than 1 m. HF RFID systems come in two forms: proximity and vicinity [35].

Proximity tags have more data storage and functionality (e.g. encryption, some processing power, and data storage and retrieval) than LF tags but require more operating power [36]. The power requirements for tag activation and operation generally limit the IZ to less than 20 cm. An HF proximity reader operates with 13.56 MHz modulated by a subcarrier at $f_c/128 = 105.9375$ kHz, $f_c/64 = 211.875$ kHz, $f_c/32 = 423.75$ kHz, or $f_c/16 = 847.5$ kHz. The tag modulates the magnetic field with a subcarrier at $f_c \pm f_c/16 = 13.56 \pm 0.8475$ MHz $= 14.4075$ and 12.7125 MHz. These subcarriers are modulated at the reader data rate.

Vicinity tags have less power and lower data rates than proximity tags [36]. Vicinity cards also operate at $f_c = 13.56$ MHz but have a maximum IZ that extends about 1 m. The subcarrier frequency is at 423.75 kHz with $f_c/32$. The subcarrier is then modulated with FSK or OOK modulation. The data rate is 26.48 kbps. Other data rates include $f_c/8, f_c/16, f_c/32, f_c/40, f_c/50, f_c/64, f_c/80,$ $f_c/100,$ and $f_c/128$. They are used in inventory control and theft deterrence systems.

Unlike LF tags, an HF tag may have anti-collision capability that allows multiple tags to simultaneously coexist in the IZ without interfering with each other. HF tags cost less than LF tags due to a smaller antenna design. They come

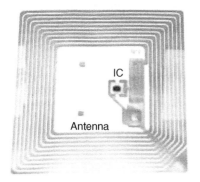

Figure 7.14 Examples of HF RFID tags.

in different sizes, some less than a centimeter in diameter. Water, biological tissues, metal, wood, and liquids have little impact on HF tag performance, but metal in close proximity to the tag detunes the antenna. HF readers are less complex and expensive than readers at higher frequencies. An HF smart card is a plastic RFID tag about the size of a credit card (54 mm × 85.5 mm × 0.8 mm). A typical HF card uses Miller encoding with subcarrier modulation at 847 kHz ($f_c/16$). HF RFID systems are the most widely used worldwide. Figure 7.14 shows two examples of HF RFID tags. The tag's memory has data to control access, collect transportation fees, store medical data, make purchases etc. It can be refreshed in order to add money, change access, or modify data.

7.6.2 Far Field

Far-field systems use dipoles instead of loops to transmit and receive signals at much higher frequencies and data rates and over larger distances compared to the NFC systems (Figure 7.15). Far-field tags modulate the backscattered field by changing the antenna impedance. Sending the stored data to the transistor gate switches the load connected to the dipole and creates a modulated backscattered signal. Any impedance mismatch generates a backscattered field from the reader signal. The transistor does not need to completely short the antenna to allow some received energy to continue to power the tag. Figure 7.16 is an example of a far-field tag for paying tolls on highways in Colorado, USA.

7.6.2.1 Multiple Readers in an Interrogation Zone

Multiple readers interfere with each other when their IZs overlap. Since near-field signals decay much faster than far-field signals ($1/r^3 \ll 1/r^2$), NFC readers have very small IZs and generally do not interfere with each other.

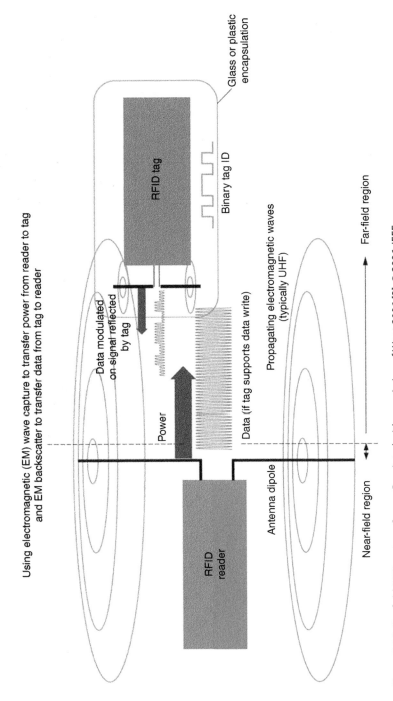

Figure 7.15 Far-field RFID system. *Source:* Reprinted with permission of Want 2006 [2]. © 2006, IEEE.

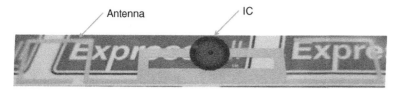

Figure 7.16 UHF passive RFID tag for paying highway tolls.

UHF and microwave readers, on the other hand, have large IZs that often require interference countermeasures such as [37]:

- Physical isolation
- RF absorber
- Shielding
- Reduced transmit power
- Frequency hopping
- Band separation

The type of countermeasure depends on the number and closeness of readers in the environment.

Reader environments fall into one of the three categories:

1. A *single reader* broadcasting in an available channel is susceptible to interference from nonreader sources. Shielding and making appropriate changes to the signal strength and antenna gain mitigate the interference.
2. A *multiple reader* environment has more channels than readers. Multiplexing reduces interference between signals.
3. A *dense reader* environment has more readers than channels, so interference abounds. Readers and tags have separate channels to prevent strong reader signals from overpowering the weaker tag signals. Dense reader mode occurs when more than 50 (North America) and 10 (Europe) readers have overlapping IZs. No two readers should transmit at the same time and on the same frequency when their IZs overlap.

Two common approaches to operating in a dense reader mode are frequency hopping (Chapter 5) and listen before talk (LBT). A UHF reader in North America supports frequency hopping over 50, 500 kHz channels between 902 and 928 MHz (915 MHz band) and resides less than 0.04 seconds in any one channel. Two readers are unlikely to operate at the same frequency within this wide bandwidth. Frequency hopping does not work for the narrower UHF bands in Europe and Japan that cannot handle as many hops [38]. In LBT, a reader checks if the transmitting channel is free. If the channel is busy, the reader tries another channel. Other approaches to dense reader mode include

- Allocate tags and readers to different channels.
- Assign a time slot to each reader.
- Dynamically assign time slots.

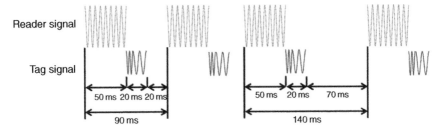

Figure 7.17 Synchronization of reader and tag signals.

Singulation means a reader picks out a tag with a specific serial number from a group of tags in its IZ [39]. The reader needs an anti-collision protocol (e.g. multiplexing or aloha) that prevents devices from interfering with each other. Tree walking, the most common approach to singulation, asks all tags with a serial number that starts with either a 1 or 0 to respond. If multiple tags respond, then the reader asks for all tags with a serial number that starts with 01 to respond. This process continues until the reader finds the desired tag (the only one still responding).

Multiple readers that simultaneously operate in the same IZ avoid mutual interference by synchronizing their transmit and receive signals as shown in Figure 7.17 [40]. In LF wireless synchronization, the reader transmits for 50 ms then the tag sends data after detecting an end of the charge burst. The reader listens for 20 ms before sending the next 90 ms signal. If the reader detects a signal from another reader it waits (backs off) 70 ms in order for the read cycle to finish. The initial reader always starts its next cycle after this 70 ms delay, so that it does not constantly back off and never reads any tags. The worst case synchronization time is 70 ms, while the worst case cycle time is 140 ms.

A reader starts communication with a passive tag by transmitting a continuous wave (CW) signal that powers the tag. Once the tag IC has enough power, it either transfers data to the reader (TTF protocol – Tag Talk First) or answers a reader query (RTF protocol – Reader Talk First) The choice of protocol depends on the number of tags within the reader's range. The two protocols are incompatible, because a TTF tag in an RTF reader IZ disrupts RTF tag communication. The TTF protocol quickly identifies a solo tag in the IZ. The reader only transmits a CW signal when not communicating with tags, so it reduces interference with other wireless systems.

7.6.2.2 Backscatter Communication

Figure 7.18 has a simplified diagram of a passive far-field RFID tag and a monostatic reader [41]. The reader generates a carrier signal 1 that passes through the circulator and bandpass filter (BPF) 2 to the transmitting antenna 3. From there

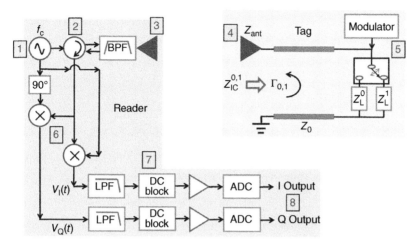

Figure 7.18 Model for the monostatic reader-tag RFID system.

the signal goes through the channel and encounters multipath, attenuation, reflection, and diffraction. It arrives at the tag antenna 4 then travels through the transmission line to the modulating circuit 5. The tag modulator uses the stored data to toggle the switch between two impedances: Z_L^0 that represents a binary zero and Z_L^1 that represents a binary one. The reflection coefficient between the antenna and IC is

$$\Gamma_{0,1} = \frac{Z_L^{0,1} - Z_{ant}^*}{Z_L^{0,1} + z_{ant}} \tag{7.4}$$

Selecting $Z_L^1 = \infty$ (open) results in $\Gamma_0 = 1$, while $Z_L^0 = 0$ (short) results in $\Gamma_0 = -1$. Load switching modulates the backscattered field with the required data. This modulated signal reflects back to the antenna 4. The antenna transmits the backscattered signal to the reader 3. The reader receives the signal through the receive antenna (different than transmit antenna for a bistatic system). Next, the signal passes through the BPF before the circulator directs it toward the receiving circuitry 2. Half the signal goes to the I channel where it is demodulated with the carrier 6. The other half goes to the Q channel where it is demodulated with the carrier shifted by 90° 6. Both signals then pass through a LPF, DC block, amplifier, and ADC 7 to get the I and Q output 8. More details are given for some of these steps in the following paragraphs.

The reader transmits a carrier signal that activates tags in the IZ before sending a query requesting tags to respond with their identification. In order to read a passive tag there must be sufficient power transferred to the tag in order to power the IC. As with any wireless system, calculating the link budget requires knowledge of the channel effects. The power delivered to the tag IC is given by [41]

$$P_{tag} = \underbrace{P_{reader}G_{reader}}_{Reader} \underbrace{\frac{\lambda^2}{(4\pi r)^2}}_{Space\ loss} \underbrace{\frac{G_{tag}\delta_p T_{load}}{\eta_{ob}L_{block}F_{tag}}}_{Tag} \tag{7.5}$$

where

P_{reader} = reader transmit power
G_{reader} = gain of reader antenna in direction of tag
G_{tag} = gain of tag antenna in direction of reader
δ_p = polarization loss factor
T_{load} = power transmission coefficient
η_{ob} = on object gain penalty
L_{block} = blockage loss
F_{tag} = fade margin

The tag size, orientation, angle, and placement impact the read range. Steep angles of incidence reduce the read range due to lower antenna gain and polarization mismatch. Maximum gain and polarization efficiency occur when the tag and reader antennas face each other. If the reader antenna and tag antennas are linearly polarized, then δ_p ranges from 0 to 1 depending on the orientations of the antennas. In order to avoid fades due to antenna orientation, a circularly polarized reader and a linear polarized tag antenna insures that $\delta_p = 0.5$.

The power delivered to the IC depends upon the match between the antenna and the IC. Some of the power reflects while the rest goes to the IC based on the transmission coefficient ($0 \leq T_{load} \leq 1$) [41]:

$$T_{load} = \frac{4Re\{Z_{ant}\}Re\{Z_L^{0,1}\}}{Re\{Z_{ant} + Z_L^{0,1}\}^2 + Im\{Z_{ant} + Z_L^{0,1}\}^2} \tag{7.6}$$

The input impedance of a tag antenna in free space differs from the tag attached to an object. This impedance difference increases as frequency increases. Low dielectric constant objects have much less impact on the tag antenna input impedance compared to water or metal. Metal-mount tags have a built-in, low dielectric between the tag and the metal object [40].

Figure 7.19 plots P_{tag} vs. separation distance for tags in free space, mounted on cardboard, and mounted on aluminum at 915 MHz [41]. This plot emphasizes that the Friis transmission formula in free space ($\delta_p = 1$, $T_{load} = 1$, $\eta_{ob} = 1$, $L_{block} = 1$, $F_{tag} = 1$) does not adequately calculate the link budget. Mounting the tag on aluminum causes over 30 dB of additional loss compared to the same tag mounted on cardboard. Setting the tag threshold at $P_{tag} = -12$ dBm (dotted horizontal line in Figure 7.19) results in an IZ of 2 m for the tag mounted on cardboard, while the tag on aluminum has an IZ much less than 1 m.

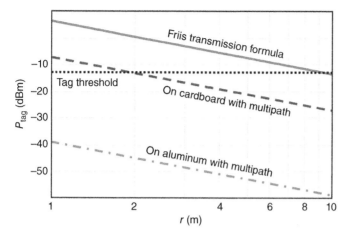

Figure 7.19 The power uplink budget for tags mounted on different materials as a function of separation distance.

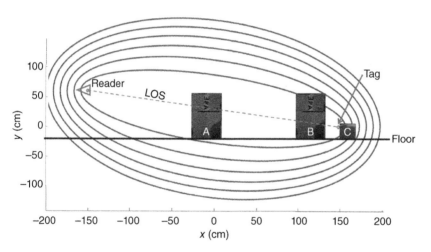

Figure 7.20 Reader-tag communications link with two boxes (A and B) blocking the LOS signal to box C.

Figure 7.20 models a situation where the RFID tag on box C lies behind boxes A and B. The tag and reader are at the foci of the Fresnel ellipsoids (Chapter 5). Box A is in zone 0 and blocks the LOS. Box B also blocks the LOS but extends up into zone 2. As a rule of thumb, when the Fresnel zone number is small (on the order of 1–2 or less), diffraction is important, and the received intensity is a complex function of position, with no well-defined shadow region [42]. When the obstacle subtends many Fresnel zones (>3–5), the tag lies in a fairly well-defined shadow and cannot communicate with the reader. Fresnel

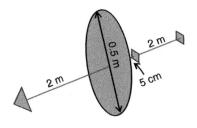

Figure 7.21 Tags in the shadow of a disk.

diffraction theory or a numerical modeling tool calculate the attenuation due
to diffraction.

Example
A reader illuminates a 1-m diameter disk that is 2 m away (Figure 7.21). A tag is
placed behind the disk at (a) 5 cm and (b) 2 m. Determine which Fresnel zone
the edge of the disk lies at 915 MHz and comment on the impact of the tag
receiving the reader signal.

Solution
Use (5.31) assuming that the reader and tag are at the foci of the Fresnel
ellipsoid.

$$r_n = \sqrt{\frac{n\lambda d_t d_r}{d_t + d_r}} \Rightarrow n = \frac{r_n^2(d_t + d_r)}{\lambda d_t d_r}$$

$$\lambda = \frac{3 \times 10^{10}}{915 \times 10^6} = 32.8 \text{ cm}$$

(a) $n = \dfrac{50^2(200 + 5)}{32.8(200)(5)} = 15.6$ Tag is in deep shadow and probably does not detect signal.

(b) $n = \dfrac{50^2(200 + 200)}{32.8(200)(200)} = 0.76$ Tag has good chance of detecting signal.

The backscattered power at the reader due to a semi-passive or chipless tag
is given by the radar range equation [41].

$$P_{\text{reader}} = \frac{P_t G_{\text{reader}} G_{\text{tag}} c^2 \sigma_{\text{rcs}}}{(4\pi)^2 f_c^2 r^4} L \tag{7.7}$$

where σ_{rcs} is the radar cross section (RCS) of the tag. A reasonable estimate
for our purposes assumes a modulated backscatter power level of 1/3 of the
absorbed power. Passive tag IC power requirements of tens or hundreds of
microwatts far exceed the receiver sensitivity. Figure 7.22 is the power received
at the reader when for a tag mounted on cardboard and aluminum at 915 MHz

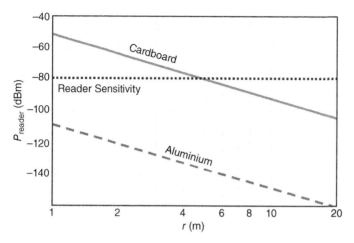

Figure 7.22 Power received at the reader when for a tag mounted on cardboard and aluminum at 915 MHz

[41]. If the reader threshold is $P_{reader} = -80$ dBm, then the tag mounted on cardboard has an IZ of about 5 m while the tag mounted on aluminum has an IZ less than 1 m.

7.6.2.3 Chipless Tags

Chipless tags dramatically reduce the cost of a tag by removing the most expensive part: the silicon IC [43]. Some chipless tags use plastic or conductive polymers instead of silicon-based microchips. Other chipless tags use materials that backscatter the reader signal in a unique signature. Figure 7.23 lists the categories of chipless RFID tags. Companies are experimenting with embedding RF reflecting fibers in paper to prevent unauthorized photocopying of certain documents. Reflective tattoo inks identify farm animals.

Delay-line-based chipless tags have microstrip discontinuities after a delay line [44]. A short interrogation pulse from a reader (1 ns) reflects from carefully placed discontinuities in the microstrip line and results in reflections as shown in Figure 7.24. The length of the delay-line between the discontinuities determines the time delay between the reflections.

One approach to a chipless tag places the RF circuit between receive and transmit antennas [45]. The circuit has cascaded resonators (Figure 7.25) that resonate at specific frequencies in order to create bandstop filters that cause an attenuation and phase jump at these frequencies. The reader interprets the spectrum of the tag signal as bits. The tag signal resonant frequencies correspond to a code. The chipless tag in Figure 7.26 consists of five second-order Piano curves that produce five peaks in the RCS [46]. These peaks represent a bit sequence.

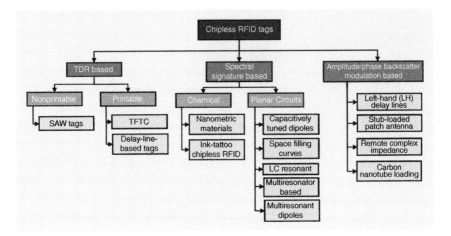

Figure 7.23 Categories of chipless RFID tags. *Source:* Reprinted with permission of Preradovic and Karmakar [43]. © 2010, IEEE.

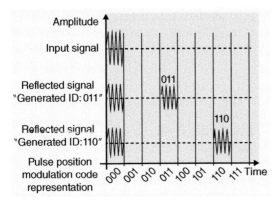

Figure 7.24 Interrogation and coding of delay-line-based chipless tag. *Source:* Reprinted with permission of Preradovic and Karmakar [43]. © 2010, IEEE.

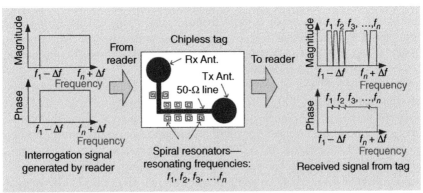

Figure 7.25 Diagram of the operation of a multiresonator-based chipless RFID tag. *Source:* Reprinted with permission of Tedjini et al. 2013 [45]. © 2010, IEEE.

Figure 7.26 Five-bit piano-curve-based tag and tag radar cross section spectral signature. *Source:* Reprinted with permission of Preradovic and Karmakar [43]. © 2010, IEEE.

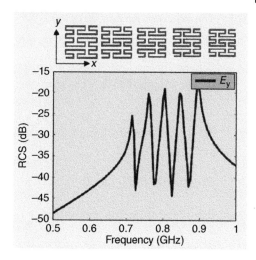

Figure 7.27 Operating principle of left-hand-delay-line based chipless RFID tag. *Source:* Reprinted with permission of Preradovic and Karmakar [43]. © 2010, IEEE.

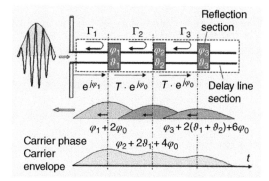

Figure 7.27 shows the operation of a left-hand (LH) delay line chipless RFID tag [47]. The interrogation pulse propagates through the periodic LH delay lines. The backscattered pulse from the discontinuities is coded with the reflected signal phase relative to a reference phase. The reflected signals have equal amplitude envelopes but their phases due to Γ_1, Γ_2, and Γ_3 are ϕ_1, ϕ_2, and ϕ_3. This approach encodes data with a higher order modulation scheme (e.g. QPSK) that increases throughput at the expense of a higher SNR.

Problems

7.1 Assume the 433 MHz tag IC has a threshold power level of -10 dBm. The reader transmits 30 dBm through an antenna with 6 dB gain. Assume the tag antenna is isotropic and $\eta_{PCE} = 0.3$. What is the maximum extent of the IZ?

7.2 The barcode data is 001234567890x where x is the checksum digit. Find x.

7.3 What is the range of the power conversion efficiency for a Schottky diode Dickson multiplier that has four stages?

7.4 Assume the 865 MHz reader has a threshold power level of −30 dBm. The reader antenna has a 6 dB gain. The tag antenna has a gain of 0 dB. Let $L = 1$.

7.5 Plot the PIE coding for the bits [1 1 0 1].

7.6 Plot the FM0 coding for the bits [1 1 0 1].

7.7 A binary data stream has the bits [1 1 0 0]. Plot the baseband symbol, square wave modulator, and Miller encoding with $M = 2$.

7.8 If $f_c = 125$ kHz and the tag bit rate is $f_c/16$, then find the data rate and the time needed to receive 128 bits.

7.9 Find the reflection coefficients for a 1 and 0 bit when $Z_{ant} = 45 + j20\,\Omega$, $Z_L^0 = 50\,\Omega$, and $Z_L^1 = 10\,\Omega$.

7.10 If the tag must receive −10 dBm, then design a reader give a tag with $G_{tag} = 0$ dB, $\delta_p = 0.5$, $T_{load} = 0.5$, $\eta_{ob} = 0.5$, $L_{block} = 1.0$, and $F_{tag} = 1.0$. Assume that the frequency is 3 GHz and the IZ extends to 10 m.

7.11 What fraction of the power gets delivered to the IC if $Z_{ant} = 45 + j20\,\Omega$, $Z_L^0 = 50\,\Omega$, and $Z_L^1 = 10\,\Omega$.

References

1 Violino, B. (2005). What is RFID? *RFID J.*, http://www.rfidjournal.com/articles/view?1339 (accessed 19 June 2019).

2 Want, R. (2006). An introduction to RFID technology. *IEEE Pervasive Comput.* 5 (1): 25–33.

3 http://www.cryptomuseum.com/covert/bugs/thing/index.htm (accessed 20 July 2018).

4 Stockman, H. (1948). Communication by means of reflected power. *Proc. IRE* 36 (10): 1196–1204.

5 https://www.smithsonianmag.com/innovation/history-bar-code-180956704/ (accessed 20 July 2018).

6 Woodland, N.J. and Silver, B. (1952). Classifying apparatus and method. US Patent 2,612,994, 7 October 1952.

7 https://electronics.howstuffworks.com/gadgets/high-tech-gadgets/upc.htm (accessed 20 July 2018).

8 Cardullo, M.W. and Parks, W.L. (1973). Transponder apparatus and system. US Patent 3,713,148, 23 January 1973.

9 Roberti, M. (2005). The history of RFID technology. *RFID J.*, http://www .rfidjournal.com/articles/view?1338 (accessed 19 June 2019).

10 Walton, C. (1983). Portable radio frequency emitting identifier. US Patent 4,384,288.

11 Lumpkins, W. (2015). RFID: an evolution of change, from World War II to the consumer marketplace. *IEEE Potentials* 34 (5): 6–12.

12 https://autoid.mit.edu/about-lab (accessed 27 January 2018).

13 Hunt, V.D., Puglia, A., and Puglia, M. (2007). *RFID A Guide to Radio Frequency Identification*. Hoboken, NJ: Wiley.

14 https://www.rfidjournal.com/glossary/?T (accessed 8 June 2018).

15 The Beginner's Guide to RFID Systems. https://www.atlasrfidstore.com/rfid-beginners-guide/ (accessed 26 April 2018).

16 http://rfid4u.com/rfid-basics-resources/how-to-select-a-correct-tag-frequency/ (accessed 30 January 2018).

17 https://en.wikipedia.org/wiki/Radio-frequency_identification#cite_note-Sen09-12 (accessed 11 June 2018).

18 http://www.hightechaid.com/standards/18000.htm (15 November 2018).

19 Sattlegger, K. and Denk, U. (2014). *Navigating your way through the RFID jungle*. Texas Instruments White Paper.

20 WEBINAR SERIES #4 EPC/RFID STANDARDS AND RFID - STUFF YOU NEeD TO KNOW," webinar presentation by GS1 US, 2012. (accessed 28 January 2018).

21 https://www.epc-rfid.info/ (accessed 22 January 2018).

22 https://rfid4u.com/rfid-basics-resources/dig-deep-rfid-tags-construction/, (accessed 13 December 2018).

23 Sanghera, P., Thornton, F., Haines, B. et al. (2007). *How to Cheat at Deploying and Securing RFID*. Burlington, MA: Syngress Publishing.

24 http://rfid4u.com/rfid-basics-resources/rfid-printers-encoders/ (accessed 30 January 2018).

25 Tran, L.-G., Cha, H.-K., and Park, W.-T. (2017). RF power harvesting: a review on designing methodologies and applications. *Micro Nano Syst. Lett.* 5 (1): 1–16.

26 Shan, H., Peterson, J. III,, Hathorn, S., and Mohammadi, S. (2018). The RFID connection: RFID technology for sensing and the internet of things. *IEEE Microw. Mag.* 19 (7): 63–79.

27 Dickson, J.F. (1976). On-chip high-voltage generation in MNOS integrated circuits using an improved voltage multiplier technique. *IEEE J. Solid-State Circuits* 11 (3): 374–378.

28 Griffin, J. *The Fundamentals of Backscatter Radio and RFID Systems Part I*. Pittsburgh, PA: Disney Research.

29 http://www.rfidjournal.com/blogs/experts/entry?11226 (accessed 20 July 2018).

30 https://blog.atlasrfidstore.com/active-rfid-vs-passive-rfid (accessed 13 December 2018).

31 Griffin, J. *The Fundamentals of Backscatter Radio and RFID Systems Part II*. Pittsburgh, PA: Disney Research.

32 Dobkin, D.M. (2013). *The RF in RFID: UHF RFID in Practice*, 2e. Oxford, UK: Elsevier.

33 Sanghera, P., Thornton, F., Haines, B. et al. (2007). *How to Cheat at Deploying and Securing RFID*. Burlington, MA: Elsevier Inc.

34 Recknagel, S. (2011). Low-frequency RFID in a Nutshell. Texas Instruments Application Report, SWRA284.

35 Novotny, D.R., Guerrieri, J.R., Francis, M. and Remley, K. (2008) HF RFID electromagnetic emissions and performance. 2008 IEEE International Symposium on Electromagnetic Compatibility, Detroit, MI (pp. 1–7).

36 Finkenzeller, K. (2003). *RFID Handbook: Fundamentals and Applications in Contactless Smart Cards and Identification*. Wiley.

37 http://rfid4u.com/rfid-basics-resources/dig-deep-dense-reader-mode-and-anti-collision/ (accessed 30 January 2018).

38 Violino, B. (2005). The basics of RFID technology. *RFID J.*

39 http://rfidsecurity.uark.edu/downloads/slides/mod04_lesson07_slides.pdf (accessed 26 April 2018).

40 Texas Instruments Technology (2004). LF Reader Synchronization. Texas Instruments Application Report, SCBA019 (11-06-26-001).

41 Griffin, J.D. and Durgin, G.D. (2009). Complete link budgets for backscatter-radio and RFID systems. *IEEE Antennas Propag. Mag.* 51 (2): 11–25.

42 Dobkin, D.M. (2008). *The RF in RFID Passive UHF RFID in Practice*. Burlington, MA: Elsevier.

43 Preradovic, S. and Karmakar, N.C. (2010). Chipless RFID: bar code of the future. *IEEE Microwave Mag.* 11 (7): 87–97.

44 Shretha, S., Vemagiri, J., Agarwal, M., and Varahramyan, K. (2007). Transmission line reflection and delay-based ID generation scheme for RFID and other applications. *Int. J. Radio Freq. Identif. Technol. Appl.* 1 (4): 401–416.

45 Tedjini, S., Karmakar, N., Perret, E. et al. (2013). Hold the chips: chipless technology, an alternative technique for RFID. *IEEE Microwave Mag.* 14 (5): 56–65.

46 McVay, J., Hoorfar, A., and Engheta, N. (2006). Space-filling curve RFID tags. 2006 IEEE Radio and Wireless Symposium Digest, San Diego, CA (17–19 January 2006), pp. 199–202.

47 Caloz, C. and Itoh, T. (2004). Transmission line approach of left-handed (LH) materials and microstrip implementation of an artificial LH transmission line. *IEEE Trans. Antennas Propag.* 52 (5): 1159–1166.

8

Direction Finding

Direction finding (DF), also known as angle of arrival (AOA) or direction of arrival (DOA) estimation, locates radio frequency (RF) signals in the environment. Pointing the antenna main beam in the direction of a signal produces the highest received signal level but has limited resolution due to the antenna beamwidth which is inversely proportional to aperture size. Strong signals entering the sidelobes overwhelm a weaker signal received by the main beam, leading to poor AOA estimates. Main beams slowly decrease away from the peak, so small angle changes in the signal produce small changes in the antenna output. For instance, Figure 8.1 shows that the antenna pattern changes by 3 dB over an 11.8° angular range. This range typically defines an antenna's resolution (3 dB beamwidth).

Consequently, most approaches to DF locate signals using nulls rather than peaks in the antenna patterns. Nulls have steep sides, so small changes in angle produce large changes in output power that a receiver easily detects. A signal entering a null produces zero antenna output. The angular change of the difference pattern null in Figure 8.1 for a 3 dB change in antenna pattern is about 0.4°. At −20 dB, the null width is 2.0°. Compare that null angular accuracy with the 12.8° beamwidth of the sum pattern.

This chapter starts by presenting some relatively simple approaches to DF that use the main beam or one null. The rest of the chapter covers sophisticated digital signal processing algorithms that precisely locate multiple signals using nulls.

8.1 Direction Finding with a Main Beam

An antenna has a maximum output power when the antenna main beam points directly at the signal. The elevation and azimuth of the antenna pointing direction determines the angular location of the signal.

Wireless Communications Systems: An Introduction, First Edition. Randy L. Haupt.
© 2020 John Wiley & Sons, Inc. Published 2020 by John Wiley & Sons, Inc.

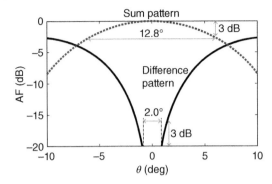

Figure 8.1 Antenna pattern null vs. peak for DF.

8.1.1 Array Output Power

The power received by an antenna equals the signal power times the antenna gain in the direction of the signal. Maximum antenna output occurs when the signal enters the antenna's main beam. Sidelobes and nulls significantly attenuate signals incident on them. The array output of an N element linear array (Figure 8.2) with element spacing d due to M signals incident on the array from M angles at θ_m is given by [1]

$$S_{out} = \sum_{m=1}^{M} \sum_{n=1}^{N} w_n (s_m e^{jk(n-1)d \sin \theta_m} + v_n) = \mathbf{w}^T (\mathbf{As} + \mathbf{n}) \tag{8.1}$$

where

$$\mathbf{s} = [s_1 \; s_2 \; \cdots \; s_M]^T = \text{signal amplitude vector}$$

$$\mathbf{w} = [w_1 \; w_2 \; \cdots \; w_N]^T = \text{element weight vector}$$

$$\mathbf{A} = \begin{bmatrix} 1 & 1 & \cdots & 1 \\ e^{jkd \sin \theta_1} & e^{jkd \sin \theta_2} & \cdots & e^{jkd \sin \theta_M} \\ \vdots & \vdots & \ddots & \vdots \\ e^{jk(N-1)d \sin \theta_1} & e^{jk(N-1)d \sin \theta_2} & \cdots & e^{jk(N-1)d \sin \theta_M} \end{bmatrix}$$

$$= \text{array steering matrix}$$

$$\mathbf{n} = [v_1 \; v_2 \; \cdots \; v_N]^T = \text{element noise vector}$$

The signal amplitude vector contains the relative amplitudes of the M signals incident on the array. Column m in the array steering matrix contain the phase differences at the elements due to a signal incident from θ_m. Each element has independent random noise with a complex value v_n and average power σ_{noise}^2, assuming additive white Gaussian noise (AWGN).

Figure 8.2 Diagram of a DF array.

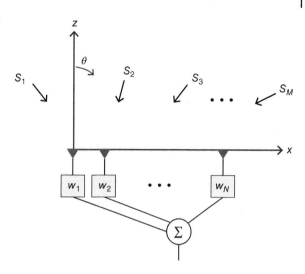

The average array output power is proportional to the expected value of the amplitude of the array output signal squared:

$$P = E[\mathbf{s}^\dagger \mathbf{s}]$$
$$= E[|\mathbf{w}^T(\mathbf{As} + \mathbf{n})|^2]$$
$$= E[\mathbf{w}^\dagger(\mathbf{As} + \mathbf{n})(\mathbf{As} + \mathbf{n})^\dagger \mathbf{w}]$$
$$= \mathbf{w}^\dagger \mathbf{C} \mathbf{w} \tag{8.2}$$

where $E[\cdot]$ is the expected value and "$\dagger$" is the complex conjugate transpose. The covariance matrix, $\mathbf{C}$, is defined by

$$\mathbf{C} = E[(\mathbf{As} + \mathbf{n})(\mathbf{As} + \mathbf{n})^\dagger]$$
$$= E[\mathbf{Ass}^\dagger\mathbf{A}^\dagger] + E[\mathbf{ns}^\dagger\mathbf{A}^\dagger] + E[\mathbf{Asn}^\dagger] + E[\mathbf{nn}^\dagger]$$
$$= \mathbf{A}E[\mathbf{ss}^\dagger]\mathbf{A}^\dagger + E[\mathbf{ns}^\dagger]\mathbf{A}^\dagger + \mathbf{A}E[\mathbf{sn}^\dagger] + E[\mathbf{nn}^\dagger]$$
$$= \mathbf{C}_s + \mathbf{C}_{s-\text{noise}} + \mathbf{C}_{\text{noise}-s} + \mathbf{C}_{\text{noise}} \tag{8.3}$$

and

$\mathbf{C}_s$ = signal covariance matrix
$\mathbf{C}_{s-\text{noise}}$ = signal–noise covariance matrix
$\mathbf{C}_{\text{noise}-s}$ = noise–signal covariance matrix
$\mathbf{C}_{\text{noise}}$ = noise covariance matrix.

The expected value operator applies to time-varying quantities, so the elements of the covariance matrix describe the cross-correlation between the signals at all of the elements. For instance, row m and column n describes how close the signal at element m looks like the signal at element n.

Uncorrelated signal and noise have $\mathbf{C}_{s-noise} = 0$ and $\mathbf{C}_{noise-s} = 0$ leaving only the signal and noise covariance matrixes

$$\mathbf{C} = \mathbf{C}_s + \mathbf{C}_{noise} \tag{8.4}$$

The noise covariance matrix diagonal elements equal the noise variance, σ^2_{noise}, and the off-diagonal elements equal zero, because the noise at one element is uncorrelated with the noise at another element

$$\mathbf{C}_{noise} = \begin{bmatrix} \sigma^2_{noise} & 0 & 0 \\ 0 & \ddots & 0 \\ 0 & 0 & \sigma^2_{noise} \end{bmatrix} \tag{8.5}$$

When no noise is present, then $\mathbf{C} = \mathbf{C}_s$. When no signal is present, then $\mathbf{C} = \mathbf{C}_{noise}$.

8.1.2 Periodogram

Scanning the antenna main beam maps the signal locations by recording output power as a function of angle. A plot of the output power vs. angle obtained from (8.2) is called a periodogram [2]. A periodogram due to one signal is the same as the antenna pattern. Periodograms have three problems with locating signals:

1. *Resolution*: The antenna beamwidth is inversely proportional to the aperture size in wavelengths. Two or more signals within the main beam beamwidth appear as one signal – they cannot be resolved. Consequently, fine resolution requires a large electrical aperture.
2. *Sidelobes*: Strong signals entering a sidelobe cannot be distinguished from weaker signals entering the main beam in the periodogram. Low sidelobes mitigate this problem but produce a wider main beam, hence degraded resolution.
3. *Accuracy*: As noted in Figure 8.1, the error in the angular location of a signal detected in a main beam is approximately $\pm\theta_{3dB}/2$.

The following example highlights these problems and sets the stage for digital beamforming DF.

Example
Plot the periodogram of an eight-element uniform linear array of isotropic elements with $d = \lambda/2$ spacing when signals of equal power are incident at $\theta_1 = 30°$, $10°$, $0°$, and $-60°$. Ignore noise.

Solution
The output power is given by

$$P(\theta) = \mathbf{w}^\dagger \mathbf{C}_s \mathbf{w} \tag{8.6}$$

Figure 8.3 Polar plot of the periodogram of an eight-element uniform array when four equal power signals are incident at $\theta_1 = 30°$, $10°$, $0°$, and $-60°$.

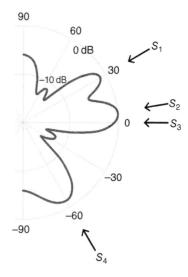

In order to steer the beam to θ_s, the weights, **w**, must include a steering phase. Since the array is uniform, the weights are given by

$$w_n = e^{-jk(n-1)d\sin\theta_s} \tag{8.7}$$

Figure 8.3 shows a polar plot of (8.6) for $0 \le \theta \le 180°$. The periodogram has peaks in the directions of $-60°$ and $30°$, so these signals are resolved. A single peak at $5°$ is due to the two signals at $0°$ and $10°$. Since they are separated by less than a beamwidth, they cannot be resolved. Note that a weak response also appears at $90°$ even though no signal is present. Signals entering sidelobes create that ghost signal when the array main beam is swept toward $90°$.

8.1.3 Wullenweber Array

A circular array of N elements scans $360°$ in azimuth by having an N_a subset of contiguous elements active at a time. Figure 8.4 shows the main beam pointing normal to the center of the N_a active elements connected to the feed network. The remaining $N - N_a$ elements have no connection to the feed network. To steer the beam from active elements 1 to N_a by $360°/N$, element 1 is disconnected from the feed network and element $N_a + 1$ is connected to the feed network. Performing this switching N times completes a $360°$ azimuth scan of the main beam.

If the circular array lies in the x–y plane, then its array factor is

$$\text{AF} = \sum_{n=1}^{N_a} w_n e^{jkr_c\cos(\phi-\phi_n)} \tag{8.8}$$

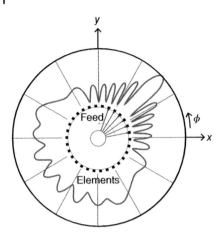

Figure 8.4 Diagram of a circular array.

where

r_c = radius of circular array

ϕ_n = angular location of element n.

Since the array lies on a curved surface, the elements need a nonlinear phase compensation to form a coherent main beam in the beam pointing direction ϕ_s.

$$w_n = e^{-kr_c \cos(\phi_s - \phi_n)} \tag{8.9}$$

Hans Rindfleisch invented the circular Wullenweber array during WW II for HF DF [3]. Figure 8.5 shows an example of an AN/FRD-10 Circularly Disposed Antenna Array (CDAA) at Gandor, NL, Canada.

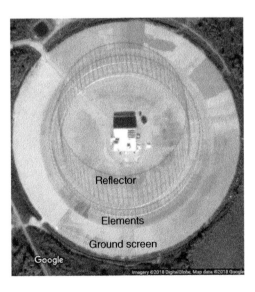

Figure 8.5 Wullenweber array near Gandor, NL, Canada. *Source:* Imagery ©2018 DigitalGlobe. Map data ©2018 Google.

8.2 Direction Finding with a Null

As already shown, a null has greater angular accuracy than a main beam, so pointing a null at a signal until the output goes to zero accurately locates the signal. A loop antenna has a null perpendicular to the plane of the loop (Chapter 4). The HF DF loop antenna in Figure 8.6 has a motorized base that mechanically steers the antenna in azimuth. Even small antennas have sharp nulls for precisely locating a signal.

An array difference pattern has a sharp null at boresight rather than a peak. Steering this null in the direction of a signal eliminates the contribution of that signal to the array output. A uniform difference pattern results when half the elements have a 180° phase shift or $w_n = 1$ for $n \leq N/2$ and $w_n = -1$ for $n > N/2$, assuming N is even. A closed form expression for a uniform difference array with an even number of elements is given by

$$\text{AF} = \frac{1 - \cos\left(\frac{N\psi}{2}\right)}{j \sin\left(\frac{\psi}{2}\right)} \tag{8.10}$$

Figure 8.6 HF DF loop antenna.

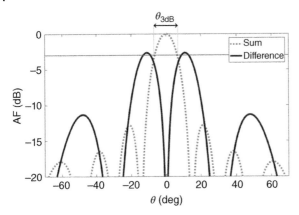

Figure 8.7 Uniform sum and difference patterns.

Bayliss found a low sidelobe taper for difference patterns that resembles the Taylor taper for sum patterns [4] in which $\bar{n} - 1$ sidelobes of equal height next to the main beam, while the rest decrease away from boresight. Simultaneous sum and difference patterns in a monopulse radar enable detection and location of targets.

Example

Plot the sum and difference patterns for an eight-element uniform array with $\lambda/2$ spacing.

Solution

The sum pattern weights are

$$\mathbf{w} = \begin{bmatrix} 1 & 1 & 1 & 1 & 1 & 1 & 1 & 1 \end{bmatrix}$$

and the difference weights are

$$\mathbf{w} = \begin{bmatrix} 1 & 1 & 1 & 1 & -1 & -1 & -1 & -1 \end{bmatrix}$$

Substitute these weights into the array factor formula to calculate the array factors. Figure 8.7 shows the two patterns that are normalized to the peak of the sum pattern (N). The sum pattern 3 dB beamwidth is 12.8°. Note how narrow the null appears compared to the main beam. A small angular change near the null of the difference pattern produces a large change in the output power. This approach minimizes the output power to determine the signal location rather than the periodogram that attempts to maximize the signal output power.

8.3 Adcock Arrays

The original Adcock array had four monopoles on the corners of a square (Figure 8.8) [5]. The elements along the x-axis combine out of phase so that

Figure 8.8 Diagram of a four monopole Adcock array.

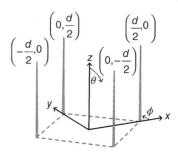

the resulting array factor has peaks at ϕ equals $0°$ and $180°$ and nulls at $90°$ and $270°$. The x-axis array factor is written as

$$AF_x(\theta, \phi) = 2j \sin \left(k\frac{d}{2} \sin \theta \cos \phi \right) \tag{8.11}$$

In contrast, the elements along the y-axis combine out of phase and in array factor peaks at $90°$ and $270°$ and nulls at $0°$ and $180°$. The y-axis array factor is given by

$$AF_y(\theta, \phi) = 2j \sin \left(k\frac{d}{2} \sin \theta \sin \phi \right) \tag{8.12}$$

where d is the separation between the elements along the x- and y-axis. AF_x and AF_y provide an estimate of the azimuth and elevation angles of a signal arriving from (θ_s, ϕ_s) [6].

$$\tan \phi_s \approx \frac{AF_y(\theta_s, \phi_s)}{AF_x(\theta_s, \phi_s)} = \frac{\sin \left(k\frac{d}{2} \sin \theta_s \sin \phi_s \right)}{\sin \left(k\frac{d}{2} \sin \theta_s \cos \phi_s \right)} \tag{8.13}$$

$$\cos \theta_s \approx \frac{1}{kd} \sqrt[4]{AF_x^2(\theta_s, \phi_s) + AF_y^2(\theta_s, \phi_s)} \tag{8.14}$$

A circular Adcock array provides better estimates for the AOA by placing additional element pairs on opposite sides of a circle with a center at the origin of the x–y axes [2]. Figure 8.9 shows an example of an eight-element Adcock

Figure 8.9 Diagram of an eight monopole circular Adcock array.

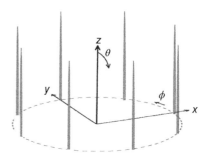

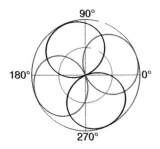

Figure 8.10 The array factors for the eight-element Adcock array.

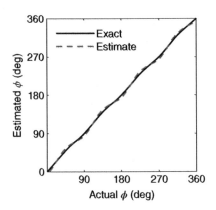

Figure 8.11 A graph of the calculated vs. estimated azimuth angles for the eight-element Adcock array.

array. The outputs of two adjacent elements add to get a single output. These four outputs are then combined using (8.11) and (8.12). If the circle has a radius of 0.25λ with eight elements, then the two array factors appear in Figure 8.10 and the azimuth angle estimate is shown in Figure 8.11. This small radius produces a small error in the AOA estimate. Increasing the radius in terms of wavelength also increases the error in the AOA estimate.

8.4 Eigenbeams

Since the covariance matrix in (8.4) describes the relationship between the signals at all the elements, its eigenvectors and eigenvalues contain important signal information. An eigen-decomposition of the covariance matrix in (8.4) results in

$$\mathbf{C} = \mathbf{Q}\mathbf{\Lambda}_\lambda\mathbf{Q}^{-1} = \underbrace{\sum_{m=1}^{M} \lambda_m \mathbf{Q}[:,m]\mathbf{Q}^\dagger[:,m]}_{\mathbf{C}_s} + \underbrace{\sigma_{\text{noise}}^2 \mathbf{I}_N}_{\mathbf{C}_N} \tag{8.15}$$

where

$\mathbf{Q} = N \times N$ matrix whose columns are the eigenvectors

$$\mathbf{\Lambda}_\lambda = \begin{bmatrix} \lambda_1 & 0 & \cdots & 0 \\ 0 & \lambda_2 & \ddots & \vdots \\ \vdots & \ddots & \ddots & 0 \\ 0 & \cdots & 0 & \lambda_N \end{bmatrix}$$

$= N \times N$ matrix whose diagonal are the N eigenvalues

λ_n = eigenvalue associated with the eigenvector in column n

$\mathbf{I}_N = N \times N$ identity matrix

M uncorrelated signals incident on an N element array ($N > M$) produces M signal eigenvalues and $N - M$ noise eigenvalues. Large eigenvalues ($\lambda_n \gg \sigma^2_{noise}$) correspond to signals while small eigenvalues ($\lambda_n \approx \sigma^2_{noise}$) correspond to noise. Noise eigenvalues are independent of θ and approximately equal the noise variance. Signal eigenvalues, on the other hand, are a function of signal direction and are proportional to the power of the signals.

Let eigenvector n from $\mathbf{Q}$ equal the array weight vector $\mathbf{w} = \mathbf{Q}[:, n]$. Substitute $\mathbf{w}$ into the equation for the array factor from Chapter 4 to get eigenbeam n [7]

$$\text{EB}_n(\theta) = \sum_{n=1}^{N} \mathbf{Q}[:, n]e^{jk(n-1)d \sin \theta} \tag{8.16}$$

Figure 8.12 graphs the 20 eigenbeams associated with a 20-element linear array ($\lambda/2$ spacing) and two signals having the same normalized power of 1 W incident at $\theta = -50°$ and $\theta = 30°$ with $\sigma^2_{noise} = 0.09$ W. The signal eigenbeams have main beams that point in the directions of the signals (Figure 8.12a). Note that the two signal eigenvectors differ and produce unique eigenbeams. All 18 noise eigenbeams have nulls pointing at $\theta = -50°$ and $\theta = 30°$ (Figure 8.12b). Any

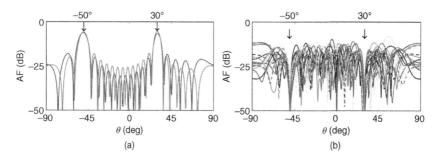

Figure 8.12 Eigenbeams for a 20-element array with two equal power signals incident at $\theta = -50°$ and $\theta = 30°$. (a) Signal eigenbeams and (b) Noise eigenbeams.

of the 20 eigenvectors have main beams or nulls pointing in the directions of signals. The signal eigenbeams use eigenbeam maxima, while the noise eigenbeams use eigenbeam minima.

Example

Plot the eigenbeams of a four-element uniform linear array with $\lambda/2$ spacing when a 1 V/m plane wave is incident at 30° and $\sigma^2_{noise} = 0.16$ W/m².

Solution

First form the covariance matrix:

$$C = \begin{bmatrix} 0.4957 & 0.0061 + j0.3316 & -0.3246 & 0.0015 - j0.3316 \\ 0.0061 - j0.3316 & 0.4953 & 0.0025 + j0.3316 & -0.3289 \\ -0.3246 & 0.0025 - j0.3316 & 0.4865 & 0.0022 + j0.3316 \\ 0.0015 + j0.3316 & -0.3289 & 0.0022 - j0.3316 & 0.4870 \end{bmatrix}$$

Next, find the eigenvectors and eigenvalues of the covariance matrix using MATLAB.

```
[eigvec,eigval]=eig(C)
```

The normalized amplitudes and phases of the eigenvectors along with the eigenvalues appear in Table 8.1. The signal eigenvalue ($\lambda_1 = 1.52$) is easy to identify, because it is much greater than the noise eigenvalues. The signal

Table 8.1 Eigenvectors and eigenvalues for a four-element uniform array with a 1 V/m plane wave incident at −50° and a 1 V/m plane wave incident at 30°.

Eigenvectors							
Signal		Noise					
Q[:, 1]		Q[:, 2]		Q[:, 3]		Q[:, 4]	
|Q[:, 1]|	∠Q[:, 1]	|Q[:, 2]|	∠Q[:, 2]	|Q[:, 3]|	∠Q[:, 3]	|Q[:, 4]|	∠Q[:, 4]
0.9964	0°	0.4272	0°	0.7380	0°	1.0000	0°
1.0000	90°	0.3754	78.41°	1.0000	169.23°	0.7576	21.89°
0.9877	180°	0.9469	176.65°	0.5372	322.34°	0.7216	18.22°
0.9922	270°	1.0000	271.41°	0.6712	190.87°	0.4911	−4.60°

Eigenvalues			
Signal		Noise	
λ_1	λ_2	λ_3	λ_4
1.52	0.22	0.20	0.19

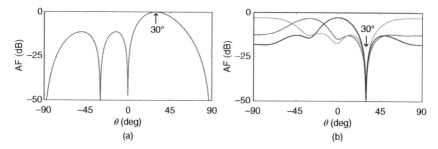

Figure 8.13 Plot of eigenbeams for the four-element uniform array with 1 V/m plane wave incident at 30°. (a) Signal eigenbeams and (b) Noise eigenbeams.

eigenbeam associated with the array weights Λ_1 has a peak at 30° due to the linear progressive phase shift between elements. Setting the steering phase equal to the phase difference between adjacent elements gives a beam pointing direction of

$$\delta_s = kd \sin \theta_s = \frac{360°}{\lambda} \frac{\lambda}{2} \sin \theta_s \Rightarrow \theta_s = \sin^{-1}\left(\frac{90°}{180°}\right) = 30°$$

Figure 8.13 is a plot of the four eigenbeams. The three noise eigenbeams have nulls at 30°.

8.5 Direction Finding Algorithms

As noted in the previous example, the covariance matrix contains information about the direction and relative power of the signals incident on the array. An eigenanalysis of the covariance matrix reveals the signal locations and strengths. This section explains how to exploit the covariance matrix to find the location of signals.

8.5.1 Capon's Minimum Variance

Capon's method (also called minimum variance spectral estimation or the maximum likelihood method) calculates an estimate of the signal power as a function of angle [8]. This approach minimizes the output power while forcing the signal power to remain constant:

$$\text{Minimize} : \mathbf{w}^\dagger \mathbf{C}\mathbf{w}$$
$$\text{Subject to} : \mathbf{w}^T \mathbf{A}\mathbf{s} = 1 \tag{8.17}$$

Thus, the output power is minimized, except in the direction of the received signals. The analytical solution to (8.17) is

$$\mathbf{w} = \frac{\mathbf{C}^{-1}\mathbf{A}}{\mathbf{A}^\dagger \mathbf{C}^{-1}\mathbf{A}} \tag{8.18}$$

which produces the spectrum given by

$$P(\theta) = \frac{1}{A^{\dagger}(\theta)C^{-1}A(\theta)} \tag{8.19}$$

Capon's method does not work well with correlated signals, because they might add to zero and avoid detection. It also requires computing the covariance matrix inverse which may be slow and error prone.

Example
An eight-element uniform array with $\lambda/2$ spacing has signals with amplitude 1.0 V/m incident at $\theta_1 = 30°$, $10°$, $0°$, and $-60°$. Find the Capon spectrum when $\sigma^2_{noise} = 0.1$ W/m^2.

Solution
The covariance matrix is calculated to be

$$
\begin{array}{llllllll}
3.91 - j0.00 & 0.98 + j1.07 & 1.06 + j1.65 & 0.59 - j0.85 & 1.28 + j1.74 & 0.50 + j0.68 & -1.78 + j0.52 & 0.99 - j1.64 \\
0.98 - j1.07 & 4.20 + j0.00 & 0.96 + j1.08 & 1.28 + j1.65 & 0.72 - j1.03 & 1.38 + j1.92 & 0.66 + j0.51 & -1.80 + j0.50 \\
1.06 - j1.65 & 0.96 - j1.08 & 4.11 + j0.00 & 1.03 + j1.23 & 1.17 + j1.70 & 0.59 - j0.70 & 1.40 + j1.98 & 0.54 + j0.74 \\
0.59 + j0.85 & 1.28 - j1.65 & 1.03 - j1.23 & 4.19 - j0.00 & 1.07 + j0.99 & 1.15 + j1.65 & 0.81 - j0.99 & 1.27 + j1.77 \\
1.28 - j1.74 & 0.72 + j1.03 & 1.17 - j1.69 & 1.07 - j0.99 & 4.21 + j0.00 & 0.95 + j1.28 & 1.30 + j1.60 & 0.66 - j0.88 \\
0.50 - j0.68 & 1.38 - j1.92 & 0.59 + j0.70 & 1.15 - j1.65 & 0.95 - j1.28 & 3.98 - j0.00 & 0.98 + j1.01 & 0.97 + j1.60 \\
-1.78 - j0.52 & 0.66 - j0.51 & 1.40 - j1.98 & 0.81 + j0.99 & 1.30 - j1.60 & 0.98 - j1.01 & 4.28 - j0.00 & 1.03 + j1.09 \\
0.99 + j1.64 & -1.80 - j0.50 & 0.54 - j0.74 & 1.27 - j1.77 & 0.66 + j0.88 & 0.97 - j1.60 & 1.03 - j1.09 & 3.89 + j0.00
\end{array}
$$

and the array steering matrix is

$$
A =
\begin{bmatrix}
-1.00 + j0.10 & 1.00 + j0.00 & -0.33 + j0.94 & 0.71 - 0.71 \\
0.87 - j0.50 & 1.00 + j0.00 & 0.21 + j0.98 & -0.71 - 0.71 \\
-0.59 + j0.81 & 1.00 + j0.00 & 0.68 + j0.73 & -0.71 + 0.71 \\
0.21 - j0.98 & 1.00 + j0.00 & 0.96 + j0.27 & 0.71 + 0.71 \\
0.21 + j0.98 & 1.00 + j0.00 & 0.96 - j0.27 & 0.71 - 0.71 \\
-0.59 - j0.81 & 1.00 + j0.00 & 0.68 - j0.73 & -0.71 - 0.71 \\
0.87 + j0.50 & 1.00 + j0.00 & 0.21 - j0.98 & -0.71 + 0.71 \\
-1.0 - j0.10 & 1.00 + j0.00 & -0.33 - j0.94 & 0.70 + 0.71
\end{bmatrix}
$$

Substitute the inverse of **C** and **A** into (8.19) and plot the result. In terms of MATLAB, the key command is

```
Pc(ic)=1/abs(A(:,ic).'* inv(C) *conj(A(:,ic)));
```

where ic indicates an angle. Figure 8.14 shows that Capon's spectrum has sharp lines at the signal angles. Unlike the periodogram, it distinctly separates the signals at $0°$ and $10°$.

Figure 8.14 Plot of the Capon power spectrum of an eight-element array with four signals present.

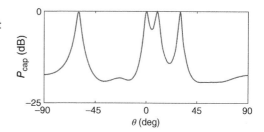

8.5.2 Pisarenko Harmonic Decomposition

Pisarenko harmonic decomposition (PHD) finds the power spectrum due to M sinusoidal signals incident on an array with AWGN present. As seen from the eigenbeam example, the noise eigenvectors approximately equal the noise variance. The eigenvector corresponding to the smallest eigenvalue ($\lambda_{min} \approx \sigma^2_{noise}$) minimizes the mean squared error of the array output with the constraint that the norm of the weight vector equals one. In this case, the power spectrum is [9]

$$P(\theta) = \frac{1}{|\mathbf{A}^\dagger(\theta)\mathbf{Q}[:, min]|^2} \tag{8.20}$$

where $\mathbf{Q}[:, min]$ is the eigenvector that corresponds to the minimum eigenvalue λ_{min}. Since the noise eigenvector is orthogonal to all the signal eigenvectors, the denominator of (8.20) goes to zero in the directions of all signals, producing peaks in the spectrum. PHD serves as a starting point for more sophisticated approaches that are less sensitive to noise. An interesting note is that (8.20) is actually the inverse of the eigenbeam power pattern corresponding to λ_{min}. In other words, (8.20) is the magnitude squared of the upside down array factor.

Example
An eight element uniform array with $\lambda/2$ spacing has signals with amplitude 1.0 V/m incident at $\theta_1 = 30°$, 10°, 0°, and $-60°$. Find the PHD spectrum when $\sigma^2_{noise} = 0.1$ W/m^2.

Solution
In order to make use of (8.20), the eigenvectors and eigenvalues of the covariance matrix must be found. The MATLAB commands are

```
[eigvec,eigval]=eig(C)
Pphd(ic)=1/abs(A(:,ic).'*eigvec(:,ii))^2
```

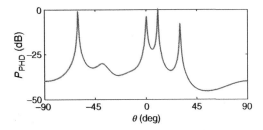

Figure 8.15 Plot of the PHD power spectrum of an eight-element array with four signals present.

where ii points to the eigenvector corresponding to the minimum eigenvalue. Figure 8.15 shows the very sharp lines in the PHD spectrum at the signal angles. It easily separates the signals at 0° and 10°. If the reader looks at Figure 8.15 upside down, it looks like an array factor with peaks at −90° and 53° and nulls in the directions of the signals!

8.5.3 MUSIC Algorithm

MUSIC (*MU*ltiple *SI*gnal *C*lassification) resembles PHD but replaces the minimum eigenvector in the power spectrum estimate with an average of the noise eigenvectors of the covariance matrix [10]. Since all of the noise eigenbeams have nulls in the direction of the signals, any of them work for DF. Averaging them makes MUSIC more robust to noise compared to PHD that only uses one of them. The MUSIC spectrum is given by

$$P(\theta) = \frac{1}{|\mathbf{A}^\dagger(\theta)\mathbf{Q}[:, M+1 : N]|^2} \tag{8.21}$$

where $\mathbf{Q}[:, M+1:N]$ are the eigenvectors corresponding to the noise (eigenvalues equal σ^2_{noise}). MUSIC requires uncorrelated or at most mildly correlated signals. The MUSIC algorithm accurately estimates the number and strengths of signals as well as their AOA for a calibrated array and uncorrelated signals [11].

In practice, no exact demarcation between signal and noise eigenvectors exists. For an unknown number of signals, the eigenvalues only estimate the number of signals present. An actual antenna array generates an approximation to the real covariance matrix. The noise eigenvalues associated with this approximate covariance matrix are close in value but not equal. The ratio of the geometric mean to their arithmetic mean of the noise eigenvalues measures the closest of the eigenvalues.

Example

An eight-element uniform array with $\lambda/2$ spacing has signals with amplitude 1.0 V/m incident at $\theta_1 = 30°$, 10°, 0°, and $-60°$. Find the MUSIC spectrum when $\sigma^2_{noise} = 0.1$ W/m^2.

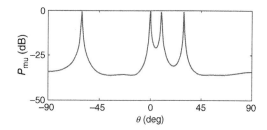

Figure 8.16 Plot of the MUSIC power spectrum of an eight-element array with four signals present.

Solution

The MUSIC spectrum in Figure 8.16 is similar to the corresponding PSD spectrum.

8.5.4 Root MUSIC

Capon's method, PHD, and MUSIC estimate the power spectrum. Peaks in the spectrum correspond to signals and their locations. A more practical algorithm called root-MUSIC, estimates the AOAs based on the roots of the array polynomial rather than the power spectrum [12]. Taking the z-transform of the denominator of (8.21) results in

$$\mathbf{A}^\dagger(\theta)\mathbf{Q}[:, M+1:N]\mathbf{Q}^\dagger[:, M+1:N]\mathbf{A}(\theta) = \sum_{m=0}^{N-1}\sum_{n=0}^{N-1} z^n C_{mn} z^{-m}$$

$$= \sum_{\ell=0}^{2N-2} c_\ell z^\ell \qquad (8.22)$$

where

$z = e^{jknd\sin\theta}$

$c_\ell = \sum_{n-m=\ell} C_{mn} =$ sum of ℓth diagonal elements of matrix $\mathbf{Q}[:, M+1:N]\mathbf{Q}^\dagger[:, M+1:N]$.

The polynomial roots come in pairs: z_m and $1/z_m^*$. One root lies inside the unit circle while the other lies outside the unit circle. Both roots have the same phase information which corresponds to the AOA. Roots on the unit circle are double roots because $z_m = 1/z_m^*$ and correspond to signals. Double roots result, because (8.22) is a power pattern. The power pattern equals the amplitude squared of the array factor. An array polynomial with $N-1$ roots has a power pattern polynomial with $2(N-1)$ roots.

Of the $2N-2$ roots of the power pattern, only roots on or very close to the unit circle correspond to the poles of the MUSIC spectrum and indicate the presence of signals. The phase of the roots of the polynomial (AOAs) in (8.22) are found from

$$\theta_m = \sin^{-1}\left(\frac{\arg(z_m)}{kd}\right) \qquad (8.23)$$

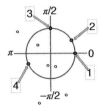

Figure 8.17 Unit circle representation of all the roots found using root MUSIC. Roots corresponding to signals are numbered in boxes.

The algorithm ignores spurious roots not close to the unit circle. Root MUSIC outperforms MUSIC in a low signal to noise ratio (SNR) environment.

Example

An eight element uniform array with $\lambda/2$ spacing has signals with amplitude 1.0 V/m incident at $\theta_1 = 30°$, $10°$, $0°$, and $-60°$. Find the location of the signals using the root MUSIC algorithm when $\sigma^2_{\text{noise}} = 0.1\ \text{W}/\text{m}^2$.

Solution

The coefficients in (8.22) are

$$c_\ell = -0.14 - j0.24, 0.33 - j0.11, -0.22 - j0.13, -0.45 + j0.52, -0.13$$
$$- j0.78, -0.62 + j1.18, -0.77 + j0.95\ 4.00, -0.77 - j0.95,$$
$$- 0.62 - j1.18, -0.13 + j0.78, -0.45 - j0.52, -0.22 + j0.13, 0.33$$
$$+ j0.11, -0.14 + j0.24$$

The $2N - 2$ roots of the array polynomial in (8.22) are listed in Table 8.2. A unit circle plot of the roots appears in Figure 8.17. Signals correspond to the double roots that have the same phase and are near the unit circle. One of the roots is outside and one inside the unit circle. The AOAs are found from the phases via

$$\theta_m = \sin^{-1}\left(\frac{\psi_m}{kd}\right) = \sin^{-1}\left(\frac{\psi_m}{\pi}\right) = -60.02°, 0.06°, 9.99°, 29.9° \quad (8.24)$$

The AOAs are slightly off due to the noise.

8.5.5 Maximum Entropy Method

The maximum entropy method (MEM), also known as the all poles model or the autoregressive model, uses the poles of the rational function model given by [13, 14]

$$P(\theta) = \frac{1}{|A^\dagger(\theta)C^{-1}[:, n]|^2} \quad (8.25)$$

where $C^{-1}(:, n)$ is the nth column of the inverse covariance matrix. The selection of n gives slightly different results.

Example

An eight-element uniform array with $\lambda/2$ spacing has signals with amplitude 1.0 V/m incident at $\theta_1 = 30°$, $10°$, $0°$, and $-60°$. Find the MEM spectrum when $\sigma^2_{\text{noise}} = 0.1\ \text{W}/\text{m}^2$.

Table 8.2 Root MUSIC roots for the eight-element array with four signals incident.

Root	Magnitude	Phase	Signal
1	2.04	−103.5°	No
2	1.77	−69.1°	No
3	1.61	147.4°	No
4	1.01	31.2°	Yes
5	0.99	31.2°	
6	1.01	89.8°	Yes
7	0.99	89.8°	
8	1.01	−155.9°	Yes
9	0.99	−155.9°	
10	1.01	0.2°	Yes
11	0.99	0.2°	
12	0.62	147.4°	No
13	0.56	−69.1°	No
14	0.49	−103.5°	No

Roots near the unit circle correspond to signals and come in pairs at the same angle (one inside and one outside the unit circle).

Figure 8.18 Plot of the MEM power spectrum of an eight-element array with four signals present.

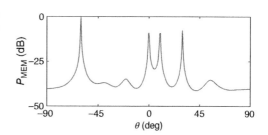

Solution

The resulting MEM spectrum is shown in Figure 8.18.

8.5.6 ESPRIT

ESPRIT (*E*stimation of *S*ignal *P*arameters via *R*otational *I*nvariance *T*echniques) applies stereoscopy to AOA estimation [15]. The algorithm begins by dividing an N-element uniform linear array into two overlapping subarrays, each having $N - 1$ elements and sharing $N - 2$ elements as shown in Figure 8.19. The shared elements are called matched pairs. Each subarray

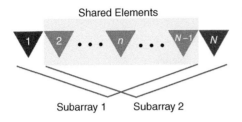

Shared Elements

Figure 8.19 An ESPRIT array has two overlapping subarrays.

Subarray 1   Subarray 2

has one element that the other subarray does not have. The M signals at the elements in the two subarrays are written as

$$\text{Subarray } 1 : \mathbf{X}_1 = \mathbf{A}[1 : N-1, :]\mathbf{s} + \mathbf{n}$$
$$\text{Subarray } 2 : \mathbf{X}_2 = \mathbf{A}[2 : N, :]\mathbf{s} + \mathbf{n} = \mathbf{A}[1 : N-1, :]\mathbf{\Phi}_s\mathbf{s} + \mathbf{n} \qquad (8.26)$$

where the $M \times M$ diagonal matrix, $\mathbf{\Phi}_s$, is given by

$$\mathbf{\Phi}_s = \begin{bmatrix} e^{jkd\sin\theta_1} & 0 & \cdots & 0 \\ 0 & e^{jkd\sin\theta_2} & 0 & 0 \\ \vdots & 0 & \ddots & \vdots \\ 0 & 0 & \cdots & e^{jkd\sin\theta_M} \end{bmatrix} = \begin{bmatrix} z_1 & 0 & \cdots & 0 \\ 0 & z_2 & 0 & 0 \\ \vdots & 0 & \ddots & \vdots \\ 0 & 0 & \cdots & z_M \end{bmatrix} \qquad (8.27)$$

Finding $\mathbf{\Phi}_s$ requires knowing $\mathbf{X}_1$ and $\mathbf{X}_2$.

The first subarray eigenvector matrix equals a matrix $\mathbf{\Psi}$ times the second subarray eigenvector matrix.

$$\mathbf{Q}[1 : N-1, 1 : M] = \mathbf{\Psi}\mathbf{Q}[2 : N, 1 : M] \qquad (8.28)$$

Solving (8.28) for $\mathbf{\Psi}$ produces an estimate of $\mathbf{\Phi}_s$. The eigenvalues of $\mathbf{\Psi}$ estimate z_m. The final step solves for the AOA estimates:

$$\theta_m = \sin^{-1}\left(\frac{\arg(\lambda_m^\Psi)}{kd}\right) \qquad (8.29)$$

where λ_m^Ψ =eigenvalues of $\mathbf{\Psi}$.

Example

An eight-element uniform array with $\lambda/2$ spacing has signals with amplitude 1.0 V/m incident at $\theta_1 = 30°$, $10°$, $0°$, and $-60°$. Find the location of the signals using the ESPRIT algorithm when $\sigma_{noise}^2 = 0.1 \text{ W/m}^2$.

Solution

The signal covariance matrix is given by (Figure 8.20)

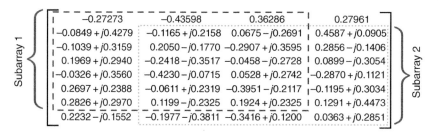

Figure 8.20 Esprit covariance matrix.

Use (8.28) to compute $\mathbf{\Psi}$

$$\mathbf{\Psi} = \begin{bmatrix} 0.5960 - 0.2807i & -0.4668 - 0.1233i & 0.2495 + 0.5460i \\ -0.0760 - 0.3009i & 0.2589 - 0.7582i & 0.1903 + 0.1133i \\ -0.2309 - 0.5797i & 0.1916 + 0.2367i & -0.6670 + 0.3268i \\ -0.1525 + 0.0351i & -0.2582 - 0.1944i & -0.0379 + 0.1709i \end{bmatrix}$$

$$\begin{bmatrix} -0.0297 + 0.3218i \\ -0.3489 - 0.4410i \\ -0.2628 - 0.2593i \\ 0.7402 - 0.4018i \end{bmatrix}$$

The eigenvalues of ψ are

$$\lambda^{\Psi} = \begin{bmatrix} 0.92 + j0.41 & -0.00 - j0.99 & 1.00 - j0.01 & 0.85 - j0.53 \end{bmatrix} \qquad (8.30)$$

Substituting the eigenvalues into (8.29) yields an estimate of the AOAs.

$$\theta_m = \begin{bmatrix} -59.86° & 0.21° & 10.17° & 30.08° \end{bmatrix}$$

8.5.7 Estimating and Finding Sources

Many DF algorithms need to know the number of signals incident on the array. One approach to estimating the number of signals sets a threshold for noise eigenvalues. Eigenvalues above the threshold belong to signals while the rest belong to noise. One algorithm for estimating M is [16]

1. Estimate the covariance matrix from K time samples.
2. Find and sort the eigenvalues $(\lambda_1 > \lambda_2 > \cdots > \lambda_N)$.
3. Find M that minimizes

$$K(N-M)\ln\left[\frac{\sum_{n=M+1}^{N} \lambda_n}{(N-M)\left(\prod_{n=M+1}^{N} \lambda_n\right)^{\frac{1}{N-M}}}\right] + f(M,N) \qquad (8.31)$$

where

$$f(M,N) = \begin{cases} M(2N-M) & \text{Akaike's information criterion} \\ 0.5M(2N-M)\ln N & \text{minimum description length} \end{cases}$$

Problems

8.1 Derive (8.10).

8.2 Use a periodogram to demonstrate the effect of separation angle between two sources using an eight-element uniform array with $\lambda/2$ spacing when $\theta_1 = -30°, 10°, 20°$ and $\theta_2 = 30°$.

8.3 An eight-element uniform array with $\lambda/2$ spacing has three signals incident upon it: $s_1(-60°)=1$, $s_2(0°)=2$, and $s_3(10°)=4$. Find the Capon spectrum.

8.4 An eight-element uniform array with $\lambda/2$ spacing has three signals incident upon it: $s_1(-60°)=1$, $s_2(0°)=2$, and $s_3(10°)=4$. Find the MEM spectrum.

8.5 An eight-element uniform array with $\lambda/2$ spacing has three signals incident upon it: $s_1(-60°)=1$, $s_2(0°)=2$, and $s_3(10°)=4$. Find the MUSIC spectrum.

8.6 An eight-element uniform array with $\lambda/2$ spacing has three signals incident upon it: $s_1(-60°)=1$, $s_2(0°)=2$, and $s_3(10°)=4$. Find the location of the signals using the root MUSIC algorithm.

8.7 An eight-element uniform array with $\lambda/2$ spacing has three signals incident upon it: $s_1(-60°)=1$, $s_2(0°)=2$, and $s_3(10°)=4$. Estimate the incident angles using ESPRIT.

References

1 Haupt, R.L. (2015). *Timed Arrays Wideband and Time Varying Antenna Arrays*. Hoboken, NJ: Wiley.
2 Haupt, R.L. (2010). *Antenna Arrays: A Computational Approach*. Hoboken, NJ: Wiley.
3 Frater, M.R. and Ryan, M. (2001). *Electronic Warfare for the Digitized Battlefield*. Norwood, MA: Artech House.

4 Bayliss, E.T. (1968). Design of monopulse antenna difference patterns with low sidelobes. *The Bell System Technical Journal* 47: 623–650.

5 Adcock, F. (1917). Improvement in means for determining the direction of a distant source of electromagnetic radiation. British Patent 1304901919.

6 Baghdady, E.J. (1989). New developments in direction-of-arrival measurement based on Adcock antenna clusters. In: *Proceedings of the IEEE Aerospace and Electronics Conference*, 1873–1879. Dayton, OH, May 22–26: IEEE.

7 Monzingo, R.A., Haupt, R.L., and Miller, T.W. (2011). *Introduction to Adaptive Antennas*, 2e. Scitech Publishing.

8 Capon, J. (1969). High-resolution frequency-wavenumber spectrum analysis. *Proceedings of the IEEE* 57 (8): 1408–1418.

9 Pisarenko, V.F. (1973). The retrieval of harmonics from a covariance function. *Geophysical Journal International* 33 (3): 347–366.

10 Schmidt, R. (1986). Multiple emitter location and signal parameter estimation. *IEEE Transactions on Antennas and Propagation* 34 (3): 276–280.

11 Schmidt, R. and Franks, R. (1986). Multiple source DF signal processing: an experimental system. *IEEE Transactions on Antennas and Propagation* 34 (3): 281–290.

12 Barabell, A. (1983). Improving the resolution performance of eigenstructure-based direction-finding algorithms. In: *ICASSP '83. IEEE International Conference on Acoustics, Speech, and Signal Processing, Boston, Massachusetts, USA*, 336–339.

13 Burg, J.P. (1972). The relationship between maximum entropy spectra and maximum likelihood spectra. *Geophysics* 37 (2): 375–376.

14 Lacoss, R.T. (1971). Data adaptive spectral analysis methods. *Geophysics* 36 (4): 661–675.

15 Paulraj, A., Roy, R., and Kailath, T. (1986). A subspace rotation approach to signal parameter estimation. *Proceedings of the IEEE* 74 (7): 1044–1046.

16 Godara, L.C. (2004). *Smart Antennas*. Boca Raton, FL: CRC Press.

9

Adaptive Arrays

Adaptive arrays modify their receive and/or transmit pattern characteristics in response to the signals in the environment. The array evolves with time in order to improve signal reception. A computer algorithm dynamically adjusts the weights and switches in response to feedback, such as signal to noise ratio (SNR). This chapter divides adaptive arrays into three categories: (i) adaptive nulling, (ii) reconfigurable antennas, and (iii) beam-switching antennas. The last two topics receive brief introductions at the end of the chapter. Adaptive nulling occupies the vast majority of space.

9.1 The Need for Adaptive Nulling

As wireless users crowd the frequency spectrum, interference becomes more common. When the sidelobe gain times the interference signal becomes large enough to drop the SNR or bit error rate (BER) below an acceptable level, then an adaptive array adjusts the element weights to improve performance. An adaptive array maximizes the main beam gain in the direction of the desired signal while minimizing the antenna pattern in the directions of the interfering signals. Howells and Applebaum developed the first adaptive nulling antenna for a radar [1, 2]. Their accomplishment went unrecognized until it was declassified a number of years later. The Howells–Applebaum algorithm must know the direction of the desired signal in order avoid placing a null in that direction. Widrow developed a similar algorithm for communications systems [3]. His least mean square (LMS) algorithm serves as the basis for many adaptive antennas. This algorithm needs to know the desired signal characteristics but not its direction.

Consider the situation in Figure 9.1 in which a desired signal enters the main beam (signal 1) and interfering signals 2 and 3 enter the sidelobes of

Wireless Communications Systems: An Introduction, First Edition. Randy L. Haupt.
© 2020 John Wiley & Sons, Inc. Published 2020 by John Wiley & Sons, Inc.

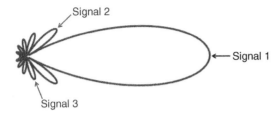

Signal 2

Signal 1

Signal 3

Figure 9.1 Three signals incident on the quiescent antenna pattern.

the quiescent pattern (pattern before adaptive algorithm places the nulls). Assume that the desired signal has an amplitude of 1 V/m and the interfering signals are 0.2 V while ignoring the noise. The desired signal received by the antenna equals the desired signal power times the gain of the main beam.

$$P_s = (1)^2 \times 1 = 1 \text{ W} \tag{9.1}$$

If the array factor sidelobe gains are -13 and -17 dB, then the received interference power is

$$P_I = (0.2)^2 \times 10^{-13/10} + (0.2)^2 \times 10^{-17/10} = 0.0028 \text{ W} \tag{9.2}$$

which leads to an signal to interference ratio (SIR) of 25.5 dB. In this case, the interfering signals have a small impact on system performance.

Now, consider the situation in which signal 2 has an amplitude of 10 V/m while the other two signals are 1 V/m. The desired received signal power is the same, but the received interference power becomes

$$P_I = (10)^2 \times 10^{-13/10} + (1)^2 \times 10^{-17/10} = 5.03 \text{ W} \tag{9.3}$$

The SIR equals -7.02 dB which means the interference overwhelms the desired signal.

Most communication systems require an SIR much larger than 0 dB, so the antenna must mitigate the interference in order to maintain the link. A low sidelobe amplitude taper improves reception of the desired signal by decreasing the received interference power relative to the received desired signal power. For instance, reducing the first sidelobe by more than 20 dB reduces the interference power below that of the desired signal. Achieving such low sidelobes, especially in small arrays, is nearly impossible and is very expensive. A static low sidelobe amplitude taper battles interference from all angles outside the main beam.

Adaptive nulling fights interference by dynamically changing the amplitude and phase weights at the elements in order to place nulls in the antenna pattern in the directions of the interfering signals. In a time varying signal environment, the adaptive algorithm continuously updates the array element weights in order to keep nulls pointing at the undesired signals as shown in Figure 9.2.

Figure 9.2 An adaptive array starts with the quiescent pattern in Figure 9.1 and places nulls in the directions of the two interfering signals while keeping the main beam pointing at the desired signal.

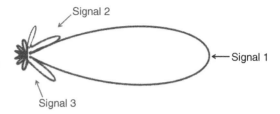

9.2 Beam Cancellation

Beam cancellation subtracts a cancellation beam (an array factor) from the quiescent array factor in order to place a null at a desired angle. When the array encounters M interfering signals coming from θ_m for $1 \leq m \leq M$, then adaptation requires steering, weighting, then subtracting M cancellation beams from the quiescent pattern [4]

$$\underbrace{\sum_{n=1}^{N} w_n e^{jk(n-1)d \sin \theta} = \sum_{n=1}^{N} a_n e^{jk(n-1)d \sin \theta}}_{\text{quiescent array factor}}$$

$$\underbrace{- \sum_{m=1}^{M} \gamma_m \sum_{n=1}^{N} b_n e^{jk(n-1)d \sin \theta} e^{-jkd(n-1)d \sin \theta_m}}_{\text{cancellation beams}} \qquad (9.4)$$

The phase shift $-kd(n-1)d \sin \theta_m$ steers the cancellation beam to the interference location θ_m. Next, γ_m weights beam m until it has the same gain as the quiescent sidelobe level at θ_m. M cancellation beams place M nulls when subtracted from the quiescent pattern if the beams are orthogonal. Beam orthogonality means that $M - 1$ beams have nulls in the direction of the peak of the Mth beam. The adapted weights in (9.4) can be found using

$$w_n = 1 - \frac{b_n}{a_n} \sum_{m=1}^{M} \gamma_m e^{-jk(n-1)d \sin \theta_m} \qquad (9.5)$$

The cancellation beam amplitude taper, b_n, controls the sidelobes and beamwidth of the cancellation beams. Usually, the cancellation beam weights either equal the amplitude taper of the quiescent array ($b_n = a_n$) or are uniform ($b_n = 1$) [5].

Example

Plot the quiescent and adapted array factors as well as the cancellation beams of an eight-element array having a 20-dB Chebyshev amplitude taper with $\lambda/2$ spacing when unwanted signals are incident at $\theta = 40°$ and $-25°$.

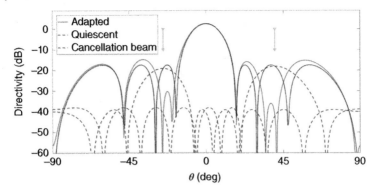

Figure 9.3 Nulls are created in the Chebyshev quiescent pattern by subtracting two cancellation beams at $\theta = 40°$ and $-25°$.

Solution

The Chebyshev weights are found from the synthesis approach in Chapter 4: $w = [0.5799\ 0.6603\ 0.8751\ 1.0000\ 1.0000\ 0.8751\ 0.6603\ 0.5799]$. The height of the sidelobes in the directions of the interfering signals are given by $\gamma = [0.0054 - j0.0825\ 0.0660 + j0.0659]$. The adapted weights calculated from (9.5) when $a_n = b_n$ are $[0.9465 + j0.0132\ 1.0289 + j0.0998\ 1.6087 - j0.1480\ 1.7607 - j0.2905\ 1.7607 + j0.2905\ 1.6087 + j0.1480\ 1.0289 - j0.0998\ 0.9465 - j0.0132]$. Figure 9.3 shows the quiescent pattern superimposed on the adapted pattern and two cancellation beams.

9.3 Optimum Weights

The error signal equals the magnitude of the difference between the desired signal (s_1) and the array output.

$$\varepsilon = |s_1 - \mathbf{w}^T(\mathbf{As} + \mathbf{n})| \tag{9.6}$$

where

$$\mathbf{s} = \begin{bmatrix} s_1 & s_2 & \cdots & s_M \end{bmatrix}^T = \text{signals}$$

$$\mathbf{w} = \begin{bmatrix} w_1 & w_2 & \cdots & w_N \end{bmatrix}^T = \text{element weights}$$

$$\mathbf{A} = \begin{bmatrix} 1 & 1 & \cdots & 1 \\ e^{jkd\sin\theta_1} & e^{jkd\sin\theta_2} & \cdots & e^{jkd\sin\theta_M} \\ \vdots & \vdots & \ddots & \vdots \\ e^{jk(N-1)d\sin\theta_1} & e^{jk(N-1)d\sin\theta_2} & \cdots & e^{jk(N-1)d\sin\theta_M} \end{bmatrix}$$

$$= \text{array steering matrix}$$

$$\mathbf{n} = \begin{bmatrix} v_1 & v_2 & \cdots & v_N \end{bmatrix}^T = \text{element noise}$$

The mean square error of (9.6) equals the expected value of the magnitude of the error squared

$$E\{\varepsilon^2\} = E\{|s_1|^2\} + \mathbf{w}^\dagger \mathbf{C}\mathbf{w} - 2\mathbf{w}^\dagger E\{s_1(\mathbf{A}\mathbf{s} + \mathbf{n})\} \tag{9.7}$$

where the covariance matrix is given by

$$\mathbf{C} = E\{(\mathbf{A}\mathbf{s} + \mathbf{n})(\mathbf{A}\mathbf{s} + \mathbf{n})^\dagger\} \tag{9.8}$$

Assuming that ε varies slowly with time, then taking the gradient of (9.7) with respect to the element weights and setting it equal to zero leads to the minimum of the mean square error

$$\nabla_w^T E\{\varepsilon^2\} = 2\mathbf{C}\mathbf{w}_{\text{opt}} - 2E\{s_1(\mathbf{A}\mathbf{s} + \mathbf{n})\}^T = 0 \tag{9.9}$$

Solving (9.9) for $\mathbf{w}$ produces the Wiener–Hopf solution [6]:

$$\mathbf{w}_{\text{opt}} = \mathbf{C}^{-1} E\{s_1(\mathbf{A}\mathbf{s} + \mathbf{n})\}^T \tag{9.10}$$

Most adaptive algorithms strive to find weights close to $\mathbf{w}_{\text{opt}}$.

9.4 Least Mean Square (LMS) Algorithm

In an adaptive algorithm, the weight vector updates by an incremental amount each time step until the interference disappears. The new weights at time sample $n + 1$ are the old weights at time sample n plus an incremental update, $\Delta\mathbf{w}[n]$

$$\mathbf{w}[n + 1] = \mathbf{w}[n] + \Delta\mathbf{w}[n] \tag{9.11}$$

Appropriate choices for $\Delta\mathbf{w}[n]$ push the weights closer to $\mathbf{w}_{\text{opt}}$. The gradient of the square of the error with respect to the weights points toward the minimum of ε^2, so it directs the weight vector toward $\mathbf{w}_{\text{opt}}$.

$$\mathbf{w}[n + 1] = \mathbf{w}[n] - \mu_0 \nabla_w^T \varepsilon^2 \tag{9.12}$$

Using a trick to rewrite the gradient in (9.12) produces

$$\mathbf{w}[n + 1] = \mathbf{w}[n] - \mu_0 \frac{\partial \varepsilon^2}{\partial \varepsilon} \frac{\partial \varepsilon}{\partial \mathbf{w}} = \mathbf{w}[n] - 2\mu_0 \varepsilon (\mathbf{A}\mathbf{s}[n] + \mathbf{n}[n]) \tag{9.13}$$

Finally, the LMS algorithm is given by [7]

$$\mathbf{w}[n + 1] = \mathbf{w}[n] + \mu\{(\mathbf{A}\mathbf{s}[n] + \mathbf{n}[n])[s_1[n] - \mathbf{w}^\dagger[n](\mathbf{A}\mathbf{s}[n] + \mathbf{n}[n])]\} \tag{9.14}$$

where μ is a step size that includes the constants resulting from taking the derivative. A large μ causes the algorithm to overshoot the optimum weights, while a small μ causes very slow convergence.

The LMS algorithm remains stable when the step size stays within bounds determined by the maximum eigenvalue (λ_{max}) of the covariance matrix [8]

$$0 \le \mu \le \frac{2}{\lambda_{max}} \tag{9.15}$$

Increasing $\lambda_{max}/\lambda_{min}$ slows the convergence speed, because (9.7) has a long, narrow valley that requires many iterations to find the minimum.

The LMS algorithm in (9.14) contains the desired signal which in practice means an approximation for the desired signal, $s_1(n)$. Approximations for $s_1(n)$ in (9.14) should [7]:

1. Be highly correlated with the desired signal and uncorrelated with the interference signals.
2. Have similar directional and spectral characteristics as those of the desired signal.

Fortunately, these conditions are not difficult to meet.

Example
An eight-element array with a uniform amplitude taper and $\lambda/2$ spacing has two interference signals: a 2 V/m signal at $\theta = 61°$ and a 4 V/m signal at $\theta = -21°$ while the desired signal is 1 V/m signal at $\theta = 0°$. Use the LMS algorithm to reduce the power received by the interfering signals while increasing the SNR.

Solution
First, form the covariance matrix then find the largest eigenvalue which is 30.6182. According to (9.15), $0 \le \mu \le 0.0653$. In order to have a slow, steady convergence, μ is chosen to be 0.0001. Figure 9.4 shows a plot of the amplitude and Figure 9.5 a plot of the phase of the adapted weights as a function of time. They seem to settle at about 0.02 ms. Random noise causes the weights to jitter. Figure 9.6 is the adapted pattern with the desired nulls superimposed on the quiescent pattern. The output signal is compared to the transmitted signal in Figure 9.7. As the weights settle down at 0.02 ms, the received signal looks like the transmitted signal. Figure 9.8 has a plot of the signal to interference

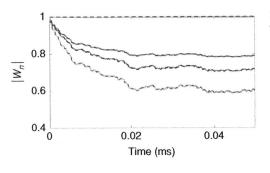

Figure 9.4 LMS amplitude weights as a function of time.

Figure 9.5 LMS phase weights as a function of time.

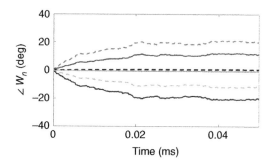

Figure 9.6 The LMS adapted pattern has nulls in the directions of the interfering signals.

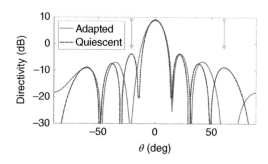

Figure 9.7 The received signal due to the LMS weights begins to track the desired signal at about 0.02 ms.

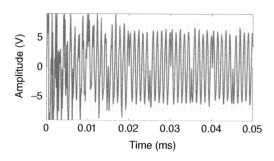

Figure 9.8 Plot of SINR as a function of time for the LMS adaptive algorithm.

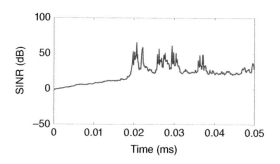

plus noise ratio (SINR) as a function of time. SINR starts at $-1.37\,dB$ and ends at $28.1\,dB$. The SINR goes above $10\,dB$ at $12.5\,\mu s$.

9.5 Sample Matrix Inversion Algorithm

In practice, the expected value needed to calculate the covariance matrix in (9.10) is replaced by an average of a finite number of signal samples. There is a tradeoff between the number of samples and the accuracy and convergence speed of the algorithm. A single sample of the covariance matrix contains too much noise to accurately represent the covariance matrix. Averaging decreases the noncoherent noise while enhancing the coherent signals. Taking more samples requires time that slows down the adaptive algorithm.

The sample covariance matrix, $\widehat{\mathbf{C}}$, averages N_s instantaneous samples of the covariance matrix

$$\widehat{\mathbf{C}} = \frac{1}{N_s} \sum_{n=1}^{N_s} (\mathbf{As}[n] + \mathbf{n}[n])(\mathbf{As}[n] + \mathbf{n}[n])^\dagger \qquad (9.16)$$

The sample matrix inversion (SMI) or direct matrix inversion (DMI) algorithm consists of calculating the sample covariance matrix in (9.16) then substituting into (9.17) to find the weights [6, 9]

$$\mathbf{w} = \widehat{\mathbf{C}}^{-1} \{ s_1[n](\mathbf{As}[n] + \mathbf{n}[n]) \}^T \qquad (9.17)$$

Example
An eight-element array with a uniform amplitude taper and $\lambda/2$ spacing has two interference signals: a $2\,V/m$ signal at $\theta = 61°$ and a $4\,V/m$ signal at $\theta = -21°$ while the desired signal is $1\,V/m$ signal at $\theta = 0°$. Use the SMI algorithm to reduce the power received by the interfering signals while increasing the SNR.

Solution
Figures 9.9 and 9.10 show plots of the amplitude and phase of the adapted weights as a function of time. The amplitudes settle at about $0.02\,ms$, while the phase settles at about $0.003\,ms$. The random noise and random variations in the interfering signals cause the weights to jitter. Figure 9.11 is the adapted pattern with the desired nulls superimposed on the quiescent pattern. The output signal is compared to the transmitted signal in Figure 9.12. The received signal looks like the transmitted signal at a very early time in the adaptation. Figure 9.13 has a plot of the SINR as a function of time. SINR starts at $0.7\,dB$ and ends at $45.6\,dB$. The SINR goes above $10\,dB$ at $0.6\,\mu s$.

Figure 9.9 SMI amplitude weights as a function of time.

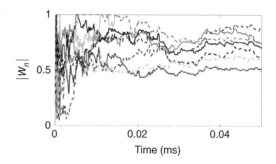

Figure 9.10 SMI phase weights as a function of time.

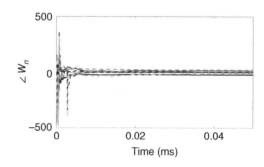

Figure 9.11 The SMI adapted pattern has nulls in the directions of the interfering signals.

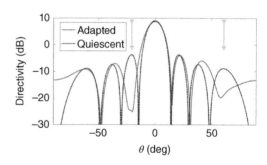

Figure 9.12 The received signal due to the SMI weights begins to track the desired signal at about 0.02 ms.

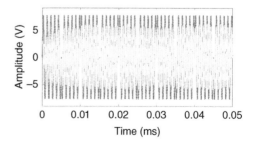

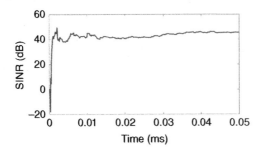

Figure 9.13 Plot of SINR as a function of time for the SMI adaptive algorithm.

9.6 Adaptive Algorithms Based on Power Minimization

The adaptive algorithms described so far require digital beamforming in order to quantify the signal at each element and form the covariance matrix. Most phased arrays, however, do not have a receiver or analog to digital convertor (ADC)/digital to analog converter (DAC) at each element. Instead, beamforming looks like the array in Figure 9.14. There are weights at each element, but the only signal available to an adaptive algorithm is the sum of all the weighted signals from the elements. As a result, only the output signal provides feedback to an adaptive algorithm that adjusts the weights. The total output power of the array is given by

$$P = |\mathbf{w}^T \mathbf{As}|^2 \tag{9.18}$$

The output power in (9.18) has the desired and interfering signals mixed together with the noise.

The desired signal power cannot be separated from the interference power in (9.18). Consequently, minimizing the total output power of the array minimizes the desired signal as well as the interference [10]. Power minimization only works as an adaptive nulling algorithm if the desired signal enters the main beam and the interference enters the sidelobes. With this assumption, small

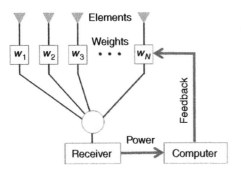

Figure 9.14 Power minimization adaptive array.

Figure 9.15 Phase shifter with four of its eight bits adaptive.

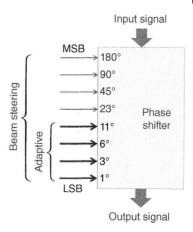

weight perturbations have a small impact on the main beam, so the desired signal remains relatively unaffected. Those same small weight perturbations place nulls in the sidelobes to reduce interference. Two approaches to constraining weights to small variations are

- *Partial adaptive nulling*: Only a subset of the elements are adaptive [11, 12].
- *Weight constraints*: Weight variations limited to small values within a specified range [13, 14].

Partial adaptive nulling has N_a adaptive elements out of a total of N elements. N_a should be large enough to form nulls in the highest sidelobes but small enough to minimize the main beam gain loss. Feedback in Figure 9.14 only goes to the N_a elements. The remaining $N - N_a$ elements have static amplitude weights and possibly beam steering phase shifters.

Making all the element weights adaptive requires weight constraints that avoid nulling the desired signal entering the main beam. A constraint limits the adaptive weights to the least significant bits (LSBs) of the amplitude and phase weights. For instance, if the phase shifters have eight bit controls for beam steering as shown in Figure 9.15, then the adaptive algorithm only has access to the four LSBs. Four bits limit the maximum adaptive phase shift to about 21° which is small enough to minimize the main beam loss but large enough to place nulls in the sidelobes. All the bits steer the beam but the adaptive algorithm only has access to the four LSBs create nulls.

9.6.1 Random Search Algorithms

Random, stochastic, or Monte Carlo search adaptive algorithms enhance the desired signal and suppress the interfering signals by searching for the appropriate array weight values [7]. For the most part, random search algorithms have slow convergence. They require minimal computations and hardware, are insensitive to discontinuities or discrete variables, and find a minimum for a multimodal cost function. Random search algorithms do not require digital beamforming arrays which makes their implementation much easier and cheaper. An adaptive array has an infinite number of weight values that reduce the interference entering the sidelobes, so a random search algorithm has good odds for successfully reducing the impact of interfering signals. Artificial intelligence and machine learning offer guided random search algorithms that control adaptive arrays [13].

Example

An eight-element uniform array has two adaptive elements: 1 and 8. Let the amplitude of these elements lie between 0 and 1, while the phase varies between $0°$ and $360°$. Find the lowest output power and the best SINR with 1000 random guesses. Assume that the array has a uniform amplitude taper and $\lambda/2$ spacing. There are two interference signals: a 2 V/m signal at $\theta = 61°$ and a 4 V/m signal at $\theta = -21°$ while the desired signal is 1 V/m signal at $\theta = 0°$.

Solution

Starting with a uniform array at trial 1, 999 random guesses are then made for the two element weights. The resulting relative output power is shown in Figure 9.16. The lowest output power of 18.0 dB is found at guess 235 compared to the 24.8 dB output power for the uniform array. The weights associated with this minimum relative output power, [$0.6482 \angle 79.5°$, 1.0, 1.0, 1.0, 1.0, 1.0, 1.0, $0.7936 \angle -40.5°$], produce the adapted array factor shown in Figure 9.17. The directivity decreases from 9.0 dB for the uniform array to 6.9 dB for the adapted pattern. The SINR associated with the 1000 weight trials is shown in Figure 9.18. The best SINR of 12.3 dB occurs at trial 235 compared to -1.4 dB for the uniform array. Only three guesses resulted in an SINR greater than 10 dB.

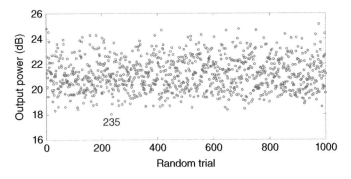

Figure 9.16 Output power for 1000 random guesses of weights.

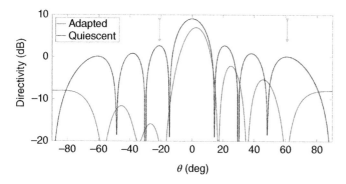

Figure 9.17 Adapted array factor superimposed on the quiescent (uniform) array factor.

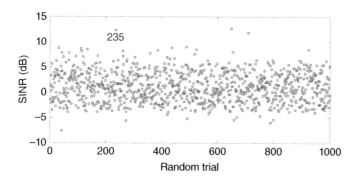

Figure 9.18 SINR for 1000 random guesses of weights.

Example

An eight-element uniform array has all the elements adaptive. Weight variation are limited to: $0.8 \leq |w_n| \leq 1.0$ and $0.0 \leq \angle w_n \leq 72°$. Find the lowest output power and the best SINR with 1000 random guesses. Assume that the array has a uniform amplitude taper and $\lambda/2$ spacing. There are two interference signals: a 2 V/m signal at $\theta = 61°$ and a 4 V/m signal at $\theta = -21°$ while the desired signal is 1 V/m signal at $\theta = 0°$.

Solution

Starting with a uniform array at trial 1, 999 random guesses are then made for the two element weights. The resulting relative output power is shown in Figure 9.19. The lowest output power of 18.6 dB is found at guess 264 compared to the 24.8 dB output power for the uniform array. The weights associated with this minimum relative output power, $[0.863 \angle 71.8°, 0.908 \angle 21.5°, 0.838 \angle 4.4°, 0.950 \angle 41.2°, 0.858 \angle 51.3°, 0.904 \angle 33.9°, 0.944 \angle 45.5°, 0.840 \angle 1.4°]$, produce the adapted array factor shown in Figure 9.20. The directivity decreases from 9.0 dB for the uniform array to 6.7 dB for the adapted pattern. The SINR

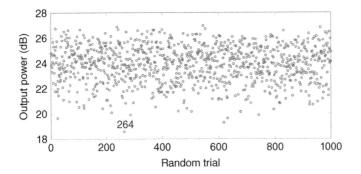

Figure 9.19 Output power for 1000 random guesses of weights.

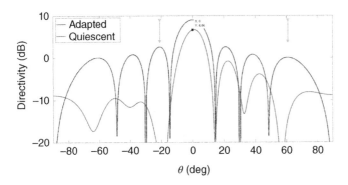

Figure 9.20 Adapted array factor superimposed on the quiescent (uniform) array factor.

associated with the 1000 weight trials is shown in Figure 9.21. The best SINR of 10.6 dB occurs at trial 264 compared to −1.4 dB for the uniform array. Only one guess resulted in an SINR greater than 10 dB.

9.6.2 Output Power Minimization Algorithms

Guessing at weights and picking the values that yield the lowest output power is a slow and inefficient way to do adaptive nulling. A better approach minimizes the calculated or measured output power of an array using numerical minimization algorithms. This approach has been successfully verified through experimental measurements as well as simulations [15, 16].

Example
An eight-element array has two adaptive elements: 1 and 8. Let the amplitude of these elements lie between 0 and 1, while the phase varies between 0° and 360°. Use the Nelder–Mead downhill simplex algorithm [17] to minimize the total output power.

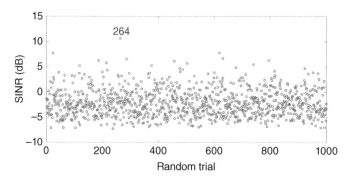

Figure 9.21 SINR for 1000 random guesses of weights.

Solution

The resulting relative output power vs. iteration is shown in Figure 9.22. The lowest output power of 16.95 dB is found after 187 iterations and 325 function evaluations compared to the 20.71 dB output power for the uniform array. The weights associated with this minimum relative output power, $[0.5819 \angle 67.6°, 1 \angle 0°, 1 \angle 0°, 1 \angle 0°, 1 \angle 0°, 1 \angle 0°, 1 \angle 0°, 0.573 \angle -65.4°]$, produce the adapted array factor shown in Figure 9.23. The directivity decreases from 9.03 for the uniform array to 6.11 dB for the adapted pattern. The SINR at the end of the optimization is 20.92 dB compared to −1.37 dB for the uniform array. This is faster and has a higher SINR than random guessing.

MATLAB commands:

```
options = optimset('Display','iter','PlotFcns',
    @optimplotfval);
[y,fval,exitflag,output] =fminsearch('partialadapf',
    rand(1,4),options)
```

The array output power is calculated in `partialadapf.m`.

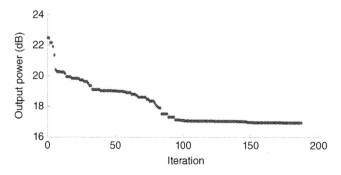

Figure 9.22 Convergence of the Nelder–Mead algorithm.

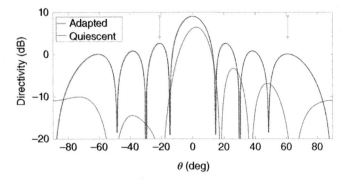

Figure 9.23 Adapted array factor calculated from the optimized weights superimposed on the quiescent (uniform) array factor.

Optimization algorithms are classified as either global or local searches. Hybrid optimization combines the two by first starting a global search that hands over its best result to a fast local optimizing algorithm. Global searches, such as genetic algorithm or particle swarm optimization guide random searches based on biological or physical processes in nature. These algorithms have been successfully applied to experimental adaptive arrays [16].

9.6.3 Beam Switching

Some array beamforming networks host multiple overlapping beams that point in slightly different directions. Figure 9.24 shows a diagram where one of four beams is selected by closing the appropriate switch. Two examples of multi-beam beamforming networks are the Rotman lens [18] and Butler matrix [19]. Each beam port receives a signal from all the elements. Path lengths differences from the elements to the beam ports cause a beam steering phase shift that points the beams in different directions. The beams overlap and cover the desired area.

9.6.4 Reconfigurable Antennas

Configuring an antenna means designing its shape, material properties, feed location, etc., so that it radiates at a desired frequency and polarization. If

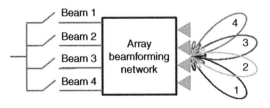

Figure 9.24 Beam switching with a multi-beam antenna.

the operating characteristics of the antenna change, then the antenna needs reconfigured to satisfy the new specifications. Reconfigurable antennas modify their performance characteristics by changing the current flow on an antenna, using mechanically movable parts, phase shifters, attenuators, diodes, tunable materials, or active materials [20].

Radio frequency (RF) switches in an antenna cause currents to flow or not flow along paths in order to change the antenna's radiation properties, as well as its impedance. Figure 9.25 shows a patch with switches that short slots when closed in order to switch between polarizations.

Reconfigurable antenna arrays modify array properties to control the antenna pattern. Figure 9.26 shows a hemispherical array for communicating with satellites. This hemispherical array follows satellites anywhere in the sky. Each triangular subarray is a planar array of antenna elements. Adjacent triangles combine to form larger arrays, so the aperture size adjusts the amount of gain need to communicate with a satellite. The active aperture moves across the dome in order to follow a satellite across the sky.

Figure 9.25 A reconfigurable slotted-patch antenna. *Source:* Haupt and Lanagan [20]. Reproduced with permission of IEEE.

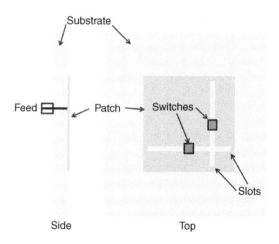

Figure 9.26 Phased array on a hemispherical surface for satellite communications. *Source:* Haupt and Lanagan [20]. Reproduced with permission of IEEE.

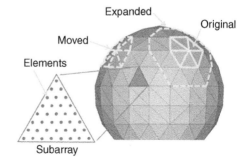

Problems

1 Plot the quiescent and adapted array factors as well as the uniform cancellation beams of an eight-element uniform array with $\lambda/2$ spacing when unwanted signals are incident at $\theta = 61°$ and $-20°$.

2 Plot the quiescent and adapted array factors as well as the 20 dB Chebyshev cancellation beams of an eight-element uniform array with $\lambda/2$ spacing when unwanted signals are incident at $\theta = 61°$ and $-20°$.

3 Plot the adapted pattern resulting from placing a null at $\theta = 16.25°$ in the array factor of a 32-element uniform array with $\lambda/2$ element spacing using random search. Four different configurations of eight adaptive elements are considered:
   (a) 1, 2, 3, 4, 29, 30, 31, 32
   (b) 13, 14, 15, 16, 17, 18, 19, 20
   (c) 1, 5, 9, 13, 17, 21, 25, 29.

4 Plot the adapted pattern resulting from placing a null at $\theta = 16.25°$ in the array factor of a 32-element uniform array with $\lambda/2$ element spacing using random search. Four different adaptive weight limits assuming every element in the array is adaptive:
   (a) $0.8 \leq a_n \leq 1.0, 0 \leq \delta_n \leq 72°$
   (b) $0.8 \leq a_n \leq 1.0, \delta_n = 0°$
   (c) $a_n = 1.0, 0 \leq \delta_n \leq 72°$
   (d) $0.5 \leq a_n \leq 1.0, 0 \leq \delta_n \leq 180°$.

5 An eight-element uniform array with $\lambda/2$ spacing has the desired signal incident at $0°$ and two interference signals incident at $-21°$ and $61°$. Use the LMS algorithm to place nulls in the antenna pattern. Assume $\sigma_{noise} = 0.01$. In MATLAB, represent the signal by $\cos(2\pi(1:K)/K)\exp(j\,rand)$ and the interference by $sign(randn(1, K))$.

6 An eight-element uniform array with $\lambda/2$ spacing has the desired signal incident at $0°$ and two interference signals incident at $-21°$ and $61°$. Use the SMI algorithm to place nulls in the antenna pattern. Assume $\sigma_{noise} = 0.01$. In MATLAB, represent the signal by $\cos(2\pi(1:K)/K)\exp(j\,rand)$ and the interference by $sign(randn(1, K))$.

References

1 Howells, P. (1976). Explorations in fixed and adaptive resolution at GE and SURC. *IEEE Transactions on Antennas and Propagation* 24 (5): 575–584.

2 Applebaum, S. (1976). Adaptive arrays. *IEEE Transactions on Antennas and Propagation* 24 (5): 585–598.

3 Widrow, B., Mantey, P.E., Griffiths, L.J. et al. (1967). Adaptive antenna systems. *Proceedings of the IEEE* 55 (12): 2143–2159.

4 Haupt, R.L. (2010). *Antenna Arrays: A Computational Approach*. Hoboken, NJ: Wiley.

5 Steyskal, H., Shore, R.A., and Haupt, R.L. (1986). Methods for null control and their effects on the radiation pattern. *IEEE AP-S Transactions* AP-34 (3): 163–166.

6 Compton, R.T. (1987). *Adaptive Antennas: Concepts and Performance*. Philadelphia, PA: Prentice-Hall.

7 Monzingo, R.A., Haupt, R.L., and Miller, T.W. (2011). *Introduction to Adaptive Antennas*, 2e. Scitech Publishing.

8 Gross, F.B. (2005). *Smart Antennas for Wireless Communications: With MATLAB*. New York: McGraw-Hill.

9 Gupta, I. (1986). SMI adaptive antenna arrays for weak interfering signals. *IEEE Transactions on Antennas and Propagation* 34 (10): 1237–1242.

10 Haupt, R.L. (2015). *Timed Arrays Wideband and Time Varying Antenna Arrays*. Hoboken, NJ: Wiley.

11 Morgan, D. (1978). Partially adaptive array techniques. *IEEE Transactions on Antennas and Propagation* 26 (6): 823–833.

12 Haupt, R.L. and Shore, R.A. (1984). Experimental partially adaptive nulling in a low sidelobe phased array. In: *Antennas and Propagation Society International Symposium*, 823–826. Boston, MA: IEEE.

13 Haupt, R.L. (1997). Phase-only adaptive nulling with a genetic algorithm. *IEEE Transactions on Antennas and Propagation* 45 (6): 1009–1015.

14 Haupt, R.L. (2010). Adaptive nulling with weight constraints. *Progress In Electromagnetics Research B* 26: 23–38.

15 Haupt, R.L. and Werner, D.H. (2007). *Genetic algorithms in electromagnetics*. Hoboken, NJ: IEEE Press: Wiley-Interscience.

16 Haupt, R.L. and Southall, H. Experimental adaptive cylindrical array. *Microwave Journal* 3: 291–296.

17 Press, W.H., and Numerical Recipes Software (Firm) (1994). *Numerical Recipes in FORTRAN*. Cambridge University Press.

18 Rotman, W. and Turner, R. (1963). Wide-angle microwave lens for line source applications. *IEEE Transactions on Antennas and Propagation* 11 (6): 623–632.

19 Butler, J. and Lowe, R. (1961). Beam-forming matrix simplifies design of electronically scanned antennas. *Electronic Design* 9: 170–173.

20 Haupt, R.L. and Lanagan, M. (2013). Reconfigurable Antennas. *IEEE Antennas and Propagation Magazine* 55 (1): 49–61.

10

MIMO

Multiple input/multiple output (MIMO) systems rely on diversity and adaptive signal processing in the transmit and receive array antennas to dramatically increase data rates and spectral efficiency [1, 2]. Diversity increases as the number of transmit and receive antenna elements increase. Both the transmit and receive arrays adapt their weights in order to emphasize productive subchannels between the transmit and receive elements in order to increase the desired signal reception.

Figure 10.1 shows four categories of wireless communication systems based on the number of elements at the transmitter (N_t) and receiver (N_r) [3]. Most communication systems have one antenna transmitting to a receive antenna $(N_t = 1, N_r = 1)$ or single input single output (SISO). The signal travels through a channel that has an impulse response given by h_{11}. SISO works well in a time invariant, high signal to noise ratio (SNR) channel. Multiple input single output (MISO) communication systems have a transmit array antenna and only one receive antenna $(N_t > 1, N_r = 1)$. A subchannel from transmit element m to the receive antenna impulse response of h_{m1}. Single input multiple output (SIMO) systems have multiple antennas at the receiver but the transmitter only has one antenna $(N_t = 1, N_r > 1)$. The subchannel impulse response from the transmit antenna to receive element n is given by h_{1n}. MIMO has an antenna array at the transmitter and receiver $(N_t > 1, N_r > 1)$. It has a subchannel impulse responses, h_{mn}, between from each of the N_t transmit antennas to each of the N_r receive antennas.

This chapter introduces the concept of MIMO and the critically important channel matrix. The channel matrix leads to finding the receive and transmit element weights that increase the channel capacity.

10.1 Types of MIMO

A MIMO communication system uses either spatial diversity or spatial multiplexing techniques to increase data transfer [4]. A transmitter with one

Wireless Communications Systems: An Introduction, First Edition. Randy L. Haupt.
© 2020 John Wiley & Sons, Inc. Published 2020 by John Wiley & Sons, Inc.

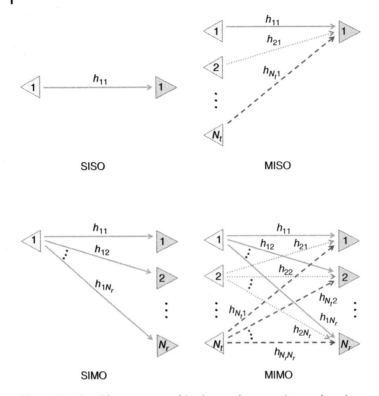

Figure 10.1 Possible antenna combinations at the transmitter and receiver.

antenna (Figure 10.2a) sends one stream of data from one location. Spatial diversity at the transmitter means that each array transmit element sends the same data signal to the receiver in order to increase the diversity of paths. Each path encounters different fading, so more paths increase the probability that one of them successfully delivers the signal. Figure 10.2b shows each of the four transmitting antennas sending the binary string "1 0 0 1". In this case, the antenna in Figure 10.2a replicates four times to form the transmit array. Placing transmit array elements far apart increases the odds that paths from one transmit element to the receiver are independent. The diversity gain for this system is $N_t = 4$, but the data rate is same as the single transmit antenna. Spatial-multiplexing at the transmitter assigns one bit in "1 0 0 1" to each of the four transmit antennas as shown in Figure 10.2c. In this case, a different data stream travels over each path or subchannel from a transmit element to a receive element. This approach increases the data rate by four but the diversity gain is zero. MIMO takes advantage of the gains provided by spatial diversity and spatial multiplexing.

Figure 10.2 Transmitting a data stream. (a) Single transmitter, (b) spatial diversity, and (c) spatial multiplexing.

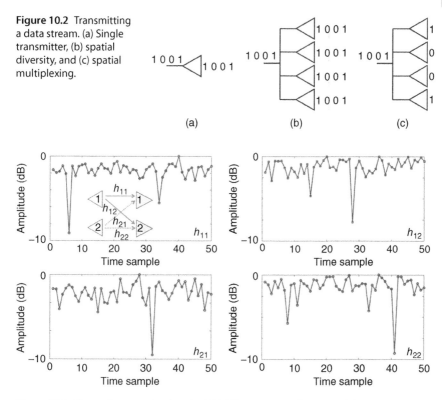

(a)　　　　　　(b)　　　　　　(c)

Figure 10.3 Channel attenuation in a 2 × 2 MIMO system as a function of time.

As an example, consider a MIMO system with an array of two elements on transmit and two elements on receive. Figure 10.3 has plots of the signal levels at the receive elements for the four paths in a 2 × 2 MIMO system over time. Each channel experiences fades at different times due to the different signal paths and a time-varying environment. A SISO system over h_{11} encounters strong signal fades at time samples 6 and 35 in which the receiver may not detect the signal. This 2 × 2 MIMO system has three other paths in which signal fades do not occur at time samples 6 and 35, so the receiver ignores the signal from the bad path and combines the signals from the good paths to optimize the signal reception. The more the transmit and receive antennas in the system (more diversity), the higher the probability that one of the $N_t \times N_r$ subchannels results in a strong signal at the receiver.

If the signal transmitted from element m, $s_{tm}(t)$, travels through a subchannel with impulse response, $h_{mn}(t)$, then the signal arriving at receive element n is given by

$$s_{rn}(t) = s_{tm}(t)^* h_{mn}(t) \tag{10.1}$$

where "*" is convolution. A MIMO system uses the impulse response from each transmit element to each receive element to increase channel capacity. Channel sounding finds the subchannel impulse responses by sending a broadband signal with a flat frequency spectrum from each transmit element to each of the receive element [5]. The broadband signal approximates an impulse function in the time domain, so the output at each receive element looks like the impulse response. Since the receiver knows the transmitted signal as well as the received signal, it uses deconvolution to find $h_{mn}(t)$ [6]. Taking the Fourier transform of (10.1) and solving for the channel transfer function produces

$$H(f) = \frac{S_r(f)}{S_t(f)} \tag{10.2}$$

The inverse Fourier transform of (10.2) results in the channel impulse response. In practice, the receiver performs the deconvolution using the z-transform. After deconvolving all the subchannel impulse responses go into a $N_r \times N_t$ channel matrix.

A simplified MIMO system model helps illustrate how to build the channel matrix. Assume that the channel does not change with time, and the signals are single frequency carriers with no modulation. The first transmit antenna sends a calibration signal to the three receive antennas. The received signals at the elements go into the three rows of column 1 in the channel matrix. Next, transmit element 2 sends a signal so that the received signal goes into the second column of the channel matrix. This process continues for transmit element 3. In this way, each transmit array element sends a signal that each receive array element records in a column of the channel matrix, **H**.

Consider a narrow band channel matrix measured at the receiver given by

$$\mathbf{H} = \begin{bmatrix} 0.0756 + j0.0848 & 0.0185 + j0.0195 & 0.2522 + j0.0651 \\ 0.0223 + j0.0032 & 0.0061 + j0.0112 & 0.1763 + j0.0230 \\ 0.0430 - j0.0131 & 0.0103 - j0.0010 & 0.0585 + j0.0266 \end{bmatrix} \tag{10.3}$$

The measured values in column 2 are about a magnitude lower than those in columns 1 and 3, so turning off transmit antenna 2 and distributing its power to transmit antennas 1 and 3 increases the signal power delivered to the receive array elements. This simple procedure is a fundamental MIMO concept.

MIMO commonly uses orthogonal frequency division multiplexing (OFDM) to split the high-speed serial data into lower-speed serial data signals for each transmit element [7]. The longer symbol periods reduce multipath time delays. As mentioned in Chapter 3, when the subcarrier spacing equals the reciprocal of the symbol period of the data signals, they are orthogonal. The resulting sinc function frequency spectra have their first nulls at the subcarrier frequencies on the adjacent channels.

10.2 The Channel Matrix

The last section introduced a simple channel matrix example. This section starts by deriving $\mathbf{H}$ from basic propagation principles. A MIMO system has N_t transmit antennas sending signals to N_r receive antennas. If an isotropic point source transmits the signal s_{tm} to a point source at a receiver r_{mn0} away in free space, then the received signal, s_{rn}, is given by

$$s_{rn}(t) = \frac{s_{tm}\left(t - \tau_{mnp}\right)}{2\sqrt{\pi}r_{mnp}} \tag{10.4}$$

The signal takes τ_{mnp} to travel along path p of length r_{mnp} from transmit element m to receive element n. Path $p = 0$ is line of sight (LOS). A SISO system with no multipath has $n = 1$, $m = 1$, and $p = 0$. The r_{mn0} path does not exist for a Rayleigh channel.

In a channel with multipath and moving transmitter and/or receiver, the signal arriving at receive array element n is [8]

$$s_{rn} = \sum_{n=1}^{N_t} \int_{-\infty}^{\infty} h_{mn}(\tau, t)s_{tm}(t - \tau)d\tau$$

$$= \sum_{n=1}^{N_t} h_{mn}(t)^* s_{tm}(t), m = 1, 2, \ldots, N_r \tag{10.5}$$

The signal transmitted by each antenna travels a different path to get to the receiver. Thus, each subchannel (e.g. from transmit antenna m to receive antenna n) has an impulse response, $h_{mn}(\tau, t)$. The variable τ represents the time delay due to the different lengths of the multipath, and the t represents the time-changing channel (Doppler). Putting (10.5) in matrix form results in

$$\mathbf{s}_r(t) = \mathbf{H}(t)^* \mathbf{s}_t(t) \tag{10.6}$$

In a flat fading channel this equation reduces to

$$\mathbf{s}_r(t) = \mathbf{H}(t)\mathbf{s}_t(t) \tag{10.7}$$

which greatly simplifies the math. Stationary transmit and receive antennas result in a static channel ($t \to 0$).

$$\mathbf{s}_r(t) = \mathbf{H}\mathbf{s}_t(t) \tag{10.8}$$

A narrow band subchannel has an inpulse response that is a complex constant. The signal transmitted from element m arrives at receive array element n via LOS and $N_p(m, n)$ single bounce multipath signals that have path lengths r_{mnp} and time of arrivals of τ_{mnp} and reflection coefficients $\Gamma_p(m, n)$. The signal in a subchannel decreases in amplitude with each reflection. After many reflections, the signal amplitude becomes very small and can be ignored.

Computer programs based on shooting and bouncing rays (SBR) (Chapter 5) eliminate all rays that fall below an amplitude threshold in order to make the computations in complex models reasonable.

$$s_{rn}(t) = \frac{s_{tm}(t - \tau_{mn0})}{2\sqrt{\pi}r_{mn0}} + \sum_{p=1}^{N_p(m,n)} \frac{s_{tm}(t - \tau_{mnp})\Gamma_p(m,n)}{2\sqrt{\pi}r_{mnp}} \tag{10.9}$$

The reflection coefficients depend on the angles of incidence and the electrical properties of the surfaces. As a result, $\mathbf{H}$ randomly changes with time when the transmitter and/or receiver moves. A carrier signal transmitted into a flat fading channel has an equivalency between time delay and phase shift by substituting $\xi = kr_{mnp} = 2\pi f\tau$

$$s_{rm}(t) = \left(\frac{e^{-jkr_{mn0}}}{2\sqrt{\pi}r_{mn0}} + \sum_{p=1}^{N_p(m,n)} \frac{\Gamma_p(m,n)e^{-jkr_{mnp}}}{2\sqrt{\pi}r_{mnp}} \right) s_{tn}(t) \tag{10.10}$$

with the channel characterized by

$$h_{mn} = \frac{e^{-jkr_{mn0}}}{2\sqrt{\pi}r_{mn0}} + \sum_{p=1}^{N_p(m,n)} \frac{\Gamma_p(m,n)e^{-jkr_{mnp}}}{2\sqrt{\pi}r_{mnp}} \tag{10.11}$$

Figure 10.4 shows a MIMO system that has LOS paths from each transmit element to each receive element as well as a one bounce multipath signal.

The relationship between the transmitted signals and the received signals in a multipath environment becomes very complex but greatly simplifies for narrow band signals:

$$\mathbf{s}_r = \mathbf{H}\mathbf{s}_t + \mathbf{n} \tag{10.12}$$

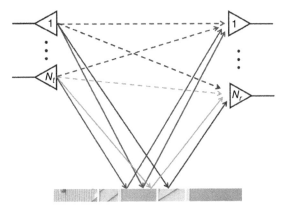

Figure 10.4 The LOS paths are dashed lines while the one bounce multipaths are solid lines.

where

$\mathbf{N} = N_t \times 1$ noise vector
$\mathbf{H} = N_r \times N_t$ channel matrix
$\mathbf{s}_t = N_t \times 1$ transmit signal vector
$\mathbf{s}_r = N_r \times 1$ transmit signal vector.

Figure 10.1 labels subchannels between the transmit array elements and the receive array elements with the appropriate subchannel impulse responses.

$$\mathbf{H} = \begin{bmatrix} h_{11} & h_{12} & \cdots & h_{1N_t} \\ h_{21} & h_{22} & & \vdots \\ & & \ddots & \\ h_{N_r 1} & \cdots & & h_{N_r N_t} \end{bmatrix} \tag{10.13}$$

Slow time variations (e.g. stationary transmitter and receiver) produce an $\mathbf{H}$ with elements that are complex constants.

Chapter 8 introduced the signal covariance matrix for direction finding which was also used for adaptive nulling in Chapter 9. Since MIMO has adaptive transmit and receive arrays, there are two signal covariance matrices of interest: one $N_t \times N_t$ matrix for transmit ($\mathbf{C}_{st}$) and one $N_r \times N_r$ matrix for receive ($\mathbf{C}_{sr}$)

$$\mathbf{C}_{sr} = E[\mathbf{s}_r^\dagger \mathbf{s}_r] \tag{10.14}$$

$$\mathbf{C}_{st} = E[\mathbf{s}_t^\dagger \mathbf{s}_t] \tag{10.15}$$

The average total transmitted power is greater than or equal to the sum of the powers radiated by all the elements:

$$P_t \geq \mathrm{tr}(\mathbf{C}_{st}) \tag{10.16}$$

where $\mathrm{tr}(\cdot)$ is the trace of a matrix or the sum of the elements along the main diagonal. For a SISO communications system, the SNR at the transmitter is defined by

$$\mathrm{SNR}_t = \frac{P_t}{\sigma_{noise}^2} \tag{10.17}$$

and at the receiver by

$$\mathrm{SNR}_r = \frac{P_t |h_{11}|^2}{\sigma_{noise}^2} \tag{10.18}$$

Experimental measurements or a computer model generates the channel matrix [9]. A switched array design has a single transmitter and a single receiver that measures $\mathbf{H}$ by sequentially connecting all combinations of transmit and receive array elements using high-speed switches. Switching times from 2 to 100 ms enable the measurement of all antenna pairs before the channel appreciably changes for most environments. Virtual arrays displace or rotate a single antenna to form an array over time. A complete channel matrix measurement takes several seconds or minutes to perform, so the channel must remain stationary over that time in order to be accurate. As a result, virtual arrays work best for fixed indoor measurements with little motion.

10.3 Recovering the Transmitted Signal Using the Channel Matrix

The simplified channel matrix in the previous section 10.2 highlights some of the basic principles of multipath channels. In reality, propagation effects in the channel impact the elements of the channel matrix in very complicated ways. The channel state information (CSI) describes how a transmitted signal changes due to the channel effects described in Chapter 5. Channel state information at the transmitter (CSIT) requires the receiver to send its version of **H** back to the transmitter. Channel state information at the receiver (CSIR) only happens when the transmitter sends a test signal for calibration. MIMO has five different cases of CSI [10]:

1. CSIT and CSIR
2. No CSIT and CSIR
3. Statistical CSIT and CSIR
4. Noisy CSI
5. No CSIT and no CSIR.

This section covers the first two cases.

CSI is either instantaneous (long term) or statistical (short term). A slow fading channel gives a system sufficient time to develop a reasonable estimate of the channel matrix. In this case, a channel matrix estimate slowly changes, so it does not need constant updating. On the other hand, a fast-fading channel matrix requires updating before the channel changes again, so channel statistics, such as the fading distribution, average channel gain, and spatial correlation replace actual measurements.

10.3.1 CSIR and CSIT

In a system with CSIR and CSIT, the receiver recovers the transmitted data by inverting the channel matrix ($N_t = N_r$) and multiplying the received signal vector (ignoring noise).

$$\mathbf{s}_t = \mathbf{H}^{-1}\mathbf{s}_r \qquad (10.19)$$

which requires an invertible $\mathbf{H}$. When $N_t \neq N_r$, a least-squares solution approximates the desired signal:

$$\hat{\mathbf{s}}_t = (\mathbf{H}^\dagger \mathbf{H})^{-1} \mathbf{H}^\dagger \mathbf{s}_r \tag{10.20}$$

The channel matrix properties (e.g. condition number and rank) predict the solution accuracy when solving (10.19) and (10.20).

Channel sounding estimates $\mathbf{H}$ by measuring the subchannel propagation characteristics. A full rank $\mathbf{H}$ has linearly independent rows and columns due to independent paths. A low matrix condition number indicates a full rank matrix (all rows and columns are linearly independent). MIMO channels with minimal multipath or a large separation distance between the transmit and receive antenna arrays have a nearly singular $\mathbf{H}$ [8]. A low-rank MIMO channel (has many linearly dependent rows and columns) behaves like a SISO channel with the same total power. A high multipath environment, on the other hand, has an $\mathbf{H}$ with high rank.

MIMO excels in a high multipath environment with no LOS signal, because $\mathbf{H}$ appears random and has a low condition number. MIMO does not work well in the presence of strong LOS signals and no multipath. As an example, a MIMO system in free space with no multipath reduces $\mathbf{H}$ to

$$\mathbf{H} = \frac{1}{2\sqrt{\pi}} \begin{bmatrix} \dfrac{e^{-jkr_{110}}}{r_{110}} & \dfrac{e^{-jkr_{120}}}{r_{120}} & \cdots & \dfrac{e^{-jkr_{1N_t0}}}{r_{1N_t0}} \\ \dfrac{e^{-jkr_{210}}}{r_{210}} & \dfrac{e^{-jkr_{220}}}{r_{220}} & & \\ \vdots & & \ddots & \vdots \\ \dfrac{e^{-jkr_{N_r10}}}{r_{N_r10}} & & \cdots & \dfrac{e^{-jkr_{N_rN_t0}}}{r_{N_rN_t0}} \end{bmatrix} \tag{10.21}$$

In this case, $\mathbf{H}$ is ill-conditioned and (10.21) has a high condition number. As a result, the calculated $\mathbf{s}_t$ has numerical errors. Increasing the element spacing in the transmit and receive arrays and/or adding multipath to the subchannels decorrelates the matrix elements and decreases the matrix condition number.

The number of data streams (transmitted packets of data) supported is less than or equal to the rank of $\mathbf{H}$, where the rank of a matrix is the maximum number of linearly independent rows or columns. Singular values are the positive square roots of the nonzero eigenvalues of $\mathbf{H}^\dagger \mathbf{H}$ and indicate which transmit subchannels deliver high-quality signals [11]. The singular value decomposition (SVD) of $\mathbf{H}$ extracts the singular values by decomposing it into

$$\mathbf{H} = \mathbf{U}\mathbf{D}\mathbf{V}^\dagger \tag{10.22}$$

$$D = \begin{bmatrix} \lambda_1 & 0 & 0 \\ 0 & \ddots & 0 \\ 0 & 0 & \lambda_{N_t} \end{bmatrix} = N_r \times N_t \text{ diagonal matrix}$$

λ_m = singular value
$U = N_r \times N_r$ column orthonormal matrix
$V = N_t \times N_t$ column orthonormal matrix.

The **V** matrix weights the data at the transmitter, while the **U** matrix weights the received signals. The singular values in **D** correspond to relative subchannel weights.

Example
Show that the singular values of **H** are the positive square roots of the nonzero eigenvalues of $H^\dagger H$ when

$$H = \begin{bmatrix} 0.0756 + j0.0848 & 0.0185 + j0.0195 & 0.2522 + j0.0651 \\ 0.0223 + j0.0032 & 0.0061 + j0.0112 & 0.1763 + j0.0230 \\ 0.0430 - j0.0131 & 0.0103 - j0.0010 & 0.0585 + j0.0266 \end{bmatrix}$$

Solution
Create an m-file that has the matrix **H** then use the commands:

```
[U,S,V] = svd(H);
[E,D] = eig(H'*H);
[S sqrt(D)]
```

to get the output:

0.3407	0	0	0.3407	0	0
0	0.0623	0	0	0.0623	0
0	0	0.0059	0	0	0.0059

The SVD decomposition provides insight into MIMO performance as shown in Figure 10.5. Substitute (10.22) into (10.12) to get

$$s_r = UDV^\dagger s_t + N \tag{10.23}$$

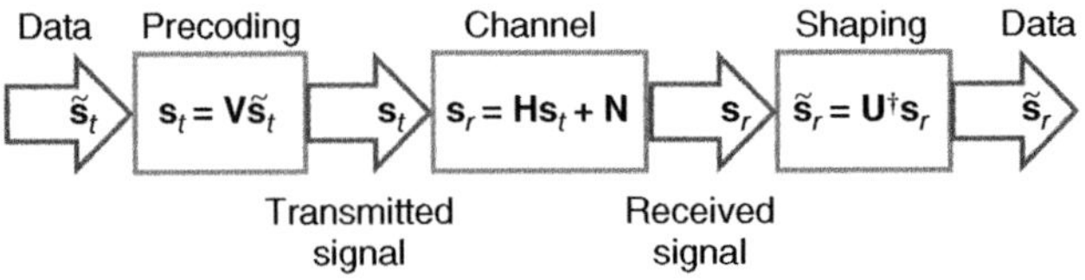

Figure 10.5 Transmitter precoding and receiver shaping.

The transmitter precodes the data signals ($\tilde{\mathbf{s}}_t$) before sending them to the transmit array elements:

$$\mathbf{s}_t = \mathbf{V}\tilde{\mathbf{s}}_t \tag{10.24}$$

Finally, shaping recovers the data ($\tilde{\mathbf{s}}_r$) at the receiver:

$$\tilde{\mathbf{s}}_r = \mathbf{U}^\dagger \mathbf{s}_r \tag{10.25}$$

Transmit precoding requires that the transmitter knows $\mathbf{V}$ (CSIT), so a system with CSIT and CSIR uses SVD.

Example
A MIMO system with two transmit elements and three receive elements has the channel matrix below, find the SVD decomposition.

$$\mathbf{H} = \begin{bmatrix} 0.8268 & 0.8558 \\ 0.6825 & 0.7192 \\ 0.5895 & 0.3907 \end{bmatrix}$$

Solution
The MATLAB command svd (H) decomposes $\mathbf{H}$:

$$\mathbf{H} = \mathbf{UDV} = \begin{bmatrix} -0.7009 & -0.2837 & -0.6544 \\ -0.5838 & -0.2989 & 0.7549 \\ -0.4098 & 0.9111 & 0.0438 \end{bmatrix} \begin{bmatrix} 1.6968 & 0 \\ 0 & 0.1417 \\ 0 & 0 \end{bmatrix}$$

$$\begin{bmatrix} -0.7187 & 0.6953 \\ -0.6953 & -0.7187 \end{bmatrix}$$

There are two independent channels with gains of 1.6968 and 0.1417. The channel with the largest singular value is the most reliable.

Alternatively, the output from the receive array can be written as a function of the singular values. Begin by substituting (10.24) into (10.23) then that result into (10.25) to get

$$\tilde{\mathbf{s}}_r = \mathbf{U}^\dagger \mathbf{UDV}^\dagger \mathbf{V}\tilde{\mathbf{s}}_t + \mathbf{U}^\dagger \mathbf{N} \tag{10.26}$$

which simplifies to [12]

$$\tilde{\mathbf{s}}_r = \mathbf{D}\tilde{\mathbf{s}}_t + \tilde{\mathbf{N}} \tag{10.27}$$

If $\mathbf{U}$ and $\mathbf{V}$ are unitary matrices (i.e. $\mathbf{U}^\dagger \mathbf{U} = \mathbf{I}_{N_r}$ and $\mathbf{V}^\dagger \mathbf{V} = \mathbf{I}_{N_t}$), then $\tilde{\mathbf{N}}$ has the same statistical properties as $\mathbf{N}$. Since $\mathbf{D}$ is a diagonal matrix of singular values, the data in a row of $\tilde{\mathbf{s}}_r$ are array weights (singular values) that multiply the data in the corresponding row of $\tilde{\mathbf{s}}_t$ plus noise. When the matrix rank is less than N_t,

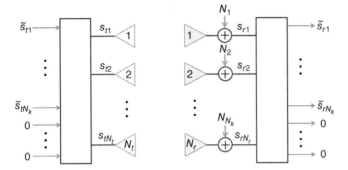

Figure 10.6 A MIMO system transmits and decodes less than or equal to N_k data streams.

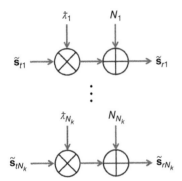

Figure 10.7 Equivalent model for Figure 10.6.

then only the data streams $\tilde{s}_{t1}$ to $\tilde{s}_{tN_k}$ get transmitted. Figure 10.6 illustrates how a MIMO system only transmits N_k data streams. The receiver only decodes up to N_k data streams. Figure 10.7 simplifies Figure 10.6 to a model of (10.27).

A time-invariant MIMO system with CSIT and CSIR has the channel capacity: [13]

$$C_{\text{CSIT-CSIR}} = E\left[\max_{\mathbf{C}_{st}: \text{tr}(\mathbf{C}_{st}) \leq P} B \log_2 \det\left(\mathbf{I}_{N_t} + \frac{1}{\sigma_{\text{noise}}^2}\mathbf{H}\mathbf{C}_{st}\mathbf{H}^\dagger\right)\right] \text{ bps}$$

(10.28)

where the maximization is over the $N_t \times N_t$ input covariance matrix $\mathbf{C}_{st}$ and $\overline{\text{SNR}}_t$ is the average SNR in the subchannels.

10.3.2 Waterfilling Algorithm

The waterfilling algorithm implements (10.26) by allocating more transmit power to higher SNR subchannels and less power to low SNR subchannels

[7]. Its name comes from refilling water glasses until they all have an equal amount of water. A glass containing water gets less water from the pitcher than a glass that has no water. In MIMO waterfilling, the amount of transmit power allocated to a subchannel is proportional to the SNR in that subchannel. Just like the empty glass gets more water, a high SNR subchannel gets more transmit power.

A MIMO subchannel has N_k subchannels associated with the N_k singular values of the SVD. Waterfilling allocates power to the subchannels up to a level of SNR_0 according to

$$P_n = \max\left[\left(\frac{1}{SNR_0} - \frac{1}{SNR_n}\right), 0\right] \tag{10.29}$$

The value of SNR_0 is chosen in order that

$$\sum_{n=1}^{N_k} P_n = P_t \tag{10.30}$$

and the subchannel SNRs are given by

$$SNR_n = \frac{\hat{\lambda}_n^2}{\sigma_{noise}^2} \tag{10.31}$$

Figure 10.8 diagrams the waterfilling process for a MIMO system with six subchannels. Subchannels with $SNR_n \leq SNR_0$ are not allocated any power, so the transmitter allocates P_t to the four subchannels according to (10.29). When P_t/σ_{noise}^2 is high, the optimum power distribution evenly allocates power across

Figure 10.8 Example of water filling with six subchannels.

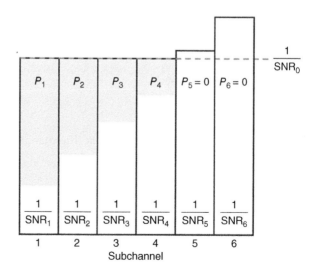

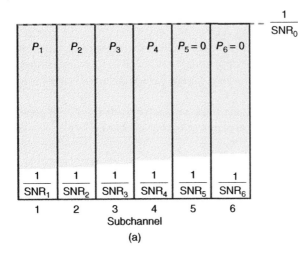

Figure 10.9 Waterfilling when $\bar{P}_t/\sigma_{noise}^2$ is high and low. (a) High SNR and (b) low SNR.

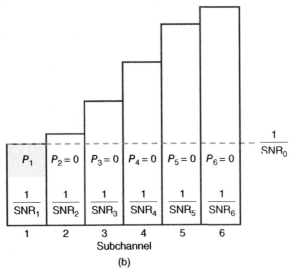

all subchannels as shown in Figure 10.9a. The channel capacity for high SNR is [14]

$$C_{hi} = N_k B \log_2 \left(\frac{P_t}{\sigma_{noise}^2} \right) \text{ bps} \tag{10.32}$$

At the other extreme, when P_t/σ_{noise}^2 is low ($P_t < 1/\lambda_2 - 1/\lambda_1$), all the power goes into the subchannel with the best SNR (Figure 10.9b). The channel capacity for low SNR is [14]

$$C_{lo} \approx B \log_2 \left(\frac{P_t}{\sigma_{noise}^2} \lambda_{max}^2 \right) \text{ bps} \tag{10.33}$$

where λ_{max} is the largest singular value.

Example
A 4×4 MIMO system has $\overline{P}_t = 1$ W and an SNR of 4 dB. Find the power allocation using waterfilling.

$$H = \begin{bmatrix} -0.7471 & 0.9534 & 0.4432 & 0.3454 \\ 0.7695 & 0.3823 & 1.0413 & 0.3896 \\ 0.7849 & 0.5910 & 0.1215 & 0.5153 \\ -0.0523 & -0.0454 & -0.2182 & 1.2589 \end{bmatrix}$$

Solution
The noise variance is $\sigma_{noise}^2 = P_t/10^{SNR/10} = 1/10^{4/10} = 0.398$
Use MATLAB to find the SVD (Table 10.1).

```
[U,D,V] = svd(H);
```

Table 10.1 Water filling example.

Subchannel	1	2	3	4
λ	1.7889	1.3038	1.2000	0.5477
$1/SNR_n = \sigma_{noise}^2/\lambda_n^2$	0.1244	0.2341	0.2764	1.3267
$1/SNR_0 = \left(P_t + \sum_{n=1}^{4} \sigma_{noise}^2/\lambda_n^2 \right)/4 = 0.7404$				
Is $1/SNR_0 - 1/SNR_n > 0$	Yes	Yes	Yes	No
set $P_4 = 0$				
$1/SNR_0 = \left(P_t + \sum_{n=1}^{3} \sigma_{noise}^2/\lambda_n^2 \right)/4 = 0.5450$				
Is $1/SNR_0 - 1/SNR_n > 0$	Yes	Yes	Yes	—
Final power allocation $\dfrac{1}{SNR_0} - \dfrac{1}{SNR_n}$	0.4206	0.3109	0.2686	—

The transmit power is allocated 42% to subchannel 1, 31% to subchannel 2, 27% to subchannel 3, and 0% to subchannel 4.

MIMO capacity equals the sum of all the subchannel capacities when the transmit power is optimally spread across the N_k subchannels [13].

$$C = \max_{P_n, \sum_{n=1}^{N_k} P_n \le P} \sum_{n=1}^{N_k} B \log_2(1 + P_n \lambda_n^2 / \sigma_{noise}^2) \text{ bps} \tag{10.34}$$

10.3.3 CSIR and No CSIT

When there is no CSIT, then the optimal transmit weights cannot be found, so setting all the transmit weights to one makes the most sense. The first step in finding the receive weights is to express the signal in terms of power [15]

$$P = \mathbf{s}_r^\dagger \mathbf{s}_r = \mathbf{s}_t^\dagger \mathbf{H}\mathbf{H}^\dagger \mathbf{s}_t = \mathbf{s}_t^\dagger \mathbf{C}_H \mathbf{s}_t \tag{10.35}$$

Now, the $N_r \times N_r$ covariance matrix $\mathbf{C}_H = \mathbf{H}\mathbf{H}^\dagger$ is square and an eigenvalue decomposition is possible:

$$\mathbf{C}_H = \mathbf{Q} \begin{bmatrix} \lambda_1 & 0 & 0 \\ 0 & \ddots & 0 \\ 0 & 0 & \lambda_{N_r} \end{bmatrix} \mathbf{Q}^{-1} \tag{10.36}$$

where the columns of $\mathbf{Q}$ are the eigenvectors and λ_n are the eigenvalues. Eigenvalue n corresponds to the received signal power level in eigenchannel (subchannel) n. The off-diagonal elements of $\mathbf{C}_H$ are the correlation between the transmitted signal streams. An increased correlation results in a decreased capacity.

When the transmitter does not know channel characteristics, each element receives an equal share:

$$P_n = P_t / N_t \tag{10.37}$$

Uncorrelated transmit elements in a random channel have an average capacity given by [8]

$$C = E \left\{ B \log_2 \left[\det \left(I_{N_t} + \frac{P_t}{N_t \sigma_{noise}^2} \mathbf{H}\mathbf{H}^\dagger \right) \right] \right\} \text{ bps} \tag{10.38}$$

MIMO capacity increases linearly with the number of elements for an equal number of transmit and receive antennas. N_t should be of the order $2N_r$ [1]. When N_t and N_r are large and $N_t > N_r$, the capacity is [8]

$$C = N_r B \log_2(1 + N_t P_t / N_r) \text{ bps} \tag{10.39}$$

As long as the ratio of N_t / N_r is constant, the capacity is a linear function of N_r.

The instantaneous eigenvalues of a random $\mathbf{C}_H$ have limits defined by [7]

$$(\sqrt{N_t} - \sqrt{N_r})^2 < \lambda_n < (\sqrt{N_t} + \sqrt{N_r})^2 \tag{10.40}$$

When N_t is much larger than N_r, all the eigenvalues cluster around N_t. Each eigenvalue is nonfading due to the high-order diversity when there are a large number of transmit elements. Thus, the uncorrelated asymmetric channel with many antennas has a very large theoretical capacity of N_r equal, constant channels with gains of N_t.

Example

For the following matricies, find (a) condition numbers (b) determine if (10.40) is true.

$$\mathbf{H}_1 = \begin{bmatrix} 0.6948 & 0.0344 & 0.7655 \\ 0.3171 & 0.4387 & 0.7952 \\ 0.9502 & 0.3816 & 0.1869 \end{bmatrix} \quad \mathbf{H}_2 = \begin{bmatrix} 1.0000 & 1.1000 & 1.2000 \\ 2.0000 & 2.0000 & 2.0000 \\ 3.3000 & 3.0000 & 3.1000 \end{bmatrix}$$

Solution

(a) $\text{cond}(\mathbf{H}_1) = 4.5408$ and $\text{cond}(\mathbf{H}_2) = 113.0466$
(b) For this channel matrix, $N_t = 3$ and $N_r = 3$, so $0 < \lambda_n < 12$
   $\mathbf{H}_1$ is a random matrix and has eigenvalues given by

   $$\text{eig}(H1' * H1).' = 0.1238 \quad 0.4018 \quad 2.5531$$

   All of these eigenvalues fall between 0 and 12
   $\mathbf{H}_2$ is not a random matrix and has eigenvalues given by

   $$\text{eig}(H2' * H2).' = 0.0035 \quad 0.0402 \quad 45.1063$$

The high condition number of $\mathbf{H}_2$ indicates that there is a correlation between rows or columns.

Problems

10.1 Find the SVD of the following channel matrices:

(a) $\begin{bmatrix} 0.2 & 0.4 & 0.8 \\ 0.7 & 0.3 & 0.4 \\ 0.5 & 0.7 & 0.2 \end{bmatrix}$, (b) $\begin{bmatrix} 0.1 & 0.5 & 0.7 & 0.9 \\ 0.2 & 0.6 & 0.4 & 0.3 \end{bmatrix}$, (c) $\begin{bmatrix} 0.4 & 0.5 \\ 0.8 & 0.2 \end{bmatrix}$,

(d) $\mathbf{H} = \begin{bmatrix} -0.6 & 0.7 & 0.3 & 0.4 \\ 0.7 & 0.4 & 0.9 & 0.5 \\ 0.7 & 0.6 & 0.1 & 0.5 \\ -0.1 & 0.0 & 0.1 & 0.8 \end{bmatrix}$

10.2 Find the eigenvalues of $H^{\dagger}H$ for the channel matrices in Problem 10.1.

10.3 Find the condition number of the matrices in Problem 10.1.

10.4 A 4×4 MIMO system has $\overline{P}_t = 1$ W W and an SNR of 5 dB. Find the power allocation using waterfilling

$$H = \begin{bmatrix} -0.6064 & 0.7356 & 0.3441 & 0.4011 \\ 0.7629 & 0.4043 & 0.9241 & 0.5388 \\ 0.7129 & 0.6188 & 0.1806 & 0.5181 \\ -0.1070 & 0.0747 & 0.0611 & 0.8884 \end{bmatrix}$$

10.5 Redo Problem 10.4 when the SNR is 20 dB.

10.6 Redo Problem 10.4 when the SNR is 2 dB.

10.7 A 3×3 MIMO system has CSIR but no CSIT. Find the capacity if SNR $= 10$ dB, $B = 1$ kHz, and

$$H = \begin{bmatrix} 0.4508 & 0.5711 & 0.3450 \\ -0.2097 & 0.4704 & 0.4510 \\ -0.6134 & -0.6382 & -0.4621 \end{bmatrix}$$

10.8 Repeat Problem 10.7 using SVD.

10.9 What is the capacity if the lowest SNR subchannel in Problem 10.7 is ignored?

10.10 Use waterfilling to allocate the power in Problem 10.7 when the total power is 1 W, the noise power is 0.1 W and the signal bandwidth is 50 kHz. What is its new channel capacity?

References

1 Winters, J.H. (1987). On the capacity of radio communication systems with diversity in a Rayleigh fading environment. *IEEE Journal on Selected Areas in Communications* 5: 871–878.

2 Foschini, G.J. (1996). Layered space-time architecture for wireless communication in a fading environment when using multi-element antennas. *Bell System Technical Journal* 1 (2): 41–59, Autumn.

3 Agilent Technologies (2008). MIMO wireless LAN PHY layer [RF] operation & measurement, Application Note 1509, 29 April 2008.

4 Rohde&Schwarz, Introduction to MIMO, Application Note 1MA102.

5 Laurenson, D. and Grant, P. (2006). A review of radio channel sounding techniques. In: *14th European Processing Conference*, Florence, Italy, 1–5. IEEE.

6 Proakis, J.G. and Manolakis, D.G. (2007). *Digital Signal Processing Principles, Algorithms, and Applications*, 4e. Upper Saddle River, NJ: Pearson Prentice Hall.

7 Bliss, D.W., Forsythe, K.W., and Chan, A.M. (2005). MIMO wireless communication. *Lincoln Laboratory Journal* 15 (1): 97–126.

8 Agilent Technologie 2010. MIMO channel modeling and emulation test challenges, Application Note, 22.

9 Jensen, M.A. and Wallace, J.W. (2004). A review of antennas and propagation for MIMO wireless communications. *IEEE Transactions on Antennas and Propagation* 52 (11): 2810–2824.

10 Molisch, A.F. (2011). *Wireless Communications*, 2e. West Sussex, UK: Wiley.

11 Andersen, J.B. (2000). Array gain and capacity for known random channels with multiple element arrays at both ends. *IEEE Journal on Selected Areas in Communications* 18 (11): 2172–2178.

12 Telatar, I. E. (1996). Capacity of multi-antenna Gaussian channels. Tech. note, AT&T Bell Lab.

13 Goldsmith, A. (2005). *Wireless Communications*. New York: Cambridge University Press.

14 Brown, T., De Carvalho, E., and Kyritsi, P. (2012). *Practical Guide to the MIMO Radio Channel with MATLAB Examples*. West Sussex, UK: Wiley.

15 Browne, D.W., Manteghi, M., Fitz, M.P., and Rahmat-Samii, Y. (2006). Experiments with compact antenna arrays for MIMO radio communications. *IEEE Transactions on Antennas and Propagation* 54 (11): 3239–3250.

11

Security

Individuals, companies, and governments demand data security from wireless networks to protect their privacy. Untethered/wireless access to a network gives users tremendous freedom of movement but requires more security measures than a wired network. To obtain access to a wired network, an unauthorized user must physically connect to a port. In contrast, an unauthorized user only needs to be within range of an antenna to connect to a wireless port. Security means protecting communication and computing services, information and data, personnel, and equipment for customers, government, and network providers [1].

The Internet consists of a mix of wired and wireless systems with many opportunities for security breaches (Figure 11.1). Wireless communication and security requirements depend on the connectivity of their fixed and mobile subsystems [2]. This chapter introduces wireless security vulnerabilities and ways to mitigate them. In order to understand the threats to wireless security and how to defend against them, this chapter starts with background about wireless networks and how devices connect to them. The second half of the chapter deals with defenses against the treats with special emphasis on encryption.

11.1   Wireless Networks

A group of devices communicate with each other through a network. A local area network (LAN) connects devices to a server over a common communications link. If the link is wireless, then the LAN is a wireless local area network (WLAN). This section introduces the basics of a wireless network.

11.1.1   Addresses on a Network

Communication between two people at a distance requires location information like an address or a telephone number. Just like people, wireless network devices find each other using their addresses on the network. The network

Wireless Communications Systems: An Introduction, First Edition. Randy L. Haupt.
© 2020 John Wiley & Sons, Inc. Published 2020 by John Wiley & Sons, Inc.

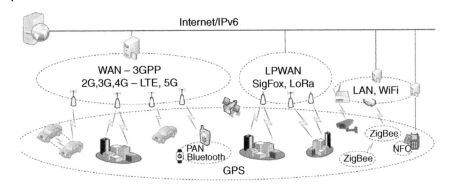

Figure 11.1 Various types of wireless networks, hierarchically connected. *Source:* Burg *et al.* [3]. Reproduced with permission of IEEE.

interface card (NIC) connects a computer to the Internet. Every NIC has a hardware address assigned by the manufacturer called an extended unique identifier (EUI) or more commonly a MAC (media access control) address [4]. A NIC converts data into a signal, then transmits it in a packet over the network. All equipment connecting to computer networks (computers, routers, servers, printers, smartphones, etc.) have a MAC (EUI) address. A 48-bit EUI (EUI-48) address has 12 hexadecimal characters broken into two 24 bit codes

$$00 \; 1f \; 19 \; \underbrace{ba \; 20 \; 39}_{} \tag{11.1}$$
$$\underbrace{}_{\text{OUI}} \quad \underbrace{}_{\text{NIC specific}}$$

This 48-bit address provides up to $2^{48} = 281, 474, 976, 710, 656$ unique values for locating all devices connected to the Internet. EUI/MAC addresses serve to direct packets from one device to another on a network. The proliferation of devices connecting to the Internet forced the adoption of an updated 64-bit EUI (EUI-64) which allows 2^{64} unique addresses.

An EUI address is either a universally administered address (UAA) or a locally administered address (LAA). Manufacturers assign unique UAAs to devices. A network administrator has the ability to replace the UAA with an LAA, so the EUI becomes the new LAA. The first 24 bits in (11.1), the organizationally unique identifier (OUI), indicate the specific vendor for that device. An "assignee" (vendor, manufacturer, or other organization) purchases an OUI from the Institute of Electrical and Electronics Engineers (IEEE) Incorporated Registration Authority. For instance, Apple has the OUI "FCFC48" [5]. Table 11.1 has an example of the OUI having a base-16 value of "ACDE48" which corresponds to the octet representation "AC-DE-48." The last row contains the binary representation.

Each device connected to a computer network has an IP (Internet protocol) address assigned by the network provider. The IP address serves as a network interface identification and a location address. Static IP addresses remain the

Table 11.1 Example OUI.

Octet identifier	0		1		2	
Hexadecimal	AC		DE		48	
Binary	1010	1100	1101	1110	0100	1000

Figure 11.2 Example of an IPv4 address.

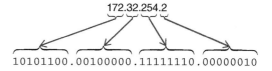

172.32.254.2

10101100.00100000.11111110.00000010

same, while dynamic IP addresses change. A data packet has an origination IP address and a destination IP address. The original Internet protocol version 4 (IPv4) address had 32 bits and is still in use (Figure 11.2). The current IPv6 has 128-bit IP addresses to accommodate the large number of devices connected to the Internet.

The Internet uses EUI addresses in combination with IP addresses to route packets from an origin to a destination – even within small LANs where two computers communicate directly with each other. FedEx and UPS operate in an analogous manner to the Internet. A FedEx or UPS worker does not carry a package directly from the sender to the recipient. Instead, the package goes to a sorting office. From there, it travels to several intermediate facilities that redirect it to the final delivery at the destination. Routers behave like the various facilities that handle the package from the origin to the destination. An IP packet temporarily stops at many different routers as it travels over the Internet. When a router receives a packet, it sends the packet to the next stop based on the destination EUI address rather than the destination IP address. It strips off the old destination EUI address (which was the router's own EUI address) and replaces it with a new destination EUI address that points to the next router along the way to the final IP address. The packet passes from router to router until it arrives at the final destination. Figure 11.3 has an example of a computer sending data to another computer via the Internet. Note how the EUI addresses in the packets change but the source and destination IP addresses do not.

11.1.2 Types of Wireless Local Area Networks

A wireless network has nodes (e.g. routers in Figure 11.3) that communicate data from a transmitter to a receiver. Figure 11.4 has diagrams of the following important network topologies for node communication:

(1) *Star network*: Nodes do not communicate with each other. All nodes directly communicate with the base station. This network increases the

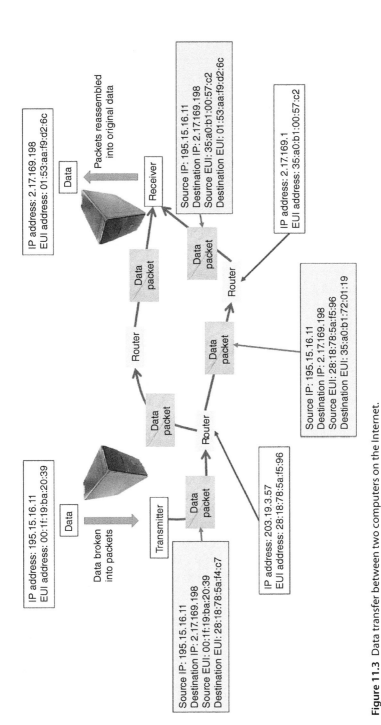

Figure 11.3 Data transfer between two computers on the Internet.

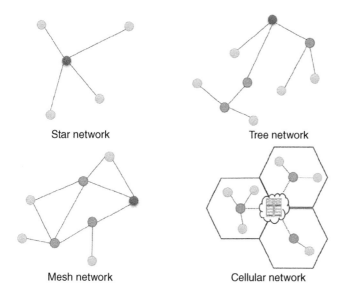

Star network   Tree network

Mesh network   Cellular network

Figure 11.4 Overview of main network topologies. *Source:* Burg *et al.* [3]. Reproduced with permission of IEEE.

risk of a network failure, increases latency, and incurs a potentially large overhead that degrades network capacity but is very simple.

(2) *Tree networks*: Nodes communicate with a designated neighboring node that forwards the data traffic to a destination. This approach extends the range of each node while routing occurs through adjacent nodes.

(3) *Mesh networks*: A flexible, robust network that allows nodes to connect to other nodes with shorter latency and more system capacity but increases routing complexity.

(4) *Cellular networks*: A cellular network is a star topology that has multiple star networks arranged in a way that minimizes the coverage overlap of the base stations. Routing data between base stations is handled via a separate network. The infrastructure is complex and costly but has a high network capacity.

A cyber-physical system (CPS) collects sensor and actuator data in order to monitor the physical environment and analyze how the changes impact their operation. The CPS then autonomously (sometimes with human-in-the-loop) influences the physical environment [6]. The IoT (Internet of things) inter-connects smart devices and connects to CPSs. Examples of CPSs include [7]: large-scale environmental systems (e.g. natural resource management), power and energy generation and distribution, transportation infrastructure, home automation, autonomous driving, personal healthcare, logistics, or industrial manufacturing. Highly distributed CPSs require wireless communications. An

OT (operational technology) system is a CPS that supports the operation of an industrial control processes. A wireless CPS has problems with latency, range, throughput, power consumption of the node, and security.

11.1.3 WLAN Examples

An autonomous vehicle has a wireless vehicle-to-infrastructure communication system to obtain traffic updates and let car manufacturers monitor vehicle status over the Internet (Figure 11.5). Vehicle-to-vehicle communication systems use high speed, reliable links that provide traffic status or link to other autonomous vehicles on the road. Cars have several independent CPSs that overlap, such as anti-lock braking, adaptive cruise control, and automated temperature control. These functions monitor actuators and sensors (e.g. tire pressure, temperature, crankshaft position, light, and collision sensors). Most CPS control and data communication occurs over wires in order to guarantee integrity and 100% availability.

In 2014, two researchers hacked into a 2014 Jeep Cherokee through a wireless entertainment and navigation system called Uconnect by knowing the vehicle's IP address [8]. They remotely gained access to a Jeep Cherokee's controller area network in order to take control of the vehicle, including shutting it down. This hack prompted Fiat Chrysler to recall 1.4 million vehicles.

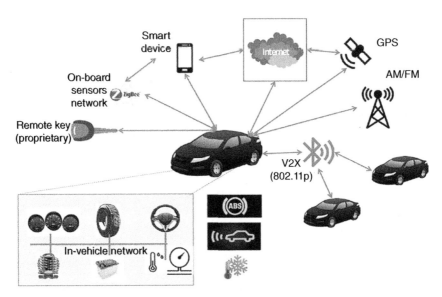

Figure 11.5 Example of wired and wireless communications in an autonomous vehicle. *Source:* Burg *et al.* [3]. Reproduced with permission of IEEE.

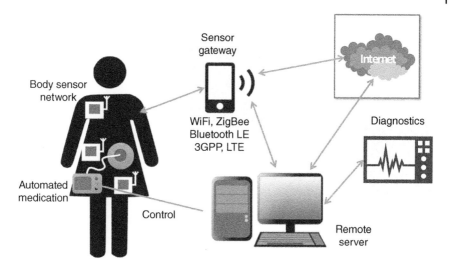

Figure 11.6 Implantable personal medical devices communicate with wireless devices that send data to the Internet. *Source:* Burg et al. [3]. Reproduced with permission of IEEE.

Figure 11.6 shows an example of wireless CPS in a personal healthcare system. Implanted medical devices perform and monitor biological functions such as heart rate and glucose level. Sensors record vital data then wirelessly transmit it to a receiver (e.g. smartphone) that in turn connects to the Internet. Health professionals anywhere in the world can monitor patient status in order to diagnose and advise patients. The system requires stringent security in order to prevent life threatening intrusions and protect patient privacy.

Figure 11.7 presents an example of a blood glucose monitoring system. A sensor mounted on the abdomen measures the blood glucose. Every five minutes, a transmitter attached to the sensor sends the data to a dedicated receiver or smart phone via Bluetooth (Appendix D). These devices store the data for 24 hours then uploads the data to a website via the Internet. The website maintains a database accessible by authorized people. Patients want this type of data secured.

Figure 11.8 portrays a smart home system that has small-scale CPSs, such as HVAC (heating, ventilation, and air conditioning). The CPSs connect to a central hub through a short-range, latency insensitive, and fault-tolerant wireless systems. Public services or energy suppliers sometimes have access to autonomous sensors distributed throughout the house. Xcel Energy offers the voluntary Savers Switch program that compensates a homeowner for letting Xcel install a switch on the air conditioners that it controls. On hot summer days, the company cycles the air conditioning every 15–20 minutes in order to conserve power usage at peak times of the day.

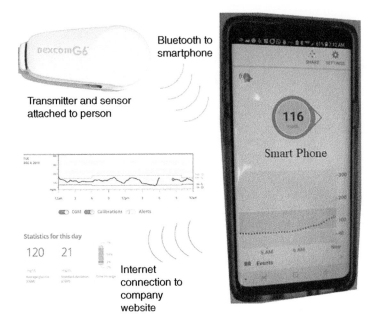

Figure 11.7 Wireless glucose monitoring system.

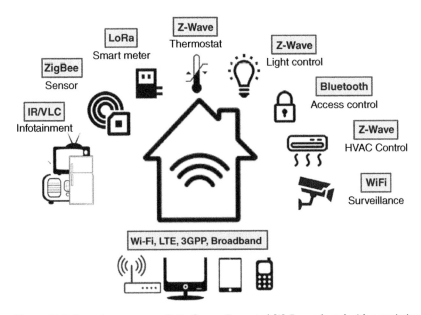

Figure 11.8 Smart home connectivity. *Source:* Burg *et al.* [3]. Reproduced with permission of IEEE.

Since wireless communication systems, like these examples, pass important information and commands, they need protection from unauthorized users. The large number of wireless devices in the home increase radio frequency interference (RFI) and potential security breaches.

11.2 Threats

Wireless devices join a WLAN through an access point (AP). The AP connects to a wired LAN typically through Ethernet. Ethernet sets the packet format that other devices on the LAN recognize, receive, and process. Figure 11.9 shows a wireless router connected to a cable provider box for Internet access through a coaxial cable. The router antennas receive the signal from a wireless device. Next, the AP inside the router transfers the data to the Internet. Routers have multiple input Ethernet ports in order to connect to devices over Ethernet cables.

A service set consists of wireless network devices that communicate with each other through the same network [9]. The three types of service sets are Basic Service Set (BSS), Independent Basic Service Set (IBSS) or ad hoc network, and Extended Service Set (ESS) [10].

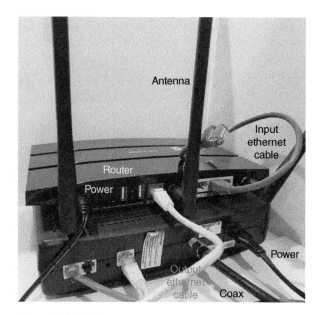

Figure 11.9 Wireless router.

A device gains access to the network through the AP by meeting the following criteria [11]:

- A matching Service Set Identifier (SSID)
- A compatible wireless data rate
- Authentication credentials.

Once the device connects to the AP, all communications associated with the device pass through the AP. An AP covers the Basic Service Area (BSA) much like a base station covers a wireless cell. The AP usually has a wired connection to a larger network. Every AP has a 32-bit SSID or network name that uniquely distinguishes it from other APs in the same physical area. The SSID acts like a password that devices need to access the WLAN.

IBSS allows multiple devices to communicate directly with each other rather than through a central device. This peer-to-peer network between devices does not require an AP, so it cannot connect to a BSS. An IBSS is ad hoc, because it does not need APs and routers.

An ESS expands the WLAN coverage area by connecting multiple BSSs through a distribution system (wired or wireless) as shown in Figure 11.10. A device that moves (roams) from one BSS to another remains connected to the LAN. When the signal gets weak, a device switches from one AP to another AP that has a better link. An ESS usually has an SSID that allows roaming from one AP to another without requiring the device to reconfigure. The ESS covers an extended service area (ESA) that consists of all the BSAs within its control.

A WLAN threat occurs when an attacker obtains unauthorized access. An attacker steals data or gains control of a system by entering an attack surface. The attack surface consists of all points where an attacker enters a system. An attack surface of an application consists of [12]

- The sum of all data/command paths that communicate with an application.

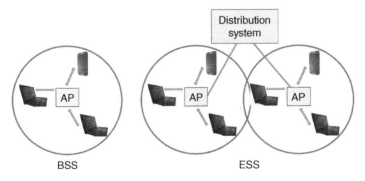

Figure 11.10 An ESS consists of multiple BSSs connected through a distribution system.

- The code that protects data/command paths, such as resource connection and authentication, authorization, activity logging, data validation, and encoding.
- All valuable application data, including secrets and keys, intellectual property, critical business data, and personal data.
- Data protection, such as encryption and checksums, access auditing, and data integrity and operational security controls.

Many different kinds of threats prey on WLANs via an attack surface. Microsoft developed a model called STRIDE that places threats into six categories [13]:

- **S**poofing is when one user poses as another user or administrator in order to fraudulently obtain authentication information, such as a username and password.
- **T**ampering means that the attacker modifies data. An example is changing an account balance.
- **R**epudiation means to delete or alter a login or transaction data. An example is deleting a purchase transaction to avoid payment.
- **I**nformation disclosure refers to stealing sensitive information (e.g. proprietary or secret data).
- **D**enial of service (DoS) attacks overwhelm a system, so that nobody can access the network.
- **E**levation of privilege occurs when an unprivileged user obtains privileged access to the entire system. The attacker bypasses system defenses and becomes trusted.

Threats attack wireless networks in seven ways [14]:

1. An **insertion attack** means that an unauthorized wireless device joins a BSS.
2. A **misconfiguration attack** occurs when software is not properly setup or updated. Devices gain access through default SSIDs or through brute force guessing an SSID.
3. Wireless sniffing tools capture the initial part of a wireless connection that includes a username and password in an **interception attack**. A device then enters the AP as a valid user.
4. **War driving attacks** result from a mobile device that searches for and then exploits WLANs.
5. A **RAP** (rogue AP) connects to a network without authorization from an administrator. RAPs have proliferated due to low-cost hardware that appears invisible to the legitimate WLAN.
6. In a **client-to-client attack**, one user attacks another user on the same BSS/ESS.
7. A **jamming attack** denies user access to a WLAN by overwhelming the AP with interfering signals.

11.3 Securing Data

A code maps one group of symbols into a new group of symbols using a codebook that contains all the mappings. For instance, the ASCII codebook maps a letter into bits. In contrast, a cipher transforms one symbol into a new symbol using an algorithm. For instance, the symbol 110 converts to 111 using the algorithm add 001 to the symbol. Codes are relatively easy to break, and the codebooks are difficult to distribute and keep secure. Consequently, ciphers have become the primary way to secure data.

11.3.1 Cryptography

Cryptography transforms (encrypts) plaintext (message or data) into ciphertext (an unreadable format that masks the content) using a cipher (algorithm). It hides the meaning of a message but does not hide the existence of the message [15]. Cryptography has five primary functions [16]:

- *Privacy/confidentiality*: Ensuring that no one reads the message except the intended receiver.
- *Authentication*: Verifying user identity.
- *Integrity*: Assuring the receiver that the received message is the same as the transmitted message.
- *Nonrepudiation*: Proof of the sender's identity.
- *Key exchange*: The sharing of crypto keys between sender and receiver.

Cryptography started over 2000 years ago when Julius Caesar invented the Caesar cipher that shifts letters in the alphabet a set number of places known to the sender and receiver (in Caesar's case three) [17]. The 25 distinct shift ciphers for a 26-letter alphabet made this code relatively easy to break. Arab scholars eventually cracked codes that simply substituted one letter for another, by noting that some letters occur more often than others in written documents. Frequency analysis of letters in the English language indicate that "e" occurs most often (see Figure 1.6). Thus, any alphabetic substitution cipher in English has the most common letter in the code representing "e."

In 1882, Frank Miller developed an unbreakable encryption method that was later called one-time pad encryption [18]. One-time pad encryption converts data into meaningless characters using a pseudo random noise (PRN) generator to determine symbol substitution. Both sender and receiver need the same substitution algorithm and PRN generator. Using PRNs to create rules for substituting one symbol for another makes the one-time pad encryption difficult to break using frequency analysis. Figure 11.11 is an example of a one-time pad encryption using PRNs generated by Google to encode the message "HELLO THERE."

Figure 11.11 Example of one-time pad encryption. *Source:* Abellán and Pruneri [18]. Reproduced with permission of IEEE.

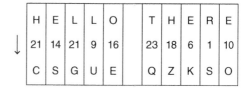

H	E	L	L	O		T	H	E	R	E
21	14	21	9	16		23	18	6	1	10
C	S	G	U	E		Q	Z	K	S	O

Figure 11.12 Enigma machine at the National Cryptologic Museum.

The famous Enigma cipher was initially developed by the Dutch for banking communications. Germans bought the patent in 1923 and created an electromechanical machine that substituted one letter for another as a message was typed (Figure 11.12). Each day, the electrical and mechanical connections for each machine were changed according to rules that were distributed once a month. Enigma had approximately 159 quintillion different settings [19]. Enigma would have been unbreakable if the users had followed proper operating procedures and spies had not passed on critical information. In 1932, Polish cryptanalysts decoded German Enigma ciphers [20]. They were successful in breaking the ciphers and producing their own Enigma machines. In 1939, Poland shared their breakthroughs with France and the United Kingdom. Alan Turing with a team of scientists discovered that a letter can be encrypted as any letter other than itself. In addition, the Germans put "Heil

Hitler" at the end of every message. That hint provided enough information to crack the code (Figure 11.12).

Cryptography creates encrypted data known as ciphertext (C_{text}) from unencrypted data known as plaintext (P_{text}) by applying mathematical transformations that converts the data into a secret code [21]. Only authorized users know the algorithm that unencrypts the ciphertext. Deciphering applies an inverse mathematical transformation to the secret code in order to recover the original data.

$$\text{Encription}: C_{text} = \mathcal{E}(P_{text}) \tag{11.2}$$

$$\text{Decription}: P_{text} = \mathcal{E}^{-1}(C_{text}) \tag{11.3}$$

where $\mathcal{E}$ = encryption and $\mathcal{E}^{-1}$ = decryption. In theory, unauthorized users cannot access the original data without knowing $\mathcal{E}^{-1}$.

A key is a random string of bits that scrambles and unscrambles data. Keys should be long, unpredictable, and unique. Three categories of data encryption based on the use of a secret key are [16]:

- Secret key cryptography (SKC) or symmetric encryption uses the same key for encryption and decryption.
- Public key cryptography (PKC) or asymmetric encryption uses one key for encryption and a second key for decryption.
- Hashing uses a mathematical transformation to irreversibly encrypt data and provide a digital fingerprint.

Figure 11.13 differentiates between the types of encryption. Note that the hash function is one-way encrypts plaintext into ciphertext that is nearly impossible to decrypt.

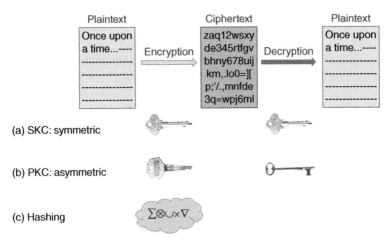

Figure 11.13 Basic functioning of the three categories of encryption.

11.3.2 Secret Key Cryptography

SKC encrypts the plaintext with a key then sends the ciphertext to the receiver. The receiver uses the same key to decrypt the ciphertext and recover the plaintext. Security depends on the difficulty of guessing the key. Examples of SKC algorithms include the triple data encryption standard (3DES) and the advanced encryption standard (AES). SKC creates either stream ciphers or block ciphers.

Stream ciphers operate on one bit of plaintext at a time with a key of pseudorandom bits to create ciphertext much longer than the plaintext. Using an unpredictable PRN generator and one-time keys improves SKC security. Stream ciphers approximate the one-time pad cipher.

Block ciphers encrypt N bits of data (block) at one time. Blocks usually contain 64, 128, or 256 bits [22]. The most important operating modes of a block cipher are [23]:

- Electronic Codebook (ECB) encrypts a plaintext block into ciphertext block using a secret key. The same plaintext block always encrypts to the same ciphertext block. ECB is susceptible to many forms of attack. A single bit error in the ciphertext causes errors in the entire block of decrypted plaintext.
- Cipher Block Chaining (CBC) performs an exclusive or of the plaintext with the previous ciphertext block before doing the encryption. In this way, two identical plaintext blocks have different encryptions. CBC protects against most brute-force, deletion, and insertion attacks. One bit error in the ciphertext, however, causes errors in the entire decrypted plaintext block as well as a bit error in the next decrypted plaintext block.
- Cipher Feedback (CFB) encrypts data into groups of bits smaller than the block size. A single bit error in the ciphertext affects the current and next block.
- Output Feedback (OFB) generates the keystream independently of both the plaintext and ciphertext bitstreams. A single bit error in OFB ciphertext generates one bit error in the decrypted plaintext.
- Counter (CTR) mode uses different keys for different blocks so that two identical blocks of plaintext will not result in the same ciphertext. Each block of ciphertext has a specific location within the encrypted message. CTR mode processes blocks in parallel – thus offering performance advantages when parallel processing and multiple processors are available – and is not susceptible to ECB's brute-force, deletion, and insertion attacks.

11.3.3 Public Key Cryptography

PKC encrypts the data with one key and decrypts it with a different key [24]. One of the keys is private and only known to the user, while the other key is public and known to others. The sender encrypts the information using the receiver's public key. The receiver decrypts the message using a private key. The receiver knows who sent the message (authentication), and the transmitter cannot deny sending the message (nonrepudiation). PKC has easy-to-compute one-way functions for creating the cipher, but the decryption inverse function is difficult to compute. In fact, SKC decrypts a message about 1000 times faster than PKC, so PKC is not used for message encryption [24]. The cipher and decipher keys are mathematically related, but knowing one key does not lead to the other key. PKC has the significant advantage over SKC of having no key distribution.

11.3.4 Hashing

Hashing transforms a character string via a hash function into a finite value called a hash [25]. The hash quickly locates an item in a database via a hash table. Two simple examples of hashes are

- Student identification numbers used to retrieve private information about a student.
- Book call numbers used to quickly locate books in a library (e.g. Library of Congress Classification).

Unlike SKC and PKC, hashing performs one-way encryption. Hashing has three steps:

1. Convert data into a hash using a hash function.
2. Store the data in a hash table.
3. Quickly retrieve the data from the table using the hash.

A hash function generates a number (hash) for an object or data string. Two equivalent objects have the same hash while two unequal objects do not. A collision occurs when two different objects have the same hash. A hash function should [26]

1. Be easy to compute.
2. Uniformly distribute storage across the hash table.
3. Avoid collisions.

Example
Map the data $D = [54\ 26\ 93\ 17\ 77\ 31]$ into a hash table with $N = 11$ slots. The hash function performs modulo 2 arithmetic mod(D,N) to get the hashes. Find the hashes for the elements in D and create the hash table.

Solution
Use the MATLAB command: hash = mod([54 26 93 17 77 31],11)

hash = 10   4   5   6   0   9

Hash	0	1	2	3	4	5	6	7	8	9	10
Data	77	—	—	—	26	93	17	—	—	31	54

Hash functions ensure that a file remains unchanged – ideal for guaranteeing data integrity. Any change made to a message causes the receiver to calculate a hash value different from the one transmitted. A good hash function does not produce the same hash value for different inputs. Small changes to the input string of the highly nonlinear hash function produce a big change in the hash. Hashing verifies a string's identity by comparing it with a securely stored string. For instance, using the last four digits of a credit card assists in verifying someone's identity. A user gains access to a network by comparing the login password to the stored hash. Hashing excels at storing passwords, because even administrators cannot decrypt the hash to gain access to the passwords. Hashing is more secure than encryption and should always be used unless the cybertext needs decrypted [25].

11.4 Defenses

Section 11.3 established the vulnerabilities associated with a wireless system. This section covers some approaches to defend wireless systems against attacks. Users expect that a message sent over a communication channel has [26]

- Authentication (the sender and receiver are accurately identified).
- Confidentiality (the message can only be understood by the receiver).
- Integrity (the message arrives unaltered).

A resilient system operates in spite of encountering unexpected inputs, subsystem failures, or environmental conditions that lie outside its operating range [27]. Fault tolerance, fault detection, and adaptation enhance resilience.

Three approaches to securing a wireless network include [28]:

1. Requiring user authentication
2. Eliminating RAPs
3. Encrypting data.

Combining all three approaches provides the best security. All security measures require continuous updating otherwise, the system becomes vulnerable to attacks.

Often times, the AP broadcasts its SSID, so it loses the basic security protection of SSID. A list of broadcasted SSIDs within range appears on a wireless device that attempts to connect to a Wi-Fi network. APs configured without an SSID allow open access to any wireless device. A first step in securing a network is to change default SSIDs and disable APs from broadcasting their SSIDs.

A second step in securing a WLAN filters EUI/MAC addresses and only allows network access to an approved EUI address. EUI filtering works best with small networks that frequently update the EUI addresses on the approved list. EUI filtering blocks a hacker who hijacked a network IP address.

SSID and EUI address filtering satisfy the first two requirements of WLAN Security. Encryption, the third defense, uses one of three protocols [29]:

- **Wired Equivalent Privacy (WEP)** is the original wireless encryption protocol designed to provide security equivalent to wired networks. WEP has many security flaws, is difficult to configure, and is easily broken.
- **Wi-Fi Protected Access (WPA)** was a replacement for WEP while the IEEE 802.11i wireless security standard was being developed (Appendix E). Most current WPA implementations use a preshared key, commonly referred to as *WPA Personal*, and the Temporal Key Integrity Protocol (TKIP) for encryption. *WPA Enterprise* uses an authentication server to generate keys or certificates.
- **Wi-Fi Protected Access version 2 (WPA2)** is based on the 802.11i wireless security standard that was finalized in 2004. WPA2 encrypts data with AES. The U.S. government uses AES to encrypt top secret information.

The IEEE 802.11 standard specifies the RC4 SKC encryption algorithm with 40-bit or 104-bit keys for WEP. Concatenating a 24-bit "initialization vector" with an encryption key produces new 64- or 128-bit keys. This key seeds a PRN generator that creates a random sequence for encrypting the data. All clients and APs using WEP on a wireless network have the same key for encrypting and decrypting data. A client authenticates through a four-step process [21]:

1. Client requests authentication to the AP.
2. AP asks the client a challenge phrase.
3. Client encrypts the challenge phrase with the shared symmetric key before transmitting it to the AP.
4. Client receives authorization when the client's response matches the AP challenge phrase.

In spite of its weaknesses, WEP provides protection for many home networks and in small networks with low security requirements [30]. EUI address filtering along with 128-bit WEP and SSID along with use of one-time keys establishes reasonable security for many applications. WEP encryption/decryption slows down data transmission.

A virtual private network (VPN) uses encryption to enable users to safely access a secure private network over a public network like the Internet. The wireless client and the wireless network must have the VPN software installed. A VPN is necessary for security on a public network accessible by anybody.

An IDS (intrusion detection system) monitors a network and identifies suspicious patterns indicative of an attack [31]. Like a burglar alarm on a house, an IDS detects an intrusion but cannot prevent or respond to an intrusion. Two main classes of IDS are rule based IDS and anomaly based IDS [2]. Rule-based IDS (also called signature based IDS) detects intrusions by comparing data with a list of signatures or patterns symptomatic of a malicious intruder. IDS has few false positives and accurately detects well-known attacks. On the down side, IDS misses attacks that do not have signatures in the intrusion database and becomes slow in high data traffic. Anomaly based IDS looks for abnormalities in data traffic patterns and "learns" patterns that correspond to threats. Anomaly detection finds new threats without using a data base, requires little maintenance, and becomes more accurate with time. On the other hand, anomaly detection has a high false alarm when learning new intruders during which it has false alarms.

Security measures are important during system design time as well as during operation (runtime) [32]. Figure 11.14 delineates design time and runtime security approaches for new and legacy systems. Design time methods verify

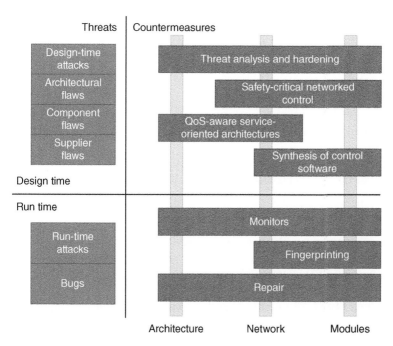

Figure 11.14 Safety and security measures at design time and runtime. *Source:* Wolf and Serpanos [32]. Reproduced with permission of IEEE.

subsystems and sets of subsystems properties. System designers scrutinize system models to ascertain attack surfaces and safety issues before testing the solution effectiveness. Runtime approaches (e.g. watchdogs, fingerprinting, repair, etc.) protect against attacks and failures during system operation. System characteristics that recognize early stage bugs and attacks can be monitored at runtime.

Problems

11.1 Find an OUI for (a) Cisco Systems, Inc., (b) Apple, Inc., (c) Dell, Inc., and (d) Agilent Technologies, Inc.

11.2 Find the binary representation of these hexadecimal numbers: (a) FC253F, (b) C8A70A, (c) 20C047, and (d) 948FEE.

11.3 Find hexadecimal representation of these binary numbers: (a) 1010010001001100110011001000, (b) 11110000100010111000100, (c) 11110000100010111000100, and (d) 110101001000000111010111.

11.4 Find the IPv4 and IPv6 addresses associated with a MAC address on a computer. These are listed in the network properties on a computer.

11.5 Convert these decimal numbers to binary: (a) 9999, (b) 365, (c) 11111, and (d) 9876.

11.6 Convert these binary numbers to decimal: (a) 111000111000111, (b) 101010101010, (c) 11010011010000010111, and (d) 1111111111.

11.7 Find the company that has the following OUI: (a) 98 : 90 : 96, (b) 08 : 00 : 20, (c) EC : AD : B8, and (d) B0 : AA : 77.

11.8 Use a Caesar cipher to encrypt the word "xray".

11.9 Write MATLAB code that will decrypt a Caesar cipher with an arbitrary shift. Use this code to decrypt (a) "BNWJQJXX" with a five letter shift and (b) "ZLUHOHVV" with a three letter shift.

11.10 Write a MATLAB function that does one-time pad encryption of the capital letters of the English alphabet. Start the function with the following code:

```
function ncode=onetime(mess,key)
```

```
nm=length(mess);
nk=length(key);
alph='ABCDEFGHIJKLMNOPQRSTUVWXYZ';
```

Find the encryption from you m-file for:
(a) mess = 'WIRELESS'; key = 'XCVBNMASDF' and
(b) mess = 'WIRELESS' key = 'XCVB'

11.11 Write a MATLAB function to count the number of times that a letter occurs in text. Distinguish between capital and lower case letters. Use a stem plot to show your results.

```
function ltrfreq=letterfreq(text)
```

11.12 Find the hashes for the data below if the hash table has 23 slots:

$$[113 \quad 117 \quad 97 \quad 100 \quad 114 \quad 108 \quad 116 \quad 105 \quad 99]$$

References

1 (2015). Security in telecommunications and information technology. In: *ITU-T, Telecommunication Standardization Bureau (TSB) Place des Nations – CH-1211 Geneva 20 – Switzerland, Sep 2015*.

2 Liu, Y. and Zhou, G. (2012). Key technologies and applications of Internet of Things. In: *Proceedings of the 5th International Conference on Intelligent Computing Technology Automation*, 197–200.

3 Burg, A., Chattopadhyay, A., and Lam, K.Y. (2018). Wireless communication and security issues for Cyber–Physical Systems and the Internet-of-Things. *Proceedings of the IEEE* 106 (1): 38–60.

4 (2017). *Guidelines for Use of Extended Unique Identifier (EUI), Organizationally Unique Identifier (OUI), and Company ID (CID)*. IEEE Standards Association.

5 https://www.adminsub.net/mac-address-finder/apple (accessed 14 August 2018).

6 Shi, J., Wan, J., Yan, H., and Suo, H. (2011). A survey of cyber-physical systems. In: *Proceedings of the International Conference on Wireless Communications and Signal Processing*, 1–6. IEEE Standards Association.

7 Khaitan, S.K. and McCalley, J.D. (2015). Design techniques and applications of cyberphysical systems: a survey. *IEEE Systems Journal* 9 (2): 350–365.

8 Miller, C. and Valasek, C. (2015). Remote exploitation of an unaltered passenger vehicle. http://illmatics.com/Remote%20Car%20Hacking.pdf (accessed 30 July 2019).

9 https://en.wikipedia.org/wiki/Service_set_(802.11_network) (accessed 1 October 2018).

10 https://www.certificationkits.com/cisco-certification/ccna-articles/cisco-ccna-wireless/cisco-ccna-wirelss-bss-a-ess (1 October 2018).

11 Chapter 1 802.11 Network Security Fundamentals (2008). *Cisco Secure Services Client Administrator Guide, Release 5.1*. Cisco Systems, Inc.

12 https://www.owasp.org/index.php/Attack_Surface_Analysis_Cheat_Sheet (1 October 2018).

13 https://docs.microsoft.com/en-us/previous-versions/commerce-server/ee823878(v=cs.20 (accessed 30 July 2019).

14 https://www.spamlaws.com/jamming-attacks.html (2 October 2018).

15 https://www.techopedia.com/definition/1770/cryptography (accessed 2 January 2019).

16 https://www.garykessler.net/library/crypto.html#intro (accessed 2 January 2019).

17 Singh, S. (1999). *The Code Book: The Science of Secrecy from Ancient Egypt to Quantum Cryptography*. New York: Anchor Books.

18 Abellán, C. and Pruneri, V. (2018). The future of cybersecurity is the quantum random number generator. *IEEE Spectrum*: 30–35.

19 https://www.scienceabc.com/innovation/the-imitation-game-how-did-the-enigma-machine-work.html (4 October 2018).

20 https://www.cia.gov/news-information/blog/2016/who-first-cracked-the-enigma-cipher.html (9 November 2018).

21 Geier, J. (2005). *Wireless Networks First-Step*. Indianapolis, IN: Cisco Press.

22 Feistel, H. (1973). Cryptography and Computer Privacy. *Scientific American* 228 (5): 15–23.

23 https://www.tutorialspoint.com/cryptography/block_cipher_modes_of_operation.htm (accessed 28 January 2019).

24 Diffie, W. and Hellman, M. (Nov 1976). New directions in cryptography. *IEEE Transactions on Information Theory* 22 (6): 644–654.

25 https://www.securityinnovationeurope.com/blog/page/whats-the-difference-between-hashing-and-encrypting (accessed 2 January 2019).

26 https://whatis.techtarget.com/definition/Confidentiality-integrity-and-availability-CIA (accessed 28 January 2019).

27 Strigini, L. (2012). Chapter 1: Fault tolerance and resilience: meanings, measures and assessment. In: *Resilience Assessment and Evaluation of Computing Systems* (eds. K. Wolter, A. Avritzer, M. Vieira and A. van Moorseled). New York: Springer.

28 https://www.netspotapp.com/wifi-encryption-and-security.html (accessed 28 January 2019).

29 Vacca, J.R. (2006). *Guide to Wireless Network Security*. New York: Springer.

30 Boncella, R.J. (2002). Wireless security: An overview. *Communications of the Association for Information Systems* 9: 269–282.

31 http://www.iup.edu/WorkArea/DownloadAsset.aspx?id=81109 (accessed 2 January 2019).

32 Wolf, M. and Serpanos, D. (Jan. 2018). Safety and security in Cyber-Physical Systems and Internet-of-Things Systems. *Proceedings of the IEEE* 106 (1): 9–20.

12

Biological Effects of RF Fields

Wireless systems expose people to a wide range of radio frequency (RF) signals. The frequency and energy of these signals determine the extent of their interaction with biological tissues. Gamma rays and X-rays have well-documented deleterious effects on humans [1]. Some interesting but nonharmful effects of electromagnetic fields occur at much lower frequencies. For instance, extremely low-frequency magnetic fields generated by high-voltage power lines disrupt the geomagnetic field orientation of cattle and roe deer while they graze [2]. Wireless communication systems typically operate at frequencies between these two extremes. Constant exposure to wireless communication signals motivates researchers to find any detrimental impact on humans. This chapter introduces the interactions of RF radiation from wireless communication systems with human biological functions.

12.1 RF Heating

Figure 12.1 divides the electromagnetic spectrum into ionizing and nonionizing radiation [3]. Ionization strips electrons from atoms and leads to tissue damage. X-rays and gamma rays are examples of ionizing radiation with high energy [4]. Gamma rays are at the pinnacle of the electromagnetic spectrum at a frequency $\geq 3 \times 10^{19}$ Hz ($\lambda \leq 10^{-11}$ m) while X-rays lie one rung below with a frequency range of $3 \times 10^{16} \leq f \leq 3 \times 10^{19}$ Hz ($10^{-11} \leq \lambda \leq 10^{-14}$ m). Wavelengths close to the size of an atom ($1 \times 10^{-10} \leq$ atom size $\leq 5 \times 10^{-10}$ m) cause resonances that ionize atoms. Nonionizing RF radiation, on the other hand, does not have enough energy to remove electrons from atoms [5].

Many research studies on RF interactions with tissue are controversial, because they result from the analysis of data collected from a human population. These studies are difficult to control and rarely result in strong correlations. Given the number of factors in any human population, even a strong correlation does not prove a causation [6].

Wireless Communications Systems: An Introduction, First Edition. Randy L. Haupt.
© 2020 John Wiley & Sons, Inc. Published 2020 by John Wiley & Sons, Inc.

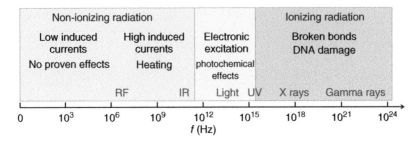

Figure 12.1 Frequency spectrum and biological effects.

RF radiation heats tissue just like microwave ovens cook food. Microwave ovens operate at 2.45 GHz which borders the cell phone frequency bands. Many RF communication systems transmitting in the microwave band, such as cell phones, heat nearby tissue. Cell phones cause health concerns like cancer, because they transmit next to the human ear and head. Human tissue does not dissipate excessive heat generated by high RF fields very well, especially in areas of low blood flow, such as the eyes [7].

Low RF radiation levels produce insignificant body heating and have no known harmful biological effects [6]. The search for nonthermal effects continues without conclusive results. In some situations (e.g. working near high-powered RF sources) appropriate limits ensure the safety of people in the vicinity. Tissue, a lossy dielectric, heats via two primary mechanisms: ionic conduction and dipolar polarization [8].

The heating conduction mechanism causes mobile charge carriers (electrons and ions) to move back and forth through the material under the influence of the microwave electric field, creating an electric current. These induced currents produce heating in the sample due to any electrical resistance resulting from charges colliding with neighboring molecules or atoms.

Human tissue primarily consists of water. Tissue with the highest water content includes muscle, skin, liver, spleen, kidney, and brain, while low water content tissues include fat, bone, teeth, nails, and hair [9]. Water molecules have a neutral charge; however, they are dipoles, because their electrons spend more time around the larger oxygen nucleus than the small hydrogen nuclei, giving the oxygen end of the molecule a negative charge. Turning off the electric field causes the water molecules to return to random orientations in the tissue as shown in Figure 12.2. Water dipoles want to align with the electric field lines in their vicinity. Their ability to adjust orientation in order to align with an applied electric field and release of thermal energy depends on frequency [10]. Low-frequency electric fields slowly change direction. The water molecules respond to the field change with no delay as shown in Figure 12.3. In contrast, Figure 12.4 shows the lack of response of the water molecules to

Figure 12.2 Random orientation of water molecules with no electric field present.

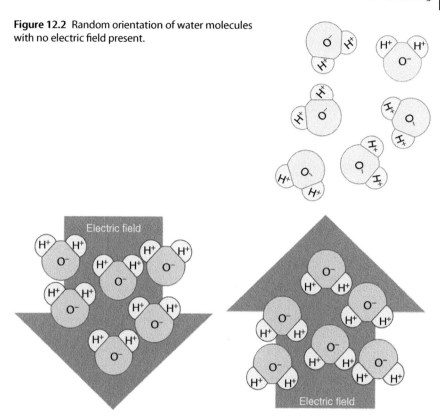

Figure 12.3 If the electric field varies too slowly, then the water molecules flip when the field changes sign.

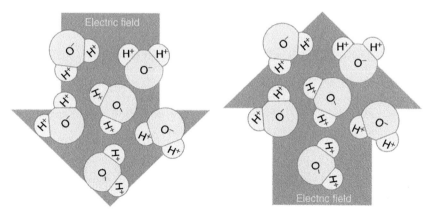

Figure 12.4 If the electric field varies too fast, then the water molecules do not have time to change orientation.

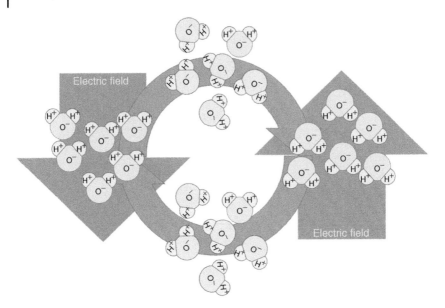

Figure 12.5 Water heating occurs when the field changes at the right speed that allows a slow reorientation of the water molecules.

a very high-frequency electric field. The field changes much faster than the water molecules have time to flip. A sweet spot in the spectrum flips the water molecules with a delay. In this case (Figure 12.5), the random molecules align with the applied electric field. Switching the field direction creates some delay in the water molecule response. In the middle of the delay, the molecules become randomly oriented again before aligning with the new direction of the electric field. This constant change between a randomized state and an aligned state produces heat. A microwave oven uses this concept to heat water molecules in food.

Lossy dielectrics have a complex dielectric constant given by

$$\varepsilon = \varepsilon' - j\varepsilon'' = \varepsilon' - j\frac{\sigma}{2\pi f} \tag{12.1}$$

where σ is the conductivity. The dielectric heating power (W) due to an electric field incident on a lossy dielectric is given by [11]

$$P = 2\pi f \varepsilon'' \tan \delta_{LF} |E|^2 \Upsilon \ \ W \tag{12.2}$$

where $\tan \delta_{LF} = \varepsilon''/\varepsilon'$ is the dielectric loss factor, and Υ is the volume of the dielectric. Even when an object has high dielectric loss, the heating efficiency for a big sample is sometimes low due to the shallow penetration depth of the microwaves. Consequently, the penetration depth quantifies the RF heating efficiency and distribution. The penetration depth (d_p) defines the distance from the surface to the point where the field strength drops by $1/e = 0.3679$. Neglecting magnetic effects, the penetration depth (same units as λ) is given by [12]

$$d_p = \frac{\lambda}{\sqrt{2\pi[(\sqrt{1 + \tan^2\delta_{LF}} - 1)]^{1/2}}} \tag{12.3}$$

Example

For muscle at 2 GHz: $\varepsilon'_r = 53.3$, $\sigma = 1.45$ S/m. Find the dielectric heating power and d_p when the electric field amplitude is 5 V/m.

Solution

$$\varepsilon'' = \frac{\sigma}{2\pi f} = \frac{1.45}{2\pi \times 2 \times 10^9} = 1.154 \times 10^{-10}$$

$$\tan\delta_{LF} = \frac{\varepsilon''}{(\varepsilon'_r \varepsilon_0)} = \frac{1.154 \times 10^{-10}}{53.3 \times 8.854\,187\,82 \times 10^{-12}} = 0.24$$

$$P = 2\pi(2 \times 10^9)(1.154 \times 10^{-10})(0.24)(5)^2 = 8.69 \text{ W}$$

$$d_p = \frac{\lambda}{\sqrt{2\pi[(\sqrt{1 + \tan^2\delta_{LF}} - 1)]^{1/2}}} = \frac{3 \times 10^8/2 \times 10^9}{\sqrt{2\pi[(\sqrt{1 + (.245)^2} - 1)]^{1/2}}}$$

$$= 0.2 \text{ m}$$

12.2 RF Dosimetry

RF dosimetry quantifies the magnitude and distribution of electromagnetic energy absorbed by biological tissue [13]. The specific absorption rate (SAR) measures the amount of RF energy absorbed by the body. RF dosimetry takes into account the shape as well as the heterogeneity of the tissues. The unit for absorbed dose of RF energy (i.e. rate of energy absorption per unit mass) is W/kg. Some factors that affect dosimetry include [14]:

- Dielectric constant
- Tissue geometry and size
- Tissue orientation and field polarization
- Field intensity and frequency
- Source configuration
- Environment
- Exposure time.

Estimates of SAR distributions in the body come from measurements in human models, in animal tissues, or from calculations.

Table 12.1 Average properties of brain, skull, and muscle tissue [16].

Frequency	Tissue	ε_r' (F/m)	σ (S/m)	ρ (kg/m³)
	Brain	68.47	0.44	1030.0
100 MHz	Skull	21.45	0.12	1850.0
	Muscle	66.19	0.73	1040.0
	Brain	46.25	0.73	1030.0
800 MHz	Skull	16.78	0.22	1850.0
	Muscle	56.21	0.93	1040.0
	Brain	45.43	0.80	1030.0
1 GHz	Skull	16.47	0.26	1850.0
	Muscle	55.74	1.01	1040.0
	Brain	43.21	1.26	1030.0
2 GHz	Skull	15.37	0.48	1850.0
	Muscle	54.17	1.51	1040.0
	Brain	39.30	3.48	1030.0
5 GHz	Skull	13.05	1.39	1850.0
	Muscle	50.13	4.24	1040.0

The National Council on Radiation Protection and Measurements (NCRP) defines SAR as the time derivative of the incremental energy absorbed by an incremental mass contained in a volume element of a given density [15]

$$SAR = \frac{1}{\Upsilon} \int_{sample} \frac{\sigma(\mathbf{r})|E_{rms}(\mathbf{r})|^2}{\rho(\mathbf{r})} d\mathbf{r} \approx \frac{\sigma|E_{rms}|^2}{\rho} \ \ W/kg \quad (12.4)$$

where

σ = tissue conductivity (S/m)
E_{rms} = RMS electric field
ρ = tissue density (kg/m³)
Υ = volume.

Values of σ and ρ for brain, skull, and muscle at five frequencies are found in Table 12.1.

A rise in the tissue temperature due to RF heating contributes to SAR according to [13]

$$SAR = \frac{c_p \Delta T}{t} \ \ W/kg \quad (12.5)$$

where

c_p = specific heat (J/g °C)

ΔT = rise in temperature ($^\circ$C)
t = exposure time (seconds).

The specific heat of water is 4.186 J/g $^\circ$C. An approximate value of c_p for bone is 3.7 J/g $^\circ$C and for muscle/brain is 1.3 J/g $^\circ$C [17].

Example
If the rms electric field $= 4$ V/m, $\sigma = 150$ S/m, $\rho = 1250$ Kg/m^3, find the SAR and incident power density.

Solution
Use (12.4) to find SAR $= 1.92$ W/kg.

Incident power density $= \frac{|E_{rms}|^2}{377\,\Omega} = 0.042$ W/m^2.

Whole-body exposure means that the incident field has a relatively uniform amplitude over the entire biological object [9]. The IEEE standard for whole-body average specific absorption rate (WBSAR) limits occupational exposure to 0.4 W/kg and public exposure to 0.08 W/kg [18]. The effect of body size and shape on WBSAR has been examined for plane wave exposure [19]. An individual's height determines the maximum RF energy absorbed at frequencies with wavelengths on the order of a person's height. The SAR at this whole body resonance frequency increases as height decreases [20].

The IEEE standard [21] establishes the whole-body maximum permissible exposure (MPE) and time averaged exposure limits for electric fields and magnetic fields (Table 12.2). The IEEE's MPE limits are spatially averaged over the whole body under two circumstances: occupational/controlled and general population/uncontrolled [13]. Occupational/controlled limits apply to people exposed in a workplace provided; those people know about the potential for exposure and have the ability to control their exposure. Limits for occupational/controlled exposure apply when an individual passes through a location with occupational/controlled limits provided the individual knows about the potential for exposure. General population/uncontrolled exposures apply to the general public or workers not fully aware of the potential for exposure or cannot exercise control over their exposure.

Example
Plot the power density limits in Table 12.2 up to 1000 GHz.

Solution
The data in Table 12.2 was entered into an m-file then graphed as a log–log plot in Figure 12.6.

Table 12.2 Limits for maximum permissible exposure (MPE) [22].

Frequency (MHz)	Power density (W/m^2)	Averaging time (min)
Limits for occupational/controlled exposures		
0.1–1.0	9 000	6
1.0–30	9 000/f^2	6
30–300	10	6
300–3 000	$f/30$	6
3 000–300 000	5	6
Limits for general population/uncontrolled exposure		
0.1–1.34	1 000	30
1.34–30	1 800/f^2	30
30–400	2.0	30
400–2 000	$f/200$	30
2 000–100 000	10	30

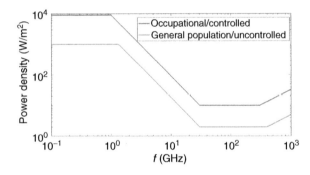

Figure 12.6 Power density limits for occupational/controlled and general population/uncontrolled exposures for up to 1000 GHz.

12.3 RF Radiation Hazards

After Hertz invented the antenna, people became interested in the biological effects of RF radiation [23]. In the late 1880s, d'Arsonval investigated the influence of RF on cells [24]. Researchers knew that shorter-wave RF induced heating in tissue. Some experiments on humans were done in the 1920s and 1930s followed by experiments with monkeys [25]. Many years later, a 42-year-old man was working 10 feet in front of a radar antenna [26]. Within seconds, he felt a sensation of heat that became intolerable in less than a minute. He moved away from the antenna, and an hour later, he was in a state of mild shock and died in a little over a week. As a result, safety procedures were developed for workers in the vicinity of high power radars. This episode startled people and motivated safer operating procedures around high power RF. Standards became prevalent and international organizations were formed to encourage the study of RF effects on humans.

12.3.1 Base Stations

A base station antenna radiates on the order of several tens of watts. The radiated RF fields for rooftops near base stations concern people working or living near them [20]. A rooftop highly attenuates signals and protects people in a building from exposure to high field levels.

12.3.2 Cell Phones

Cell phone radiation penetrates approximately 2 cm into the brain at 1800–1900 MHz [27]. For cell phones held against the ear, the SAR drops off rapidly for the regions of the brain away from the antenna and is negligible for the rest of the human body except for the hand.

12.3.3 Medical Tests

Magnetic resonance imaging (MRI) potentially causes adverse effects in some patients, including a potential for uncomfortable exposure to acoustic noise, heating, and sensory disturbances (in particular vertigo). However, all of these effects are reversible and can be prevented or ameliorated. Insufficient evidence prevents drawing firm conclusions about long-term health effects. Theory suggests that no permanent damage results if scanners are operated in line with existing guidelines that limit the exposure of patients to static and RF fields during MRI procedures [9]. The RF guidelines avoid excessive elevation of core body temperature or local temperature in the head, trunk, or extremities by restricting the SAR.

Healthcare facilities use high power RF for the ablation of tumors and in diathermy for deep tissue heating [6]. Ablation radiates 200 W at 915 and 2450 MHz and 500 W at 500–750 kHz. Both methods aim to heat the target tissues to 65–98 °C. Thermal ablation of a number of tumor types, including liver, breast, thyroid, and prostrate, has a success rate of 95% for a single treatment, with a three-year overall patient survival rate of 90%. While this technology is minimally invasive and cause only the local ablation of target tissues with minimal damage to overlying structures or surrounding tissues, there are concerns about possible collateral damage to normal structures adjacent to the desired zone of ablation.

12.4 Modeling RF Interactions with Humans

Tomographic medical imaging techniques created three-dimensional computer models (called voxel models, tomographic models, or phantoms) based on the human anatomy [28]. The models match the actual dimensions of organs.

The National Library of Medicine (NLM) Visible Human Project (VHP) created publicly available complete, anatomically detailed, three-dimensional representations of a human male cadaver and a human female cadaver [29]. The 15 GB Visible Man data set became available in 1994 and the 40 GB Visible Woman in 1995. The data sets serve as (i) a reference for the study of human anatomy, (ii) public-domain data for testing medical imaging algorithms, and (iii) a test bed and model for the construction of network-accessible image libraries. Figure 12.7 shows a section through the Visible Human Male – head, including cerebellum, cerebral cortex, brainstem, and nasal passages.

A computational electromagnetics simulation segmented the phantom in Figure 12.8. A cell phone transmitting 1 W through a PIFA antenna next to the right ear of the phantom heats the tissue close by. Figure 12.8 shows the SAR distribution averaged over 10-g of contiguous tissue at 900 MHz induced in the entire head as well as a plane cut through maximum average SAR [30].

Researchers use specialized laboratory test equipment called Dosimetric Assessment SYstem (DASY) for conducting SAR measurements. The equipment consists of a "phantom" (human or box), precision robot, RF field sensors, and mobile phone holder [31]. The phantom contains tissue simulating liquids that represents the electrical properties of human tissue. The robot moves the probe through the simulating liquid and measures the SAR vs. position in the phantom.

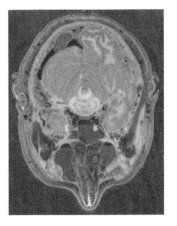

Figure 12.7 Male human head from the VHP. *Source:* www.nlm.nih.gov.

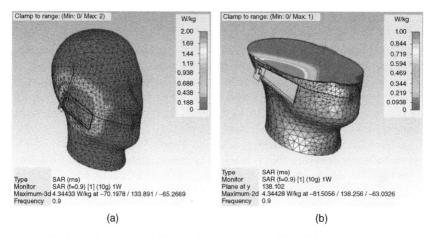

Figure 12.8 The 10-g averaged SAR distributions in the SAM model at 900 MHz: (a) the 3-D surface SAR distribution and (b) the 2-D SAR distribution in a cutting plane. *Source:* Zhao *et al.* [30]. Reproduced with permission of IEEE.

The change in temperature in a dielectric due to an RF field is expressed as

$$\Delta T = \frac{P_a t}{\text{mass} \times c_p} \,^\circ\text{K}$$

(12.6)

where

P_a = power absorbed (W)
t = exposure time (seconds)
mass= mass of object (g)
c_p = specific heat (J/(kg °C).

The following procedures measure SAR inside a head phantom [31]:

- Position the handset against the phantom body and switched to full power.
- The precision robot moves the RF probe throughout the phantom head measuring the radio signal level in the head phantom.
- Convert the measured data into SAR (W/kg).
- The full test is conducted at all operating frequencies and using different phone positions.
- The maximum level measured is recorded as the SAR value against the head.

The following procedures measure SAR inside a body (box) phantom [31]:

- Position the handset against the phantom body and switched to full power.
- The precision robot moves the RF probe throughout the phantom body measuring the radio signal level in the body near the phone.
- Convert the measured data into SAR (W/kg).
- The maximum level measured is recorded as the SAR value against the body.

12.5   Harmful Effects of RF Radiation

At mobile phone frequencies, most of the energy absorbed by the skin and other superficial tissues produce a negligible temperature rise in the brain or any other organs of the body. A person's ear receives the most RF radiation from a smart phone. Researchers have found no basis for concerns over auditory perception and on acoustic evoked potentials as well as auditory functions of the cochlea or auditory brainstem responses [32].

The Surveillance, Epidemiology, and End Results (SEER) Program of the National Cancer Institute (NCI) tracks US cancer statistics [33]. SEER data shows even as cell phone use in the United States significantly increased, the age-adjusted incidence of brain cancer remained unchanged. Wide spread use of mobile phones did not start until the early 1990s, so epidemiological studies only assess cancers that occur since then. Animal studies have not shown an increased cancer risk due to long-term exposure to RF signals.

The International Agency for Research on Cancer (IARC) launched a large study, Interphone, that looked for links between use of mobile phones and head and neck cancers in adults [34]. The data from 13 participating countries found no increased risk of glioma or meningioma (brain tumors) with mobile phone use of more than 10 years. People who reported the highest 10% of cumulative hours of cell phone use had some indications of an increased risk of glioma, although increasing risk with increasing the duration of use did not increase the risk. No increase in risk of acoustic neuroma (benign tumor on auditory nerves) was found. The researchers concluded that biases and errors question these conclusions and prevent a causal interpretation. This study led IARC to classify RF fields as possibly carcinogenic to humans (Group 2B). In other words, there is a credible causal association, but chance, bias or confounding cannot be ruled out with reasonable confidence.

Several studies investigated the impact of RF radiation from cell phones on brain electrical activity, cognitive function, sleep, heart rate, and blood pressure. No evidence of adverse health effects were found from field levels that do not cause tissue heating [27]. Some people claim to have electromagnetic hypersensitivity (EHS) or are sensitive to electromagnetic fields, particularly radiation from cell phones. The World Health Organization (WHO) concluded after many studies [35]: "Whatever its cause, EHS can be a disabling problem for the affected individual. EHS has no clear diagnostic criteria and there is no scientific basis to link EHS symptoms to electromagnetic field exposure. Further, EHS is not a medical diagnosis, nor is it clear that it represents a single medical problem."

Problems

12.1   A 1 GHz plane wave is incident on a dielectric with $\varepsilon' = 60\varepsilon_0$ and $\sigma = 1\,\text{S/m}$. Find (a) dielectric loss factor, (b) penetration depth, and (c) dielectric heating power if the electric field amplitude is 1 V/m.

12.2 Plot the penetration depth into the brain for $100 \leq f \leq 5000$ MHz.

12.3 Plot the penetration depth into the skull for $100 \leq f \leq 5000$ MHz.

12.4 Plot the penetration depth into the muscle for $100 \leq f \leq 5000$ MHz.

12.5 Calculate SAR for Table 12.1.

12.6 Plot SAR for brain, muscle, and skull from 100 MHz to 5 GHz when a 5 V/m electric field is present.

12.7 You walk within 10 m of a base station that is radiating 10 W through an antenna with 10 dB of gain at 2 GHz. Calculate the SAR in the brain.

12.8 How much exposure time is needed to raise the following tissue 1 °C: (a) brain $c_p = 3.6$ J/g °C, (b) skull $c_p = 2.0$ J/g °C, and (c) muscle $c_p = 3.4$ J/g °C.

12.9 The permittivity and conductivity of many different tissues as a function of frequency are available at https://itis.swiss/virtual-population/tissue-properties/database/dielectric-properties. Pick a tissue not listed in Table 12.1. Plot (a) penetration depth vs. frequency and (b) SAR vs. frequency for a 1 V/m electric field.

References

1 Mahaffey, J. (2015). *Atomic Accidents A History of Nuclear Meltdowns and Disasters; From the Ozark Mountains to Fukushima*. New York: Pegasus Books.
2 Burda, H., Begall, S., Červený, J. et al. (2009). Extremely low-frequency electromagnetic fields disrupt magnetic alignment of ruminants. *Proceedings of the National Academy of Sciences* 106 (14): 5708–5713.
3 Flores-McLaughlin, J., Runnels, J., and Gaza, R. (2017). Overview of non-ionizing radiation safety operations on the International Space Station. *Journal of Space Safety Engineering* 4 (2): 61–63.
4 http://physicscentral.com/explore/action/radiationandhumans.cfm (17 November 2018).
5 https://www.fda.gov/radiation-emittingproducts/radiationemittingproductsandprocedures/homebusinessandentertainment/cellphones/ucm116282.htm (accessed 17 November 2018).
6 (2006). *Framework for Developing Health-Based EMF Standards*. Switzerland: World Health Organization Press, World Health Organization Press.
7 Cleveland, R.F. Jr. and Ulcek, J.L. (1999). *Questions and Answers About Biological Effects and Potential Hazards of Radiofrequency Electromagnetic*

Fields, 4e. Federal Communications Commission Office of Engineering & Technology, OET Bulletin 56.

8 Baker-Jarvis, J. and Kim, S. (2012). The interaction of radio-frequency fields with dielectric materials at macroscopic to mesoscopic scales. *Journal of Research of the National Institute of Standards and Technology* 117: 1–60.

9 Chou, C.K., Bassen, H., Osepchuk, J. et al. (1996). Radio frequency electromagnetic exposure: tutorial review on experimental dosimetry. *Bioelectromagnetics* 17 (3): 195–208.

10 https://www.microdenshi.co.jp/en/microwave (accessed 17 November 2018).

11 https://www.ncbi.nlm.nih.gov/pmc/articles/PMC5502878 (accessed 17 November 2018).

12 Johnk, C.T.A. (1988). *Engineering Electromagnetic Fields and Waves*, 2e. New York: Wiley.

13 https://www.gpo.gov/fdsys/pkg/CFR-2011-title47-vol1/xml/CFR-2011-title47-vol1-sec1-1310.xml (accessed 18 November 2018).

14 http://www.who.int/peh-emf/meetings/04_Chou.pdf (accessed 19 November 2018).

15 https://www.everythingrf.com/rf-calculators/sar-rf-exposure-calculator (accessed 11 December 2018).

16 https://www.fcc.gov/general/body-tissue-dielectric-parameters (11 December 2018).

17 Giering, K., Lamprecht, I., and Minet, O. (1996). Specific heat capacities of human and animal tissues. In: *SPIE Proceedings Volume 2624, Laser-Tissue Interaction and Tissue Optics*, SPIE = The International Society of Optics and Photonics, 188–197.

18 IEEE Standard C95.1 2005. *IEEE Standard for Safety Levels with Respect to Human Exposure to RadioFrequency Electromagnetic Fields, 3 kHz to 300 GHz*, IEEE.

19 Hirata, A., Fujiwara, O., Nagaoka, T., and Watanabe, S. (2010). Estimation of whole-body average SAR in human models due to plane-wave exposure at resonance frequency. *IEEE Transactions on Electromagnetic Compatibility* 52 (1): 41–48.

20 Gandhi, O.P. (1929). Dosimetry—the absorption properties of man and experimental animals. *Bulletin of the New York Academy of Medicine* 55 (11): 999–1020.

21 IEEE Std C95.1-2005 (Revision of IEEE Std C95.1-1991) (2006, 238). *IEEE Standard for Safety Levels with Respect to Human Exposure to Radio Frequency Electromagnetic Fields, 3 kHz to 300 GHz*, 1. IEEE.

22 https://www.rfsafetysolutions.com/RF%20Radiation%20Pages/IEEE_Standards.html (accessed 17 November 2018).

23 Cook, H.J., Steneck, N.H., Vander, A.J., and Kane, G.L. (1980). Early research on the biological effects of microwave radiation: 1940–1960. *Annals of Science* 37: 323–351.

24 d'Arsonval, A. (1893). 'Influence de l'électricité sur la cellule microbienné. *Arch. de physiol, norm. et path.* 5 (5): 66–69.

25 Turner, J.J. (1962). *The Effects of Radar on the Human Body*. Whippany, NJ, RM-TR-62-1, 21 Mar: U.S. Army Ordance Missle Command Liaison Office, Bell Telephone Laboratories.

26 McLaughlin, J.T. (1957). Tissue destruction and death from microwave radiation (radar). *California Medicine* 86 (5): 336–339.

27 National Research Council (2008). *Identification of Research Needs Relating to Potential Biological or Adverse Health Effects of Wireless Communication Devices*. Washington, DC: The National Academies Press.

28 Caon, M. (2004). Voxel-based computational models of real human anatomy: a review. *Radiation and Environmental Biophysics* 42 (4): 229–235.

29 https://www.nlm.nih.gov/research/visible/visible_human.html (accessed 18 November 2018).

30 Zhao, L., Ye, Q., Wu, K. et al. (2016). A new high-resolution electromagnetic human head model: a useful resource for a new specific-absorption-rate assessment model. *IEEE Antennas and Propagation Magazine* 58 (5): 32–42.

31 http://www.emfexplained.info/?ID=25584 (accessed 11 December 2018).

32 Sievert, U., Eggert, S., and Pau, H.W. (Apr 2005). Can mobile phone emissions affect auditory functions of cochlea or brain stem? *Otolaryngology Head and Neck Surgery* 132 (3): 451–455.

33 http://seer.cancer.gov (accessed 23 October 2018).

34 The INTERPHONE Study Group (2010). Brain tumour risk in relation to mobile telephone use: results of the INTERPHONE international case–control study. *International Journal of Epidemiology* 39 (3): 675–694.

35 https://www.who.int/peh-emf/publications/facts/fs296/en (accessed 18 December 2018).

Appendix A

MATLAB Tips

A.1 Introduction

If you are unfamiliar with MATLAB or need to refresh your skills, I suggest starting at https://www.mathworks.com/support/learn-with-matlab-tutorials .html

A search of the Internet will produce many books and helpful resources on MATLAB.

There are a wide range of MATLAB toolboxes dedicated to topics in wireless communications. I tried to stick to basic MATLAB in my examples and homework problems, so that readers do not have to spend a lot of money buying toolboxes. If you want to become a wireless communications MATLAB guru, then these are the most relevant MATLAB packages:

MATLAB and SIMULINK	
Signal Processing Toolbox	RF Toolbox
Phased Array Toolbox	Communications Toolbox
DSP System Toolbox	LTE Toolbox
Audio Toolbox	WLAN Toolbox
Antenna Toolbox	5G Toolbox

If you do not have money to buy MATLAB, then here are some cheap alternatives:

- There are free online calculators for many of the problems in this book.
- Use Python – it is MATLAB – like and is free.
- Other programming languages and math software tools.

Wireless Communications Systems: An Introduction, First Edition. Randy L. Haupt.
© 2020 John Wiley & Sons, Inc. Published 2020 by John Wiley & Sons, Inc.

A.2   Plotting Hint

My pet peeve with MATLAB is the default graphics. Everything is too tiny – font and line widths. I suggest that you create an m-file called startup and put it in the bin directory in MATLAB. Everytime you start MATLAB, these commands are executed. You can put additional commands in this file. Here are some helpful commands for plotting:

```
set(0,'DefaultAxesFontSize',28)
set(0,'DefaultLineLineWidth',2)
set(0,'DefaultTextFontSize',28)
set(0,'DefaultAxesLineStyleOrder','-|--|:|-.')
set(0,'DefaultAxesColorOrder',[0 0 1;1 0 0;0 0 0;0 1 0])
set(0,'DefaultAxesFontName','Times New Roman')
set(0,'DefaultTextFontName','Times New Roman')
```

These commands save you from changing these parameters every time you call MATLAB. Your graphics will look much better!

Appendix B

OSI Layers

In 1983, the International Standards Organization (ISO) developed the Open Systems Interconnection (OSI) model to define a framework for computer communications (Figure B.1). The OSI model has seven layers described below in detail. Lower layers (1–4) concern moving data around, while upper layers (5–7) deal with application-level data. A mnemonic trick for memorizing the seven layers: "Please Do Not Tell Secret Passwords Anytime."

B.1   Layer 1: Physical

The physical layer transmits signals across a communication medium. It includes but is not limited to cables; antennas; electronics; power; bit rate; point-to-point; multipoint or point-to-multipoint line configuration; network topology; serial or parallel communication; simplex, half duplex or full duplex transmission mode; modulation; line coding; synchronization; circuit switching; multiplexing; equalization; training sequences; pulse shaping; FEC; and bit interleaving.

B.2   Layer 2: Data Link

The data link layer transforms bits from the physical layer into a frame for the network layer. It provides direct node-to-node data transfer, and error correction from the physical layer. The Data Link Layer has encoding and data compression. Two sublayers in the data link are the Media Access Control (MAC) layer and the Logical Link Control (LLC) layer. In the networking world, most switches operate in layer 2.

B.3   Layer 3: Network

The network layer controls packet routing and works with IP addresses.

Wireless Communications Systems: An Introduction, First Edition. Randy L. Haupt.
© 2020 John Wiley & Sons, Inc. Published 2020 by John Wiley & Sons, Inc.

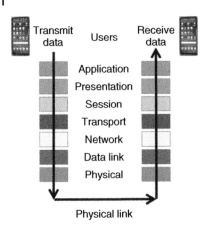

Figure B.1 OSI layers.

Physical link

B.4 Layer 4: Transport

The transport layer splits the data from the session layer into packets for delivery on the network layer and verifies that the packets arrive error free at the other end. It decides how much data to send, the data rate, destination, etc. The Transmission Control Protocol (TCP) occurs at this layer.

B.5 Layer 5: Session

The session layer establishes and manages sessions, conversions, or dialogues between two computers. Two devices communicate via a session. Functions at this layer involve setup, coordination, and termination between the applications at each end of the session. Examples include: IPv4, IPv6, and Apple Talk.

B.6 Layer 6: Presentation

The presentation layer manages the syntax and semantics of the information transmitted between two computers. It represents the preparation or translation of application format to network format, or from network formatting to application format. This layer prepares data for the application or the network. Encryption and decryption happens in this layer. Examples include ASCII, JPEG, tif, and gif.

B.7   Layer 7: Application

The application layer deals with data symbols and has several common protocols, such as file transfer, virtual terminal, and email. Most users only deal with this layer. Some examples include browsers, email, video conferencing, HTTP (HyperText Transfer Protocol), and FTP (File Transfer Protocol).

B.7 Layer 7 Application

Appendix C

Cellular Generations

First generation (1G) wireless systems were analog systems that only supported voice calls. Nippon Telephone and Telegraph (NTT) in Tokyo, Japan, Europe Nordic Mobile Telephone (NMT), and Bell labs AMPS (Advanced Mobile Phone Service) started service in the early 1980s. The desire for data communication, security, and high data rates initiated a digital trend with the second generation (2G) and continues today [1–3]. Table C.1 compares the four generations of cell phones.

Table C.1 Comparison of wireless generations [4].

Generation	1	2	3	4
Introduced	1981	1991	2000–2002	2009
Introduced	USA	Finland	Japan	South Korea
Technology	AMPS, NMT, TACS	IS-95, GSM	IMT2000, WCDMA	LTE, WiMAX
Multiple Access	FDMA	TDMA, CDMA	CDMA	CDMA
Switching type	Circuit switching	Circuit switching for Voice and Packet switching for Data	Packet switching except for Air Interface	Packet switching
Data rate	2.4–14.4 kbps	14.4 Kbps	3.1 Mbps	100 Mbps
Supports	Voice only	Voice and Data	Voice and Data	Voice and Data
Internet	None	Narrowband	Broadband	Ultra Broadband
Bandwidth	Analog	25 MHz	25 MHz	100 MHz

(Continued)

Wireless Communications Systems: An Introduction, First Edition. Randy L. Haupt.
© 2020 John Wiley & Sons, Inc. Published 2020 by John Wiley & Sons, Inc.

Table C.1 (Continued)

Generation	1	2	3	4
Operating frequencies	800 MHz	GSM: 900, 1800 MHz CDMA: 800 MHz	2100 MHz	850, 1800 MHz
Advantage	Simpler (less complex) network elements	Multimedia features (SMS, MMS), Internet access and SIM introduced	High security, international roaming	Speed, high speed handoffs, MIMO technology, Global mobility
Disadvantages	Limited capacity, not secure, poor battery life, large phone size, background interference	Low network range, slow data rates	High power consumption, Low network coverage, High cost of spectrum licence	Hard to implement, complicated hardware required
Applications	Voice Calls	Voice calls, Short messages, browsing	Video conferencing, mobile TV, GPS	High speed applications, mobile TV, Wearable devices

References

1 Vora, L.J. (2015). Evolution of mobile generation technology: 1G to 5G and review of upcoming wireless technology 5G. *International Journal of Modern Trends in Engineering and Research* 2 (10): 281–290.

2 Bhandari, N., Devra, S., and Singh, K. (2017). Evolution of cellular network: from 1G to 5G. *International Journal of Engineering and Techniques* 3 (5): 98–105.

3 Mohammad Meraj ud in Mir , Dr. Sumit Kumar (2015). Evolution of mobile wireless technology from 0G to 5G. *International Journal of Computer Science and Information Technologies* 6 (3): 2545–2551.

4 http://www.zseries.in/telecom%20lab/telecom%20generations/#.XEE83lxKhPY (accessed 17 January 2019).

Appendix D

Bluetooth

In 1994, Ericsson invented a short distance wireless technology for exchanging data between fixed and mobile devices over the ISM (Industrial, Scientific, and Medical) band 2.4–2.4835 GHz [1]. It was named after the Scandinavian king Harald Bluetooth. Bluetooth (🔵) operates at frequencies between 2402 and 2480 MHz, or 2400 and 2483.5 MHz including guard bands 2 MHz wide at the bottom end and 3.5 MHz wide at the top as shown in Figure D.1 [2]. It uses frequency-hopping spread spectrum (FHSS) at 1600 hops per second. The data packets transmit on one of 79 designated channels. Each channel has $B = 1$ MHz with 800 hops per second. Bluetooth has many updates starting with version 1 and extending to the current version 5.

Bluetooth Low Energy (BLE) occupies the same frequency band as Bluetooth using 40 $B = 2$-MHz channels and operates with significantly lower power consumption and cost. BLE enables small sensors to operate off tiny batteries for months.

The original Bluetooth used Gaussian frequency-shift keying (GFSK) modulation. Devices using GFSK operate in basic rate (BR) mode at 1 Mbit/s. Enhanced Data Rate (EDR) mode came later with $\pi/4$-DPSK modulation at 2 Mbit/s and 8-DPSK modulation at 3 Mbit/s. Bluetooth that combined the BR and EDR modes is called BR/EDR radio.

Bluetooth has a master/slave architecture for passing packets. One master communicates with up to seven slaves that synchronize to the master's clock that has a period of 312.5 μs. Two clock periods make up a slot of 625 μs, and two slots make up a slot pair of 1250 μs. The master transmits in even slots and receives in odd slots for single-slot packets. The slave receives in even slots and transmits in odd slots. Packets may be 1, 3, or 5 slots long. The master rapidly switches from one device to another checking for a slave to address.

Wireless Communications Systems: An Introduction, First Edition. Randy L. Haupt.
© 2020 John Wiley & Sons, Inc. Published 2020 by John Wiley & Sons, Inc.

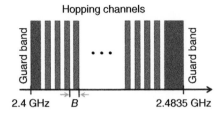

Hopping channels

Figure D.1 Bluetooth spectrum.

References

1 Morrow, R. (2002). *Bluetooth: Operation and Use*. New York: McGraw-Hill.
2 https://www.edgefxkits.com/blog/bluetooth-technology-and-its-working (accessed 3 February 2019).

Appendix E

Wi-Fi

Wi-Fi represents the IEEE 802.11* standards where * is a, b, e, f, g, h, I, j, k, n, s, u, ac, ad, af, ah, and ax [1]. A worldwide network of companies called the Wi-Fi Alliance owns the Wi-Fi registered trademark. Wi-Fi provides broadband Internet connection via access points over an area known as a hotspot. The 802.11 standards specify Wi-Fi communications at bands around 900 MHz, 2.4 GHz, 3.6 GHz, 4.9 GHz, 5 GHz, 5.9 GHz, and 60 GHz bands. Each frequency band has many channels that countries regulate in terms of allowable channels, allowed users, and maximum power levels within these frequency ranges. Bandwidth requirements increased with the evolution of 802.11 standards.

Wi-Fi Alliance started referring to Wi-Fi by number versions rather than using the very confusing IEEE standard designation that has letters that make no sense to most people. All new versions of Wi-Fi receive one higher number than the previous version [2]. Higher number are compatible with lower numbers and have higher performance. This new designation does not apply to the outdated Wi-Fi 1–3.

Table E.1 summarizes the characteristics of the different Wi-Fi releases. In the 2.4-GHz ISM band microwave ovens, cordless telephones, USB 3.0 hubs, and Bluetooth devices interfere with Wi-Fi. The United States has 11 channels in the 2.4 GHz band, while Australia and Europe have 13, and Japan 14. A Wi-Fi signal only occupies two or three channels spaced across the 2.4 GHz band in order to limit interference between channels.

CCK (complementary code keying) is a spread spectrum technique for low data rates up to 11 Mbps. 802.11a/n/ac use the 5 GHz U-NII band, which, for much of the world, offers at least 23 nonoverlapping 20 MHz channels rather than the 2.4 GHz ISM frequency band, where the channels are only 5 MHz wide. Common building materials absorb frequencies near 5 GHz, resulting in a shorter range.

802.11n uses double the radio spectrum/bandwidth (40 MHz) compared to 802.11a or 802.11g (20 MHz). This means there can be only one 802.11n network on the 2.4 GHz band at a given location, without interference to/from other WLAN traffic. 802.11n can also be set to limit itself to 20 MHz bandwidth

Wireless Communications Systems: An Introduction, First Edition. Randy L. Haupt.
© 2020 John Wiley & Sons, Inc. Published 2020 by John Wiley & Sons, Inc.

Table E.1 Characteristics of Wi-Fi 1–Wi-Fi 6 [2, 3].

Generation	Standard	Date	Max. rate	Frequency (GHz)	Channel width (MHz)	Modulation	Multi-plexing
1	802.11b	1999	11 Mbps	2.4	20	CCK	
2	802.11a	1999	54 Mbps	5	20	BPSK, QPSK, 16-QAM, 64-QAM	OFDM
3	802.11g	2003	54 Mbps	2.4	20	CCK, DSSS	OFDM
4	802.11n	2009	600 Mbps	2.4 or 5	20 or 40	CCK, DSSS	OFDM
5	802.11ac	2014	6.93 Gbps	5.8	20, 40, and 80	BPSK, QPSK, 16-QAM, 64-QAM, 256-QAM	OFDM
6	802.11ax	2019	14 Gbps	2.4 and 5	20, 40, 80, and 160	BPSK, QPSK, 16-QAM, 64-QAM, 256-QAM, 1024-QAM	OFDMA

to prevent interference in dense community. Wi-Fi 4 was first to use multiple input/multiple output (MIMO).

Salient features of Wi-Fi 6 include [4]:

- Uplink and downlink orthogonal frequency division multiple access (OFDMA) increases efficiency and lowers latency for high demand environments
- Multi-user multiple input, multiple output (MU-MIMO) allows more data to be transferred at one time, enabling access points (APs) to handle larger numbers of devices simultaneously
- Transmit beamforming enables higher data rates at a given range to increase network capacity
- quadrature amplitude modulation mode (1024-QAM) increases throughput for emerging, bandwidth-intensive use cases
- Frequency Division Multiple Access or OFDMA. The Wi-Fi access point can talk to more devices at once.

References

1 https://www.electronics-notes.com/articles/connectivity/wifi-ieee-802-11/standards.php (accessed 17 January 2019).

2 Kastrenakes, J. (2018). Wi-Fi now has version numbers, and Wi-Fi 6 comes out next year. https://www.theverge.com/2018/10/3/17926212/wifi-6-version-numbers-announced (accessed 10 March 2018).

3 https://ccm.net/contents/802-introduction-to-wi-fi-802-11-or-wifi (accessed 14 February 2019).

4 IEEE 802.11ax: The Sixth Generation of Wi-Fi, Cisco Technical White Paper. https://www.cisco.com/c/dam/en/us/products/collateral/wireless/white-paper-c11-740788.pdf (accessed 14 February 2019).

Appendix F

Software-Defined Radios

Radios existed long before computers and software, so the idea of replacing some of the hardware in a radio with a computer and software did not occur until late in the twentieth century. Software-defined radio (SDR) alters the transmit/receive waveform in software rather than hardware. Traditional hardware functions, such as modulation, carrier frequency, or coding, now become lines of code.

F.1 SDR Basics

The benefits of SDR are:

- *Flexibility*: Requires no hardware changes when switching between functions
- *Interoperability*: Works with old and new systems that have appropriate software
- *Ease of upgrade*: Adds new features and advances through software updates
- *Efficiency*: Supports many different radios
- *Higher-level interfaces*: Operates through GUI and network interfaces.

The ultimate goal of SDR technology is to have a radio that communicates at any frequency, bandwidth, modulation, and data rate through software updates rather than hardware changes. The ideal SDR is [1]

1. *Multiband*: Operates on two or more bands either sequentially or simultaneously.
2. *Multicarrier or multichannel*: Simultaneously operates at more than one frequency within the same band or in two different bands at the same time.
3. *Multimode*: The ability to switch modulation schemes.
4. *Multirate*: The ability to process different data rates.
5. *Variable bandwidth*: Channel bandwidth determined by digital filters.

Wireless Communications Systems: An Introduction, First Edition. Randy L. Haupt.
© 2020 John Wiley & Sons, Inc. Published 2020 by John Wiley & Sons, Inc.

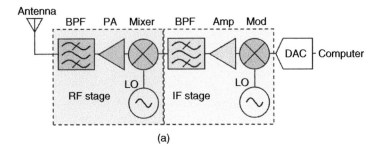

(a)

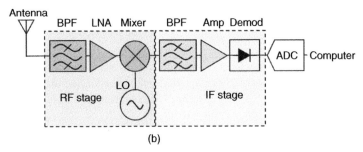

(b)

Figure F.1 Block diagram of a two-stage hardware radio transmitter and receiver. (a) Transmitter and (b) receiver

The SDR Wireless Innovation Forum (WINNF) defines five tiers of radios based on which parts are configurable [2].

Tier 0: All functions in Figure F.1 are hardware. No functions are software reconfigurable.

Tier 1: A radio with limited functions controlled by software, such as power levels and interconnections, but not mode or frequency. All functions in Figure F.1 are performed in hardware with some software control.

Tier 2: A radio with a hardware-based RF front end but software control of frequency, modulation and waveform generation/detection, wide/narrowband operation, and security (Figure F.2).

Tier 3: Nearly all functions performed in software. The analog to digital convertors (ADCs) and digital-to-analog converters (DACs) are as close to the antenna as possible (Figure F.3).

Tier 4: Full programmability and supports a broad range of functions and frequencies. With many electronic items such as cellphones having many different radios and standards a software definable multifunction phone would fall into this category (Figure F.3).

Figure F.2 An IF SDR transmitter and receiver eliminates the IF stage. (a) Transmitter and (b) receiver

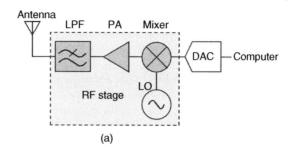

(a)

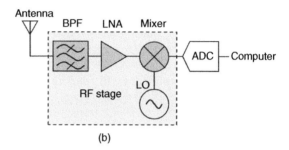

(b)

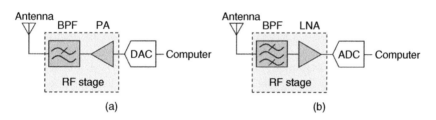

(a)   (b)

Figure F.3 SDR DUC transmitter and DDC receiver. (a) Transmitter and (b) receiver

F.2 SDR Hardware

Figure F.1 is a simplified example of a typical two-stage transmitter and receiver. The radio frequency (RF) and IF (intermediate frequency) stages are hardware. The DAC in the transmitter (Figure F.1a) sends an analog baseband signal to the IF stage. The IF stage upconverts the baseband frequency to an intermediate frequency. This signal is amplified then passed through a bandpass filter (BPF). The IF stage output goes to the RF stage input where it is upconverted to RF, amplified, and filtered before heading to the antenna. The receiver has a reverse process in which the receive antenna sends its signal to the RF stage where it is filtered, amplified, and downconverted to an IF (Figure F.1b). The IF stage then filters, amplifies, and demodulates the IF signal. An ADC at the output of the IF stage converts the IF output to binary format for input to a computer. The RF and IF stages are designed to interface the ADC input and DAC output with the antenna.

The antenna and ADC specifications restrict SDR performance [1]. Most antennas have a narrow bandwidth that limits multiband operation. Systems operating over a wide bandwidth require a matching circuit that adapts to the operating frequency changes. The antenna gain also changes with frequency, so the antenna must adapt its size as the frequency changes. The ADC and DAC have an upper frequency limit. Signals above that limit require frequency downconvertion or upconversion.

Figure F.2a shows a transmitter that inputs the bits from the DAC into an RF stage without need for the IF stage. The RF stage is necessary, because the DAC output frequency is too low. Once the IF signal is up converted, then it passes through a power amplifier and BPF before going to the antenna. Figure F.2b is the receiver without an IF stage.

A direct upconversion (DUC) transmitter (Figure F.3a) passes digital data through a power amplifier and BPF without the need for upconversion. A direct downconversion (DDC) receiver (Figure F.3b) passes the RF signal through a BPF and low noise amplifier (LNA) before the input of the ADC without the need of an IF stage.

The mass-produced DVB-T TV tuner dongle based on the RTL2832U chipset serves as the basis for the RTL-SDR [3]. The RTL maximum sample rate is 3.2 MS/s (mega samples per second). This rate drops samples, though, so 2.4 MS/s is more realistic [4]. The ENOB is 7 bits. These dongles were intended for TV, so they have a 75 Ω input impedance. Newer dongles have 50 Ω SMA connectors. Figure F.4 is an example of an RTL SDR with a monopole antenna plugged into the USB port of a laptop computer.

F.3   SDR Software

The Software Communications Architecture (SCA) specifies how the hardware and software in an SDR interface with each other. It uses CORBA (Common Object Request Broker Architecture) which enables software components written in multiple computer languages and running on multiple computers to work together. The advantages of SCA are [2]:

- Software modules written by different sources are compatible.
- Reusing modules significantly reduce costs.

For amateurs, GNU radio serves the same purpose as SCA and is supported by a community of developers [5]. GNU radio is free and distributed under the terms of the GNU General Public Licence (GPL). It can be used with readily

Figure F.4 An RTL SDR.

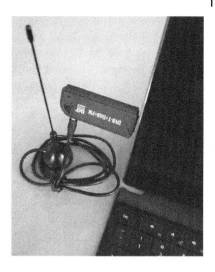

available low-cost external RF hardware to create software-defined radios, or without hardware in a simulation-like environment. All of the code is copyright of the Free Software Foundation.

MATLAB serves as another avenue for interfacing with RTL-SDR [6].

F.4 Cognitive Radio

Cognitive radio is an SDR capable of altering its operating behavior based on an awareness of its environment. The benefits of CR include [7]

- Optimize use of spectrum
- Organize interoperability
- Map locations of units, rank candidates for dispatch (nearest, equipment and training, ready to move)
- Reconfigure networks to meet current needs
- Respond to priority structures
- Reach hidden nodes.

References

1 Fette, B.A. (2007). Basics of software defined radio, part 1. *EE Times*.
2 Mitola, J. III. Software radios: Survey, critical evaluation and future directions. *IEEE Aerospace and Electronic Systems Magazine* 8 (4): 25–36.
3 Laufer, C. *The Hobbyist's Guide to the RTL-SDR*, 4e.
4 https://www.rtl-sdr.com/about-rtl-sdr (accessed 17 January 2019).

5 https://www.gnuradio.org/about (accessed 9 January 2019).

6 Stewart, R.W., Barlee, K.W., Atkinson, D.S.W., and Crockett, L.H. (2017). *Software Defined Radio Using MATLAB & Simulink and the RTL-SDR.* Glasgow, Scotland: University of Strathclyde.

7 Cook, P.G. (2007). Introduction to software defined radio, cognitive radio and the SDR Forum. http://www.npstc.org/download.jsp?tableId=37& column=217&id=1280&file=IO-Cook-0802145%20NPSTCa.pdf (accessed 25 June 2019).

Index

Wireless Communications Systems: An Introduction, First Edition. Randy L. Haupt.
© 2020 John Wiley & Sons, Inc. Published 2020 by John Wiley & Sons, Inc.

Printed and bound by CPI Group (UK) Ltd, Croydon, CR0 4YY

12/03/2024

14469437-0001